D0079182

BIOENGINEERING OF THE SKIN

Skin Biomechanics

DERMATOLOGY: CLINICAL & BASIC SCIENCE SERIES

Series Editor Howard I. Maibach, M.D.

Published Titles:

Pesticide Dermatoses
Homero Penagos, Michael O'Malley, and Howard I. Maibach

Hand Eczema, Second Edition
Torkil Menné and Howard I. Maibach

Dermatologic Botany
Javier Avalos and Howard I. Maibach

Dry Skin and Moisturizers: Chemistry and Function
Marie Loden and Howard I. Maibach

Skin Reactions to Drugs
Kirsti Kauppinen, Kristiina Alanko, Matti Hannuksela, and Howard I. Maibach

Contact Urticaria Syndrome
Smita Amin, Arto Lahti, and Howard I. Maibach

Bioengineering of the Skin: Skin Surface, Imaging, and Analysis
Klaus P. Wilhelm, Peter Elsner, Enzo Berardesca, and Howard I. Maibach

Bioengineering of the Skin: Methods and Instrumentation
Enzo Berardesca, Peter Elsner, Klaus P. Wilhelm, and Howard I. Maibach

Bioengineering of the Skin: Cutaneous Blood Flow and Erythema
Enzo Berardesca, Peter Elsner, and Howard I. Maibach

Bioengineering of the Skin: Water and the Stratum Corneum
Peter Elsner, Enzo Berardesca, and Howard I. Maibach

Human Papillomavirus Infections in Dermatovenereology
Gerd Gross and Geo von Krogh

The Irritant Contact Dermatitis Syndrome
Pieter van der Valk, Pieter Coenrads, and Howard I. Maibach

Dermatologic Research Techniques
Howard I. Maibach

Skin Cancer: Mechanisms and Human Relevance
Hasan Mukhtar

Skin Cancer: Mechanisms and Human Relevance
Hasan Mukhtar

Protective Gloves for Occupational Use
Gunh Mellström, J.E. Walhberg, and Howard I. Maibach

Pigmentation and Pigmentary Disorders
Norman Levine

Nickel and The Skin: Immunology and Toxicology
Howard I. Maibach and Torkil Menné

DERMATOLOGY: CLINICAL & BASIC SCIENCE SERIES

BIOENGINEERING OF THE SKIN

Skin Biomechanics

Edited by
Peter Elsner
Enzo Berardesca
Klaus-P. Wilhelm
Howard I. Maibach

CRC PRESS

Boca Raton London New York Washington, D.C.

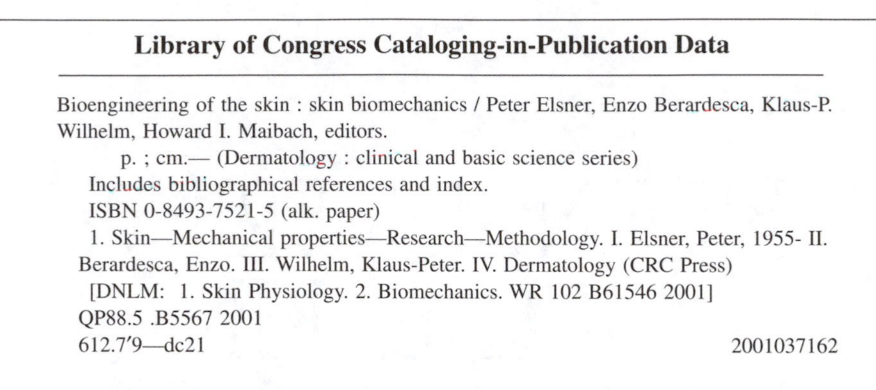

Library of Congress Cataloging-in-Publication Data

Bioengineering of the skin : skin biomechanics / Peter Elsner, Enzo Berardesca, Klaus-P. Wilhelm, Howard I. Maibach, editors.
p. ; cm.— (Dermatology : clinical and basic science series)
Includes bibliographical references and index.
ISBN 0-8493-7521-5 (alk. paper)
1. Skin—Mechanical properties—Research—Methodology. I. Elsner, Peter, 1955- II. Berardesca, Enzo. III. Wilhelm, Klaus-Peter. IV. Dermatology (CRC Press)
[DNLM: 1. Skin Physiology. 2. Biomechanics. WR 102 B61546 2001]
QP88.5 .B5567 2001
612.7′9—dc21 2001037162

Catalog record is available from the Library of Congress

Direct all inquiries to CRC Press LLC, 2000 N.W. Corporate Blvd., Boca Raton, Florida 33431.

No claim to original U.S. Government works
International Standard Book Number 0-8493-7521-5
Library of Congress Card Number 2001037162
Printed in the United States of America 1 2 3 4 5 6 7 8 9 0
Printed on acid-free paper

Series Preface

Our goal in creating the *Dermatology: Clinical & Basic Science Series* is to present the insights of experts on emerging applied and experimental techniques and theoretical concepts that are, or will be, at the vanguard of dermatology. These books cover new and exciting multidisciplinary areas of cutaneous research; and we want them to be the books every physician will use to become acquainted with new methodologies in skin research. These books can be given to graduate students and postdoctoral fellows when they are looking for guidance to start a new line of research.

The series consists of books that are edited by experts and that consist of chapters written by the leaders in a particular field. The books are richly illustrated and contain comprehensive bibliographies. Each chapter provides substantial background material relevant to the particular subject. These books contain detailed tricks of the trade and information regarding where the methods presented can be safely applied. In addition, information on where to buy equipment and helpful web sites for solving both practical and theoretical problems are included.

We are working with these goals in mind. As the books become available, the efforts put in by the publisher, the book editors, and the individual authors will contribute to the further development of dermatology research and clinical practice. The extent to which we achieve this goal will be determined by the utility of these books.

Howard I. Maibach, M.D.

Preface

The skin plays an important role in maintaining the integrity of the living organism while allowing the interaction of the organism with its environment. To fulfill these functions, mechanical stability is as important as flexibility. The mechanical properties of skin are very diverse depending on the anatomical location, and they evolve throughout life from the fetus to old age. Both genetic and acquired skin diseases modify skin biomechanics, as do intrinsic and photoaging. Since aging is so closely linked with changes of skin mechanical properties that lead to wrinkles and furrows, the desire for eternal youth leads to attempts to modify skin mechanics by a variety of interventions, including cosmeceuticals, peeling, and laser treatments.

It is within this wide scope of interests that this book gathers up-to-date information on the noninvasive assessment of skin biomechanics by modern bioengineering technology. The editors are grateful that leading investigators have shared their experiences in the development and use of standard and new techniques, their applications in dermatology, and in the testing of pharmaceutical, cosmetic, and nonfood products for safety and efficacy. The editors are indebted to all authors for the knowledge and effort they have invested in this project. At the same time, we would like to thank Ms. Barbara Norwitz and Ms. Tiffany Lane of CRC Press, Boca Raton, for their help in the publishing process.

We sincerely hope that this book will provide valuable advice to our readers and that it will stimulate them to apply bioengineering techniques skillfully in their professional settings.

Jena/Pavia/Hamburg/San Francisco, May 2001

Peter Elsner, M.D.

Enzo Berardesca, M.D.

Klaus-P. Wilhelm, M.D.

Howard I. Maibach, M.D.

The Editors

Peter Elsner, M.D., studied medicine at Julius Maximilians University, Würzburg, Germany, from 1974 to 1981 and was trained as a dermatologist and allergologist at the Department of Dermatology, Würzburg University, 1983 to 1987. He received his doctoral degree in 1981 and his lectureship in dermatology in 1987. From 1988 to 1989, he was visiting research dermatologist at the Department of Dermatology, University of California at San Francisco; and from 1991 to 1997, he was consultant and associate professor, Department of Dermatology, University of Zurich, Switzerland. Since 1997 he has served as professor and chairman, Department of Dermatology and Allergology, Friedrich Schiller University, Jena, Germany.

Dr. Elsner has published more than 200 original papers and 18 books. He is a member of more than 30 scientific societies; has served as chairman of the International Society for Bioengineering and the Skin (ISBS) and as a member of the Scientific Committee for Cosmetics and Non-Food Products (SCCNFP) of the European Commission and the European Group on Efficacy Measurement of Cosmetics and Other Topical Products (EEMCO).

Enzo Berardesca, M.D., is senior dermatologist and professor at the School of Dermatology of the University of Pavia in Pavia, Italy. Dr. Berardesca obtained his training at the University of Pavia and earned his M.D. in 1979. He served as resident and dermatologist at the Department of Dermatology, IRCCS Policlinico S. Matteo, Pavia, from 1982 to 1987, and as research assistant at the Department of Dermatology, University of California School of Medicine in San Francisco in 1987. He assumed his present position in 1988.

Dr. Berardesca has been chairman of the International Society for Bioengineering and the Skin from 1990 to 1996 and is a member of the Society for Investigative Dermatology, the European Society for Dermatological Research, the Italian Group for Research on Contact Dermatitis (GIRDCA), and the European Group for Standardization of Efficacy Measurements of Cosmetics (EEMCO group). He is currently vice chairman of the EEMCO group. He has organized several international meetings on skin bioengineering and irritant contact dermatitis in Europe.

Dr. Berardesca's current major research interests are irritant dermatitis, barrier function, and noninvasive techniques to investigate skin physiology (with particular regard to racial differences in skin function), sensitive skin, and efficacy evaluation of topical products.

He has authored five books and more than 200 papers and book chapters.

Klaus-P. Wilhelm, M.D., is president and medical director of proDERM Institute for Applied Dermatological Research, Schenefeld/Hamburg, Germany, and Lecturer of Dermatology at the Medical University of Lübeck, Germany.

Dr. Wilhelm earned his M.D. degree in 1986 from the Medical University of Lübeck and was awarded the degree of Lecturer by the same institution in 1995.

From 1988 to 1990, Dr. Wilhelm was a visiting scientist at the Department of Dermatology, University of California, San Francisco Medical School. He completed his residency at the Department of Dermatology, Medical University of Lübeck in 1993. In 1994 he founded the contract research institute proDERM in Schenefeld/Hamburg.

Dr. Wilhelm is a member of the Executive Board of the International Society for Bioengineering and the Skin and a member of the European Society for Dermatological Research, the European Contact Dermatitis Society, the German Dermatological Society, and the American Academy of Dermatology. He has received three consecutive government grants and has published more than 40 scientific papers and book chapters. His research interests include physiology of healthy and diseased skin, irritant contact dermatitis, skin pharmacology, and evaluation of bioinstrumentation techniques for the skin.

Howard Maibach, M.D., is a Professor of Dermatology at the University of California, San Francisco, and has been a leading contributor to experimental research in dermatopharmacology, and to clinical research on contact dermatitis, contact uticaria, and other skin conditions. His work on pesticides includes clinical research on glyphosate, chlorothalonil, sodium hypochlorite, norflurazon, diethyl toluamide, and isothiazolin compounds. His experimental work include research on the local lymph node assay, and the evaluation of the percutaneous absorption of atrazine, boron-containing pesticides, phenoxy herbicides, acetochlor, glyphosate, and many other compounds.

Contributors

J. Asserin
Laboratory of Engineering
and Cutaneous Biology
St. Jacques University Hospital
Besançon, France

André O. Barel
Laboratory of General
and Biochemical Chemistry
Faculty of Physical Education
and Physiotherapy
Free University of Brussels
Brussels, Belgium

Enzo Berardesca
Department of Dermatology
University of Pavia
Pavia, Italy

Undine Berndt
Department of Dermatology
Friedrich Schiller University
Jena, Germany

Jean de Rigal
Départment Evaluation
L'Oreal Recherche Appliquée et
Développement
Chevilly Larue, France

Hristo P. Dobrev
Department of Dermatology
Higher Medical Institute
Plovdiv, Bulgaria

C. Edwards
Department of Dermatology
Royal Gwent Hospital
Newport, Gwent
Wales

Peter Elsner
Department of Dermatology
Friedrich Schiller University
Jena, Germany

Vincent Falanga
Boston University
and
Department of Dermatology
and Skin Surgery
Roger Williams Medical Center
Providence, Rhode Island

Christopher J. Graves
University of Wales College of
Medicine
Cardiff, Wales

HansJörg G. Häuselmann
Center for Rheumatology
and Bone Disease
Zurich, Switzerland

T. Hermanns-Lê
Department of Dermatopathology
University Medical Center
Sart Tilman
Liège, Belgium

Karl Huber
Department of Rheumatology
and Physical Medicine
University Hospital
Zurich, Switzerland

Phillippe Humbert
Laboratory of Engineering
and Cutaneous Biology
St. Jacques University Hospital
Besançon, France

Gregor B. E. Jemec
Division of Dermatology
Department of Medicine
Roskilde Hospital
Roskilde, Denmark

Anina Knuutinen
Department of Dermatology
Oulu University Hospital and
University of Oulu
Oulu, Finland

Tina Holst Larsen
Division of Dermatology
Department of Medicine
Roskilde Hospital
Roskilde, Denmark

Howard I. Maibach
Department of Dermatology
University of California
San Francisco, California

Paul J. Matts
The Proctor & Gamble Company
Egham, Surrey
United Kingdom

Beat Michel
Department of Rheumatology
and Physical Medicine
University Hospital
Zurich, Switzerland

Aarne Oikarinen
Department of Dermatology
Oulu University Hospital
and
University of Oulu
Oulu, Finland

Gerald E. Piérard
Department of Dermatopathology
University Medical Center
Sart Tilman
Liège, Belgium

C. Piérard-Franchimont
Department of Dermatopathology
University Medical Center
Sart Tilman
Liège, Belgium

John R. Potts
Third Party Research
and Development
Cortland Manor, New York

Peter T. Pugliese
Consultant
Reading, Pennsylvania

Marco Romanelli
Department of Dermatology
University of Pisa School
of Medicine
Pisa, Italy

Claudia Rona
Department of Dermatology
University of Pavia
Pavia, Italy

Burkhart Seifert
Division of Biostatistics
Department of Social
and Preventive Medicine
University of Zurich
Zurich, Switzerland

Jørgen Serup
Department of Dermatology
University Hospital
Linköping, Sweden

D. Varchon
Applied Mechanics Laboratory
U.M.R./C.N.R.S.
Besançon, France

P. Vescovo
Applied Mechanics Laboratory
U.M.R./C.N.R.S.
Besançon, France

H. Gerhard Vogel
Standortslehre d. University
Ottobrunn, Germany

R. Randall Wickett
College of Pharmacy
University of Cincinnati Medical Center
Cincinnati, Ohio

Klaus-P. Wilhelm
proDERM Institute for Applied
Dermatological Research
Schenefeld, Germany
and
Department of Dermatology
Medical University of Lübeck
Lübeck, Germany

Wolf-Ingo Worret
Department of Dermatology
and Allergy
Technical University of Munich
Munich, Germany

H. Zahouani
Laboratory of Tribology
and Dynamic Systems
U.M.R. C.N.R.S.
Central School of Lyon
Ecully, France

Table of Contents

SECTION I Introduction

Chapter 1
Mechanical Properties of Human Skin: Biochemical Aspects 3
Aarne Oikarinen and Anina Knuutinen

Chapter 2
Mechanical Properties of Human Skin: Animal Models 17
H. Gerhard Vogel

Chapter 3
Mechanical Properties of Human Skin: Elasticity Parameters
and Their Relevance 41
Jørgen Serup

Chapter 4
Mechanical Properties of the Skin during Friction Assessment............................. 49
H. Zahouani, J. Asserin, and Phillippe Humbert

Section II Elasticity and Viscoelasticity

Part 1 General Aspects

Chapter 5
Hardware and Basic Principles of the Dermal Torque Meter 63
Jean de Rigal

Chapter 6
In Vivo Tensile Tests on Human Skin: The Extensometers 77
P. Vescovo, D. Varchon, and Phillippe Humbert

Chapter 7
Hardware and Measuring Principle: The Cutometer® 91
Undine Berndt and Peter Elsner

Chapter 8
Hardware and Measurement Principles: The Gas-Bearing Electrodynamometer and Linear Skin Rheometer 99
Paul J. Matts

Chapter 9
Hardware and Measuring Principles: The Dermaflex A 111
Jørgen Serup

Chapter 10
Hardware and Measuring Principles: The DermaLab 117
Jørgen Serup

Chapter 11
Hardware and Measuring Principles: The Dermagraph in Patients with Systemic Sclerosis and in Healthy Volunteers 123
HansJörg Häuselmann, Karl Huber, Burkhart Seifert, and Beat Michel

Chapter 12
Hardware and Measuring Principles: The Durometer 139
Marco Romanelli and Vincent Falanga

Chapter 13
Hardware and Measuring Principles: The Ballistometer 147
Peter T. Pugliese and John R. Potts

Chapter 14
Hardware and Measuring Principles: The Microindentometer 161
Christopher J. Graves and C. Edwards

Chapter 15
Standardization of Skin Biomechanical Measurements 179
R. Randall Wickett

Chapter 16
Mapping Mechanical Properties of Human Skin 187
Klaus-P. Wilhelm and Howard I. Maibach

Chapter 17
Skin Mechanics and Hydration 199
Tina Holst Larsen and Gregor B. E. Jemec

Chapter 18
Skin Tensile Strength in Scleroderma 207
Gerald E. Piérard, T. Hermanns-Lê, and C. Piérard-Franchimont

Chapter 19
Mechanical Properties in Other Dermatological Diseases 215
Hristo P. Dobrev

Part 2 Product Testing

Chapter 20
Skin Biomechanics: Antiaging Products 231
Claudia Rona and Enzo Berardesca

Chapter 21
Product Testing: Moisturizers 241
André O. Barel

Chapter 22
Antikeloidal Products 257
Wolf-Ingo Worret

Index 269

Section I

Introduction

1 Mechanical Properties of Human Skin: Biochemical Aspects

Aarne Oikarinen and Anina Knuutinen

CONTENTS

Introduction 3
Collagen 4
Elastin 8
Alterations in the Synthesis of Collagen and Elastin during Aging 10
Degradation of Collagen and Elastin 11
Function and Alteration of the Basement Membrane 12
Comments 13
References 13

INTRODUCTION

The mechanical properties of skin are due to the thickness and qualitative properties of epidermis, dermis, and subcutis. There are marked variations in these parameters in different parts of the body. During aging and in many diseases, qualitative and quantitative changes occur in epidermis and dermis. Since collagen and elastin are the major components of skin, this overview focuses on these proteins, emphasizing the synthesis, degradation, and genetic alterations that take place in them. Furthermore, certain physiological phenomena and diseases are illustrated that affect the quantity or quality of collagen and elastin and lead to alterations in the physical parameters and appearance of skin.

Human skin is composed of epidermal and dermal layers, each of which has its own functional importance. Epidermis consists mainly of keratinocytes and, to a lesser extent, melanocytes, Langerhans cells, Merkel cells, and unmyelinated axons. Dermis consists of eccrine and apocrine glands, hair follicles, veins, nerves, and a fine network of collagen fibers, elastic fibers, and other components of the extracellular matrix (ECM). ECM consists primarily of proteins and complex sugars, which form fibrillar networks and a ground substance. Collagen is an important structural component of skin connective tissue and provides the tensile strength of skin.

0-8493-7521-5/02/$0.00+$1.50

Approximately 70 to 80% of the dry weight of skin consists of collagen. The most abundant collagen types in skin are types I and III; the former accounts for 80% of the total collagen content of skin and the latter for approximately 15%.[1] The other collagen types present in skin include type IV collagen, which is abundant in the basement membrane (BM); type V collagen, which is located pericellularly; type VI collagen, which plays a role in matrix assembly and is present as microfibrils between collagen fibers; and type VII collagen, which is a structural component of anchoring fibrils.[2] Elastin accounts for only about 1 to 2% of the dry weight of skin but is important for the maintenance of skin elasticity and resilience. Glycosaminoglycans are of central importance for the maintenance of a water balance in skin, even though the quantities in ECM are small (0.1 to 0.3% of the dry weight of skin).[3,4] The BM of skin is a flexible sheetlike structure, which contains multiple different molecules.[5] Mutations in various BM components may cause variable clinical diseases, such as epidermolysis bullosa, in which the mechanical resistance of the BM is reduced.

In this overview, the synthesis and organization of collagen, elastin, and BM are elucidated. Furthermore, alterations in collagen and elastin are discussed in relation to changes in the mechanical characteristics of skin.

COLLAGEN

A characteristic feature of the collagen molecule is a triple-helical conformation of three α chains, which can be similar or dissimilar polypeptide chains. Type III collagen, for example, consists of three identical α1(III) chains encoded by a single gene, whereas type I collagen is composed of two identical α1(I) chains, which are synthesized from the same gene, and an α2(I) chain, which is synthesized from another gene.[6] Each of the polypeptide chains forms a leftward helix, and the three helical chains wrap around each other to form a right-handed superhelix, which is stabilized in the extracellular space by cross-linking between chains and molecules (Figure 1.1). The triple-helical conformation of a collagen molecule requires the presence of glycine as every third amino acid in the polypeptide chains, which results in a series of Gly-X-Y, where X and Y can be any amino acid except glycine. The other amino acids essential for the triple-helical structure are proline and 4-hydroxyproline. Formation of 4-hydroxyproline and C-terminal disulfide bonds is crucial for the formation of the triple helix. Lysine is an amino acid also commonly found in the Y position, and it serves as a site for sugar attachment when converted into hydroxylysine by a specific enzyme.[1,7,8]

Skin collagen synthesis takes place mainly in fibroblasts. The synthesis of collagen has an intracellular and an extracellular phase, both of which involve post-translational modifications crucial for the formation of stable triple-helical collagen molecules, with appropriate cross-links (Figure 1.2). Intracellular modifications include hydroxylation of proline residues in the Y position into 4-hydroxyproline and of some proline residues in the X position into 3-hydroxyproline as well as hydroxylation of lysine residues in the Y position into hydroxylysine.[7] The reactions are catalyzed by specific enzymes, prolyl-4-hydroxylase, prolyl-3-hydroxylase, and lysyl hydroxylase, respectively, in the presence of Fe^{2+}, oxygen, 2-oxoglutarate, and

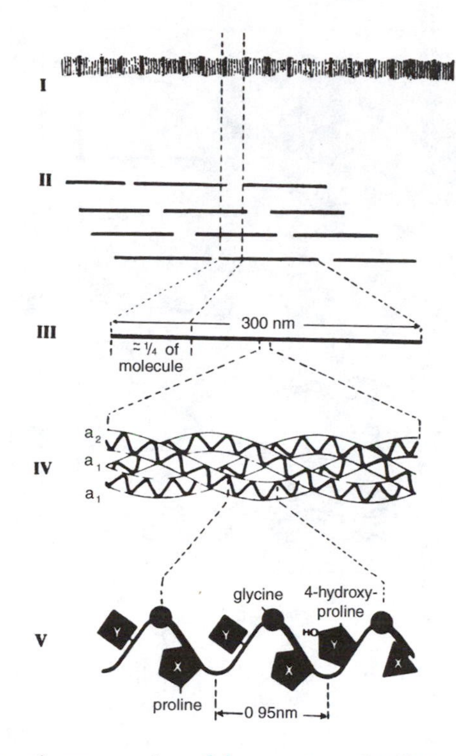

FIGURE 1.1 Schematic presentation of the structure of collagen. (I) The collagen fibers in tissues demonstrate repetitive periodicity when examined by electron microscopy. (II) The fibers consist of individual collagen molecules aligned in a quarter-stagger arrangement. (III) Each collagen molecule is approximately 300 nm long. (IV) The collagen molecules consist of three individual polypeptides, α-chains, which are twisted around each other in a right-handed, triple-helical conformation. (V) Each α-chain has a primary sequence of amino acids in a repetitive X-Y-Gly sequence. As indicated, the X position is frequently occupied by a prolyl residue and the Y position by a 4-hydroxyproline residue. The individual α-chains have a left-handed helical secondary structure with a pitch of 0.95 nm. (Modified from Prockop, D.J. and Guzman, N.A., Collagen diseases and the biosynthesis of collagen, *Hosp. Pract.*, 12, 61–68, 1977.)

ascorbate. Ascorbate is essential for the biosynthesis of collagen and acts as a cofactor in the hydroxylation of proline and lysine.[9]

Glycosylation of hydroxylysine and asparagine residues also takes place intracellularly. Both hydroxylation and glycosylation continue until the triple-helical conformation of the developing molecule is achieved. The procollagen molecules synthesized intracellularly are excreted into the extracellular space, where the large aminoterminal and carboxyterminal propeptides of the procollagens are cleaved *en block* by specific endoproteinases.[10] This cleavage of propeptides enables the initiation of fibril formation.[11] The molecular weights of the aminoterminal propeptides of type I and III procollagens (PINP and PIIINP) are 35,000 and 45,000, respectively.[10] Since procollagens and mature collagens are synthesized in a ratio of 1:1, the amount of procollagen propeptides in serum and interstitial fluid reflects

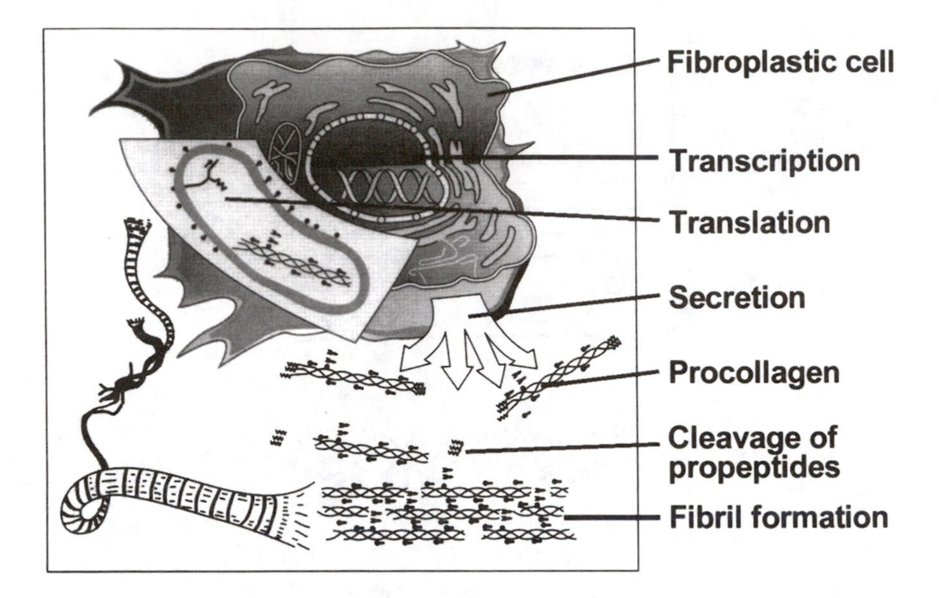

FIGURE 1.2 The intracellular and extracellular steps of the synthesis of fibrillar collagen.

the rate of ongoing collagen synthesis.[1,12,13] In adult human skin, the ratio of type I to type III collagen is approximately 5:1 to 6:1,[1] but there may be a tendency toward an increased relative amount of type III collagen in the skin of elderly individuals.[14]

In living tissues, the existing collagen fibers gradually undergo chemical reactions that lead to the formation of covalent bonds between adjacent polypeptide chains, which make the fibers less soluble and more resistant to proteolytic enzymes. The first step in this reaction sequence is enzymatic, involving oxidation of the ε-amino groups of lysine or hydroxylysine residues by the lysyl oxidase enzyme, which results in the formation of aldehydes derived from the corresponding amino acids. Two such aldehydes may, consequently, react with one another or one aldehyde may bind to another ε-amino group. Either way, cross-links connecting two polypeptide chains, i.e., bivalent cross-links, are formed. The number of bivalent cross-links peaks at some point, after which their number begins to decline, as they develop into more-complicated structures connecting three or more polypeptide chains.[15]

Diseases with disturbed collagen metabolism include acquired diseases, such as scleroderma and scleredema, in which accumulation of collagen leads to thickening and stiffening of skin,[16,17] diabetic thick skin, presenting as thickening of skin, and keloids composed of excessive amounts of collagen (Table 1.1). In scleroderma and scleredema, increased synthesis of collagen results in thickening of skin.[17,18] In diabetic thick skin, nonenzymatic glycosylation of collagen is the most likely cause of the changes observed in skin.[19] A reduced amount of collagen can be found in skin atrophy, which may be a result of normal aging, or may be induced by topical or systemic glucocorticoids, or may be caused by genetic factors, such as focal dermal hypoplasia. In steroid-induced skin atrophy, the reduced amounts of collagen mRNA and the consequently reduced synthesis of collagen induce thinning of skin, as shown in Figure 1.3.[20]

TABLE 1.1
Mechanical Properties of Skin in Various Genetic and Acquired Diseases Resulting from Changes in Collagen

Disease	Characteristics of Skin	Basic Biochemical Etiology
Steroid-induced atrophy	Thinning of skin	Reduced synthesis of type I and III collagens
Age-related atrophy	Thinning of skin	Reduced synthesis of type I and III collagens
Striae	Bluish /livid red lesions of variable size and shape	Not known, reduced collagen synthesis in steroid-induced striae
Focal dermal hypoplasia	Thinning of the dermis	Not known
Ehlers–Danlos syndrome (ED); includes at least ten different subtypes	Hyperextensible skin Thinning of skin Fragility of skin	Mutations in types I and V collagen, genes in type I and II ED, mutations in type III collagen gene in type IV ED, mutations in lysyl hydroxylase gene in type VI ED, defect in conversions of procollagen to collagen in type VII ED, defect in lysyl oxidase in type IX ED
Osteogenesis imperfecta	Thin, fragile skin	In most cases, mutations in type I collagen
Scleroderma	Thickening and stiffening of skin	Generally increased deposition of collagen
Keloids	Tumorlike thickening of skin	Increased deposition of collagen
Diabetic thick skin	Thickening and tautness of skin	Increase in nonenzymatic glycosylation of collagen

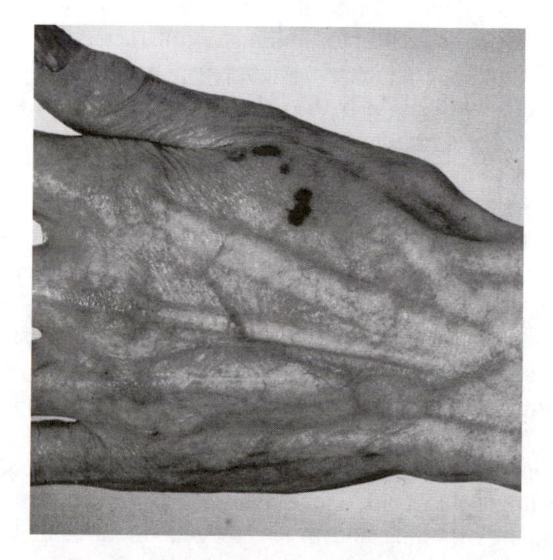

FIGURE 1.3 Skin atrophy after topical glucocorticoid treatment. Skin thickness was 0.65 mm in a steroid-treated dorsum of the hand, whereas skin thickness in age-matched controls was 1.3 mm.

Changes in collagen are also found in various hereditary conditions. These include osteogenesis imperfecta, which involves changes in tissues rich in type I collagen, such as bones, ligaments, and skin, and Ehlers–Danlos (ED) syndrome, which has a wide variety of clinical manifestations, depending on the underlying defect in collagen metabolism.[7,21,22] Several gene defects associated with collagen-related diseases have been elucidated. For example, defects in the genes encoding the proα1(I) or proα2(I) chain of type I procollagen are commonly found in osteogenesis imperfecta, mutations in types I and V collagen genes have been found in ED types I and II,[22] and mutations in type III procollagen occurs in ED type IV.[6,23] The clinical picture in ED can vary from hyperextensible skin, as illustrated in Figure 1.4, due to abnormal fibrillogenesis of collagen, to thinning of the skin, as in ED type IV (Figure 1.5). This patient presented with a markedly reduced synthesis rate of type III collagen in the skin.[24]

ELASTIN

Elastic fibers are composed of an amorphous material, elastin, which accounts for 90% of the mature fibers, and of a microfibrillar component, which consists of microfibrils, 10 to 12 nm in size, primarily located around elastin, but partly also interspersed within it.[25] Microfibrils contain several glycoproteins; of these, fibrillin has been studied in most detail.[26] Elastic fibers are assembled in dermis as a three-dimensional net. Oxytalan fibers occur perpendicular to epidermis and are connected to elaunin fibers, which run parallel to epidermis.[25]

Elastin synthesis takes place in embryonic and rapidly growing tissues and in cells derived from these. Elastin is a polypeptide approximately 70 kDa in size, which is encoded by a single copy gene found in chromosome 7.[27,28] The elastin gene encodes tropoelastin, a precursor protein for elastin. Tropoelastin is synthesized intracellularly and then excreted into the extracellular space, where cross-linking takes place.[26] A high degree of cross-linking is characteristic of elastin, and the formation of desmosines is unique to it. A copper-dependent enzyme, lysyl oxidase, is involved in the cross-linking of both collagen and elastin.[29] In the cross-links of elastin, the lysine residues present as pairs in polyalanine sequences in such a way that there are always two or three amino acids, usually alanines, between two lysine residues, thus forming sequences of Lys-Ala-Ala-Lys or Lys-Ala-Ala-Ala-Lys. These alanine-rich cross-linking domains have an α-helical conformation. In addition to the cross-linking domains, elastin has hydrophobic domains containing glycine, proline, and valine residues. The mechanisms of elastic fiber assembly are not well known, but microfibrils become visible first, after which elastin appears as an amorphous material that then coalesces and forms the core of the fiber. Most microfibrils are transferred to the outer aspect of the fiber, where they remain in mature tissue.[26,27] Abnormalities in elastic fiber morphology and assembly are seen in a number of congenital skin diseases, and specific gene defects behind genodermatosis have recently been found. Cutis laxa is a skin disease that presents in mild cases as predominant wrinkling and in severe genetic cases as widespread elastic fiber damage in skin and internal organs[30] (Table 1.2). Disturbed elastin cross-linking, due to defects in the copper metabolism and/or function of lysyl oxidase, has been suggested to cause X-linked cutis laxa.[28] A defect in the fibrillin1 gene is found in Marfan syndrome.[27,31] In pseudoxanthoma elasticum, abnormal elastin fibrillogenesis occurs by an unknown cause and results in a lax and wrinkled appearance of the skin. Anetoderma involves local degradation of elastic fibers, causing sacklike protrusions.[32] In elastoderma, conversely, local accumulation of abnormal elastic fibers leads to delayed recoil and elasticity of the skin (Figure 1.6).[33]

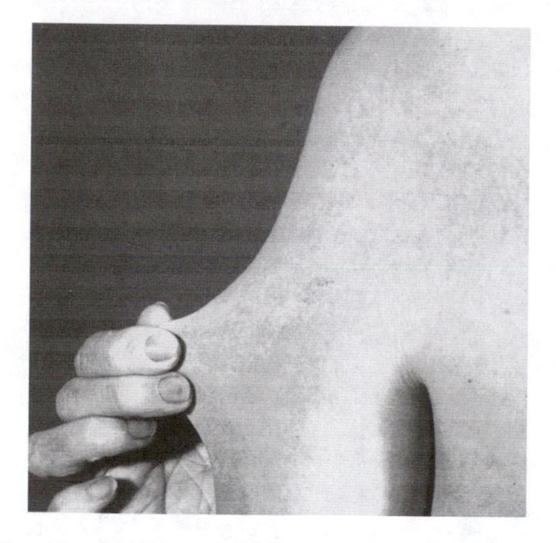

FIGURE 1.4 The skin of a patient with ED type I is hyperextensible.

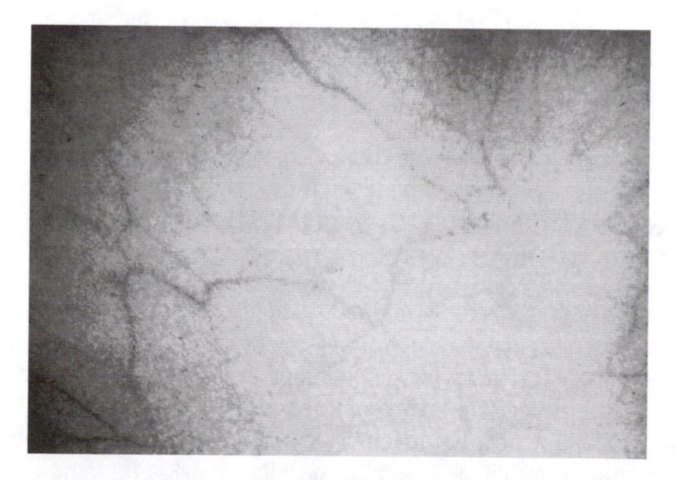

FIGURE 1.5 The skin of a 20-year-old male with ED type IV is translucent with readily visible blood vessels. The concentration of type III collagen propeptide (PIIINP) was 32.5 μg/L in the suction blisters of the patient, whereas the mean value in the controls was 106 μg/L, indicating markedly reduced synthesis of type III collagen in the patient's skin fibroblasts. Skin thickness was markedly reduced: 0.82 mm in the forearm of the patient and 1.49 mm in the controls.

TABLE 1.2
Skin Properties in Diseases Affecting Elastin

Disease	Characteristic of the Skin	Biochemical Alterations
Cutis laxa	Loose, sagging skin	Decreased amount of elastin
Anetoderma	Sacklike lesions	Local degradation of elastic fibers
Pseudoxanthoma elasticum	Lax and wrinkled skin	Accumulation of abnormal elastin
Actinic elastosis	Diffuse thickening and wrinkling of sun-exposed skin	Accumulation of abnormal elastin in dermis; elevated levels of MMPs; decrease in collagen synthesis, increase in elastin synthesis
Marfan syndrome	Hyperextensible skin, striae	Mutations in fibrillin gene

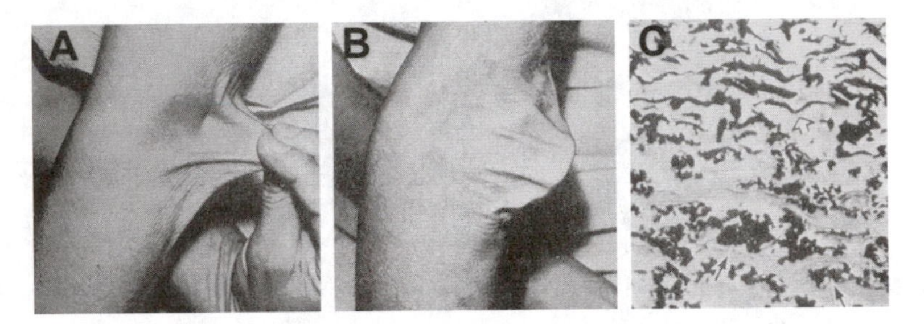

FIGURE 1.6 The right arm of a patient with elastoderma, demonstrating the laxity (A) and incomplete and delayed recoil (B) of skin. Histopathology revealed only a few normal-appearing fibers (the white arrow) and abnormal elastic structures (the black arrows) in the lower dermis (C) (Verhoeff–van Gieson stain). Modified from Kornberg, R.L. et al., Elastoderma—disease of elastin accumulation within the skin, *New Engl. J. Med.*, 312, 771–774, 1985. With permission.

ALTERATIONS IN THE SYNTHESIS OF COLLAGEN AND ELASTIN DURING AGING

Along with increasing age, skin wrinkling gradually becomes evident, especially in sun-exposed areas, such as the face. Several distinct histological features have been observed within wrinkles, including reduction of oxytalan fibers in the dermis under wrinkles, profound collagen atrophy, and decreased amounts of type IV and VII collagens at the dermoepidermal junction as well as decreased amounts of dermal chondroitine sulfates, which are essential for balanced skin hydration.[34]

Skin collagen synthesis declines with aging and as the result of such external factors as long-term sun exposure and medications, for example, D-penicillamine and topical corticosteroids.[35–37] In aging skin, collagen fibers become thicker and less soluble and the synthesis of collagen declines.[38] Skin thickness remains quite constant between 10 and 70 years of age, after which a marked decrease in skin thickness occurs.[39] Precursors of both type I and III collagens also decrease in photodamaged skin, and the degree of reduction in collagen production correlates with the amount of photodamage.[40]

The elastic properties of skin are also affected by aging. Along with increasing age, dermal elastic fibers become thicker and fragmented and oxytalan fibers appear fragmented and shortened.[41] Disintegration of elastic fibers is already seen in a minority of fibers between ages 30 and 70, but the changes become more profound after the age of 70 years, affecting a majority of the fibers.[25] As a result of the decreased number of elastic fibers in aged skin, the elastic recovery of skin decreases in elderly people.[4] Flattening of the dermo-epidermal junction is seen in both sun-exposed and sun-protected skin in elderly people.[42] Epidermal thickness declines with age in sun-protected areas, whereas sun-exposed regions develop an irregular epidermis with both thickened and atrophic regions.[43] A distinct feature of photoaged skin is a decrease in the ultrasound echogenicity of the upper dermis, which causes a subepidermal low-echogenic band.[44–46]

The ultraviolet (UV) radiation reaching the earth surface consists of UVA (320 to 400 nm) and UVB (280 to 320 nm) radiation. Shortwave UVC does not pass through the atmosphere.[47] UVA penetrates deep into tissues and has direct effects on dermal cells, including fibroblasts. UVB, on the other hand, has indirect effects on the ECM turnover by inducing the production of certain lymphokines and cytokines.[13,48] In actinic elastosis, the number of abnormal elastic fibers increases in the dermis, and the amount of collagen is reduced. *In vitro* studies have shown that the life spans of dermal fibroblasts and keratinocytes are shorter than normal in sun-exposed skin specimens.[49,50] It has also been demonstrated that elastin mRNA levels are elevated in photoaged skin, indicating transcriptional upregulation of the gene that codes elastin.[51] Reactive oxygen species activated by UV radiation are thought to play an important role in UV-induced DNA damage, cellular senescence, and aging.[47] Upon aging, the capacity to repair DNA decreases, thus increasing the risk of malignant transformations.[52]

DEGRADATION OF COLLAGEN AND ELASTIN

Three major families of proteases degrade components of the extracellular matrix. These protease families are called serine, cysteine, and metalloproteinases, and they are important in the wound healing process and in tumor invasion and metastasis.[53] Matrix metalloproteinases (MMPs) and tissue inhibitors of matrix metalloproteinases (TIMPs) regulate the degradation of collagen, elastin, and other components of ECM.[54] It has been suggested that matrix metalloproteinases could have a crucial role in the degradation of collagen in actinic elastosis, since UV radiation has been shown to rapidly induce MMPs in skin and cell cultures. MMP-1, MMP-8, and MMP-13 (collagenases 1, 2, and 3) are the principal MMPs capable of initiating the degradation of fibrillar collagens I, II, III, and V. MMP-2 and MMP-9 are important in the final degradation of fibrillar collagens. MMP-2, MMP-3, MMP-7, MMP-9, MMP-10, and MMP-12 are capable of degrading elastin.[54,55] MMP-1 degrades type III collagen at a faster rate than types I and II, whereas MMP-8 degrades type I collagen at a rate much faster than type III and, unlike MMP-1, is also important in the cleavage of type II collagen, which is abundant in cartilage.[56] The expression of collagenase (MMP-1), 92-kDa gelatinase (MMP-9), and stromelysin has been shown to increase after a single exposure to UV radiation, and the activity of MMP-1 and MMP-9 remained maximally elevated for 7 days after four UV exposures at 2-day intervals. Topical tretinoin inhibited UV-induced MMP activity but did not counteract the induction of TIMPs.[57] Abundant evidence supports the use of topical tretinoin in the treatment of early signs of actinic damage.[37,48,57–60]

FUNCTION AND ALTERATION OF THE BASEMENT MEMBRANE

Epithelial or endothelial cells and mesenchymal connective tissue are separated by a basement membrane (BM) or basal lamina, which is a flexible sheetlike structure approximately 50 to 100 nm thick.[5] Structurally, the BM consists of two layers: *lamina lucida,* which is adjacent to the basal plasma membrane of epithelial cells, and *lamina densa*, which is just below the *lamina lucida*. In skin, BM separates the epidermis from the dermis and forms a dermal–epidermal junction (DEJ). Depending on the type of tissue, the BM determines cell polarity, influences cell metabolism, organizes the proteins in adjacent plasma membranes, induces cell differentiation, and guides cell migration. The major components of BM are type IV collagen, a large heparin sulfate proteoglycan perlecan, and the glycoproteins laminin-1 and nidogen/entactin. Laminins are large flexible cross-shaped glycoproteins composed of three polypeptide chains. They bind to type IV collagen, heparan sulfate, nidogen, and cell surface laminin receptor proteins. Type IV collagen forms a network that is important in the maintenance of the mechanical stability of BM. The binding of another independent network, formed by laminin-1 on this network, is stabilized by nidogen-1 and nidogen-2. BM acts as a selective barrier to the movements of cells. BM beneath an epithelial layer prevents the fibroblasts in the underlying connective tissue from contacting the epithelial cells. The epithelial cells are linked to BM by integrins. In addition, the DEJ contains plaquelike hemidesmosomes at the surface of epithelial cells. Hemidesmosomes contain plectin, bullous pemphigoid antigen 1 (BPAG1), collagen XVII, and integrin $\alpha6\beta4$. They link the keratin cytoskeleton to laminin-5 in *lamina lucida*. Laminin-5 is linked to type VII collagen in *lamina densa*. Type VII collagen forms anchoring fibrils that firmly bind to the underlying connective tissue.[61,62]

There are several diseases, mostly inherited, in the epidermolysis bullosa (EB) disease group that affect the BM zone.[63–65] EB can be divided into four types, in which blister formation occurs at different levels. In EB simplex, the blistering is due to mutations in the keratins 5 and 14. As a consequence, keratinocytes are easily detached from each other, and blistering occurs within the epidermis.[66] In EB types where blistering occurs within the hemidesmosomes or between the basement membrane and the cell membranes, mutations may occur in type XVII collagen, plectin, or $\alpha6$ or $\beta4$ integrins. In junctional EB, the blistering takes place within *lamina lucida*, and mutations have been found in genes that code laminin 5. In dystrophic EB, the blistering takes place under BM and is due to mutations in type VII collagen.

COMMENTS

Collagen and elastin are important for the structural integrity of skin. Thus, alterations in the quantity or quality of these proteins may cause changes in the mechanical properties of skin. Diseases involving a reduced collagen content of skin are characterized by atrophic skin, readily visible blood vessels, and easy bruising, which may result in paperlike scars. In contrast, if the quantity of collagen increases, skin becomes thick and taut and skin elasticity is limited, if not completely nonexistent. There are various genetic diseases in which collagen genes or enzymes participating in collagen biosynthesis are mutated. As a result, a wide range of diseases affecting skin and blood vessels may develop. Skin may, for example, be fragile, thin, or hyperextensible. Similarly, changes in the quantity or quality of elastin cause changes in the elastic properties of skin. If elastic fibers are fragmented or reduced in quantity, skin looks old and sags. If the quantity of elastin is increased, as in pseudoxanthoma elasticum or solar damage, skin may be thickened and inelastic.

There are several methods available to elucidate various aspects of collagen and elastin. Skin biopsies are useful in the assessment of collagen quantity, different types of collagen, and the rate of collagen synthesis. Histology, immunohistochemistry, and electron microscopy (EM) can be used to investigate changes in collagen fibers. Elastin can be studied by histological analysis and EM. Gene defects are best characterized by white blood cells, from which specific gene deletions and other mutations can be analyzed.

Similarly, the integrity of the BM zone can be studied by histological, immunohistochemical, and EM methods. If one wants to look at specific mutations of the BM components, there are several methods available to characterize the mutations in most of the proteins contributing to the integrity of BM.

REFERENCES

1. Uitto, J., Olsen, D.R., and Fazio, M.J., Extracellular matrix of the skin: 50 years of progress, *J. Invest. Dermatol.*, 92 (Suppl. 4), 61–77, 1989.
2. Van den Rest, M. and Brückner, P., Collagen: diversity of the molecular and supramolecular levels, *Curr. Opin. Struct. Biol.*, 3, 430–436, 1993.
3. Oikarinen, A., Aging of the skin connective tissue: how to measure the biochemical and mechanical properties of aging dermis, *Photodermatol. Photoimmunol. Photomed.*, 10, 47–52, 1994.
4. Bernstein, E.F. and Uitto, J., The effect of photodamage on dermal extracellular matrix, *Clin. Dermatol.*, 14, 143–151, 1996.
5. Timpl, R. and Brown, J.C., Supramolecular assembly of basement membranes, *Bioessays*, 18, 123–132, 1996.
6. Prockop, D.J., Mutations in collagen genes as a cause of connective-tissue diseases, *New Engl. J. Med.*, 326, 540–546, 1992.
7. Prockop, D.J. and Kivirikko, K.I., Heritable diseases of collagen, *New Engl. J. Med.*, 311, 376–386, 1984.

8. Burgeson, R.E. and Morris, N.P., The collagen family of proteins, in *Connective Tissue Disease. Molecular Pathology of the Extracellular Matrix*, Uitto, J. and Perejda, A.J., Eds., Marcel Dekker, New York, 1987, 3–28.
9. Kivirikko, K.I. and Myllylä, R., Posttranslational enzymes in the biosynthesis of collagen: intracellular enzymes, in *Methods in Enzymology*, Cunningham, L.W. and Frederiksen, D.W., Eds., Academic Press, New York, 1982, 245–249.
10. Risteli, J., Niemi, S., Kauppila, S., Melkko, J., and Risteli, L., Collagen propeptides as indicators of collagen assembly, *Acta. Orthop. Scand.*, 66 (Suppl. 266), 183–188, 1995.
11. Prockop, D.J. and Fertala, A., The collagen fibril: the almost crystalline structure, *J. Struct. Biol.*, 122, 111–118, 1998.
12. Autio, P. and Oikarinen, A., Suction blister techniques for measurement of human skin collagen synthesis, *Skin Res. Technol.*, 3, 88–94, 1997.
13. Oikarinen, A., Basic aspects of collagen metabolism, in *Skin Cancer and UV Radiation*, Altmeyer, P., Hoffman, K., and Stucker, M., Eds., Springer-Verlag, Berlin, 1997, 77–93.
14. Lovell, C.R., Smolenski, K.A., Duance, V.C., Light, N.D., Young, S., and Dyson, M., Type I and III collagen content and fibre distribution in normal human skin during ageing, *Br. J. Dermatol.*, 117, 419–428, 1987.
15. Last, J.A., Armstrong, L.G., and Reiser, K.M., Biosynthesis of collagen crosslinks, *Int. J. Biochem.*, 22, 559–564, 1990.
16. Rodnan, G.P., Lipinski, E., and Luksick, J., Skin thickness and collagen content in progressive systemic sclerosis and localized scleroderma, *Arthritis Rheum.*, 22, 130–140, 1979.
17. Uitto, J., Bauer, E.A., and Eisen, A.Z., Scleroderma: Increased biosynthesis of triple-helical type I and type III procollagens associated with unaltered expression of collagenase by skin fibroblasts in culture, *J. Clin. Invest.*, 64, 921–930, 1979.
18. Oikarinen, A. et al., Sclerederma and paraproteinemia: enhanced collagen production and elevated type I procollagen messenger RNA level in fibroblasts grown from cultures from the fibrotic skin of a patient, *Arch. Dermatol.*, 123, 221–229, 1987.
19. Salmela, P.I., Oikarinen, A., Pirttiaho, H., Knip, M., Niemi M., and Ryhänen, L., Increased non-enzymatic glycosylation and reduced solubility of skin collagen in insulin-dependent diabetic patients, *Diabetes Res.*, 11, 115–120, 1989.
20. Oikarinen, A., Haapasaari, K.-M., Sutinen, M., and Tasanen, K., The molecular basis of glucocorticoid-induced skin atrophy: topical glucocorticoid apparently decreases both collagen synthesis and the corresponding collagen mRNA level in human skin *in vivo*, *Br. J. Dermatol.*, 139, 1106–1110, 1998.
21. Prockop, D.J., Kivirikko, K.I., Tuderman, L., and Guzman, N.A., The biosynthesis of collagen and its disorders, *New Engl. J. Med.*, 301, 77–85, 1979.
22. Myllylä, R. and Kivirikko, K.I., Collagen and collagen-related diseases, *Ann. Med.*, 33, 7–21, 2001.
23. Byers, P.H., Ehlers-Danlos syndrome: recent advances and current understanding of the clinical and genetic heterogeneity, *J. Invest. Dermatol.*, 103, 475–525, 1994.
24. Autio, P., Turpeinen, M., Risteli, J., Kallioinen, M., Kiistala, U., and Oikarinen, A., Ehlers-Danlos type IV: non-invasive techniques as diagnostic support, *Br. J. Dermatol.*, 137, 653–655, 1997.
25. Braverman, I.M. and Fonferko, E., Studies in cutaneous aging: I. The elastic fiber network, *J. Invest. Dermatol.*, 78, 434–443, 1982.
26. Rosenbloom, J., Abrams, W.R., and Mecham, R., Extracellular matrix 4: the elastic fiber, *FASEB J.*, 7, 1208–1218, 1993.

27. Christiano, A.M. and Uitto, J., Molecular pathology of the elastic fibers, *J. Invest. Dermatol.*, 103, (Suppl. 5), 53–57, 1994.
28. Debelle, L. and Tamburro, A.M., Elastin: molecular description and function, *Int. J. Biochem. Cell. Biol.*, 31, 261–272, 1999.
29. Davidson, J.M., Elastin structure and biology, in *Connective Tissue Disease. Molecular Pathology of the Extracellular Matrix*, Uitto, J. and Perejda, A.J., Eds., Marcel Dekker, New York, 1987, 29–54.
30. Hashimoto, K. and Kanzaki, T., Cutis laxa: ultrastructural and biochemical studies, *Arch. Dermatol.*, 111, 861–873, 1975.
31. Godfrey, M., From fluorescence to the gene: the skin in the Marfan syndrome, *J. Invest. Dermatol.*, 103, (Suppl. 5), 58–62, 1994.
32. Oikarinen, A.I., Palatsi, R., Adomian, G.E., Oikarinen, H., Clark, J.G., and Uitto, J., Anetoderma: biochemical and ultrastructural demonstration of an elastin defect in the skin of three patients, *J. Am. Acad. Dermatol.*, 11, 64–72, 1984.
33. Kornberg, R.L., Hendler, S.S., Oikarinen, A.I., Matsuoka, L.Y., and Uitto, J., Elastoderma — disease of elastin accumulation within the skin, *New Engl. J. Med.*, 312, 771–774, 1985.
34. Contet-Audonneau, J.L., Jeanmaire, C., and Pauly, G., A histological study of human wrinkle structures: comparison between sun-exposed areas of the face, with or without wrinkles, and sun-protected areas, *Br. J. Dermatol.*, 140, 1038–1047, 1999.
35. Oikarinen, A., Dermal connective tissue modulated by pharmacologic agents, *Int. J. Dermatol.*, 31, 149–156, 1992.
36. Autio, P., Risteli, J., Haukipuro, K., Risteli, L., and Oikarinen, A., Collagen synthesis in human skin *in vivo*: modulation by aging, ultraviolet B irradiation and localization, *Photodermatol. Photoimmunol. Photomed.*, 11, 1–5, 1994.
37. Kang, S., Fisher, G.J., and Voorhees, J.J., Photoaging and topical tretinoin. Therapy, pathogenesis, and prevention, *Arch. Dermatol.*, 133, 1280–1284, 1997.
38. Fenske, N.A. and Lober, C.W., Structural and functional changes of normal aging skin, *J. Am. Acad. Dermatol.*, 15, 571–585, 1986.
39. Escoffier, C., de Rigal, J., Rochefort, A., Vasselet, R., Lévêque, J.-L., and Agache, P.G., Age-related mechanical properties of human skin: an *in vivo* study, *J. Invest. Dermatol.*, 93, 353–357, 1989.
40. Talwar, H.S., Griffiths, C.E.M., Fisher, G.J., Hamilton, T.A., and Voorhees, J.J., Reduced type I and type III procollagens in photodamaged adult human skin, *J. Invest. Dermatol.*, 105, 285–290, 1995.
41. Gogly, B., Godeau, G., Gilbert, S., Legrand, J.M., Kut, C., Pellat, B., and Goldberg, M., Morphometric analysis of collagen and elastic fibers in normal skin and gingiva in relation to age, *Clin. Oral. Invest.*, 1, 147–152, 1997.
42. Lavker, R.M., Structural alterations in exposed and unexposed aged skin, *J. Invest. Dermatol.*, 73, 59–66, 1979.
43. Marks, R. and Edwards, C., The measurement of photodamage, *Br. J. Dermatol.*, 127 (Suppl. 41), 7–13, 1992.
44. De Rigal, J., Escoffier, C., Querleux, B., Faivre, B., Agache, P., and Lévêque, J.-L., Assessment of aging of the human skin by *in vivo* ultrasonic imaging, *J. Invest. Dermatol.*, 93, 621–625, 1989.
45. Gniadecka, M. and Jemec, G.B.E., Quantitative evaluation of chronological ageing and photoageing *in vivo*: studies on skin echogenicity and thickness, *Br. J. Dermatol.*, 139, 815–821, 1998.
46. Pellacani, G. and Seidenari, S., Variations in facial skin thickness and echogenicity with site and age, *Acta Derm. Venereol.* (Stockholm), 79, 366–369, 1999.

47. Mariéthoz, E., Richard, M.-J., Polla, L.L., Kreps, S.E., Dall'Ava, J., and Polla, B.S., Oxidant/antioxidant imbalance in skin aging: environmental and adaptive factors, *Rev. Environ. Health*, 13, 147–168, 1998.
48. Kligman, L.H., The ultraviolet-irradiated hairless mouse: a model for photoaging, *J. Am. Acad. Dermatol.*, 21, 623–631, 1989.
49. Gilchrest, B.A., Szabo, G., Flynn, E., and Goldwyn, R.M., Chronologic and actinically induced aging in human facial skin, *J. Invest. Dermatol.*, 80 (Suppl.), 81–85, 1983.
50. Gilchrest, B.A. and Yaar, M., Ageing and photoageing of the skin: observations at the cellular and molecular level, *Br. J. Dermatol.*, 127 (Suppl. 41), 25–30, 1992.
51. Bernstein, E.F., Chen, Y.Q., Tamai, K., Shepley, K.J., Resnik, K.S., Zhang, H., Tuan, R., Mauviel, A., and Uitto, J., Enhanced elastin and fibrillin gene expression in chronically photodamaged skin, *J. Invest. Dermatol.*, 103, 182–186, 1994.
52. Grossman, D. and Leffell, D.J., The molecular basis of nonmelanoma skin cancer, *Arch. Dermatol.*, 133, 1263–1270, 1997.
53. Kähäri, V.-M. and Saarialho-Kere, U., Matrix metalloproteinases in skin, *Exp. Dermatol.*, 6, 199–213, 1997.
54. Mauch, C., Regulation of connective tissue turnover by cell-matrix interactions, *Arch. Dermatol. Res.*, 290 (Suppl.), 30–36, 1998.
55. Kähäri, V.-M. and Saarialho-Kere, U., Matrix metalloproteinases and their inhibitors in tumour growth and invasion, *Ann. Med.*, 31, 34–45, 1999.
56. Jeffrey, J.J., Interstitial collagenases, in *Matrix Metalloproteinases*, Parks, W.C. and Mecham, R.P., Eds., Academic Press, New York, 1998, 15–42.
57. Fisher, G.J., Wang, Z.Q., Datta, S.C., Varani, J., Kang, S., and Voorhees, J.J., Pathophysiology of premature skin aging induced by ultraviolet light, *New Engl. J. Med.*, 337, 1419–1428, 1997.
58. Griffiths, C.E.M., Russman, A.N., Majmudar, G., Singer, R.S., Hamilton, T.A., and Voorhees, J.J., Restoration of collagen formation in photodamaged human skin by tretinoin (retinoic acid), *New Engl. J. Med.*, 329, 530–535, 1993.
59. Uitto, J., Understanding premature skin aging. (Editorial), *New Engl. J. Med.*, 337, 1463–1465, 1997.
60. Kang, S. and Voorhees, J.J., Photoaging therapy with topical tretinoin: an evidence-based analysis, *J. Am. Acad. Dermatol.*, 39, (Suppl. 2), 55–61, 1998.
61. Christiano, A.M. and Uitto, J., Molecular complexity of the cutaneous basement membrane zone. Revelations from the paradigms of epidermolysis bullosa, *Exp. Dermatol.*, 5, 1–11, 1996.
62. Burgeson, R.E. and Christiano, A.M., The dermal-epidermal junction, *Curr. Opin. Cell Biol.*, 9, 654–658, 1997.
63. Eady, R.A.J. and Dunnill, M.G.S., Epidermolysis bullosa: hereditary skin fragility diseases as paradigms in cell biology, *Arch. Dermatol. Res.*, 287, 2–9, 1994.
64. Uitto, J., Pulkkinen, L., and McLean, W.H.I., Epidermolysis bullosa: a spectrum of clinical phenotypes explained by molecular heterogeneity, *Mol. Med. Today*, 3, 457–465, 1997.
65. Pulkkinen, L. and Uitto, J., Mutation analysis and molecular genetics of epidermolysis bullosa, *Matrix Biol.*, 18, 29–42, 1999.
66. Lorden, L.D. and McLean, W.H.I., Human keratin diseases: hereditary fragility of specific epithelial tissues, *Exp. Dermatol.*, 5, 297–307, 1996.
67. Prockop, D.J. and Guzman, N.A., Collagen diseases and the biosynthesis of collagen, *Hosp. Pract.*, 12, 61–68, 1977.

2 Mechanical Properties of Human Skin: Animal Models

H. Gerhard Vogel

CONTENTS

Introduction 17
Studies *in Vitro* (*ex Vivo*) 18
  Preparation of Samples 18
  Stress–Strain Curves *in Vitro* 18
  Anisotropy of Skin 21
  Hysteresis Experiments 21
  Relaxation Experiments 22
  Isorheological Behavior 23
  Repeated Strain 24
  Creep Experiments 24
  Studies in Skin Wounds 25
  Thermocontraction 26
Studies *in Vivo* 27
  Stress–Strain Curves *in Vivo* 27
  Repeated Strain *in Vivo* 28
  In Vivo Recovery after Repeated Strain 29
Conclusions 30
References 31

INTRODUCTION

A significant question is whether animal models are suitable to predict mechanical properties in human skin. Can the results of animal studies be extrapolated to human skin, and are the basic approaches the same? Several attempts have been performed to describe the mechanical properties of skin by mathematical models (Ridge and Wright, 1964, 1965, 1966; Harkness, 1968, 1971; Hirsch and Sonnerup, 1968; Jamison et al., 1968; Viidik, 1968, 1969, 1973a,b, 1978, 1979; Frisén et al., 1969a,b; Veronda and Westman, 1970; Danielson, 1973; Soong and Huang, 1973; Wilkes

0-8493-7521-5/02/$0.00+$1.50

et al., 1973; Jenkins and Little, 1974; Lanir and Fung, 1974; Vogel, 1976, 1986; Barbanel and Evans, 1977; Barbanel et al., 1978; Lanir, 1979; Barbanel and Payne, 1981; Burlin, 1980, 1981; Fung, 1981; Sanjeevi, 1982; Potts and Breuer, 1983). Most of these authors used models derived from studies in polymers (Ferry, 1970). The simplest mechanical model analogous to a viscoelastic system is a spring combined with a dashpot, either in series (Maxwell element) or in parallel (Voigt or Kelvin element). Combinations of these elements were used to explain the mechanical phenomena in connective tissue, such as stress-strain behavior, relaxation and mechanical recovery, hysteresis, and creep phenomena (Jamison et al., 1968; Frisén et al., 1969a,b; Hirsch and Sonnerup, 1968; Vogel, 1976a, 1993a,b; Riedl and Nemetscheck, 1977; Vogel and Hilgner, 1979a; Viidik, 1968, 1969, 1973, 1978, 1979). Larrabee (1986), Larrabee and Sutton (1986), and Larrabee and Galt (1986) reviewed the theoretical and experimental mechanics of skin and soft tissue and proposed a mathematical model of skin deformation based on the finite-element method. A finite-element-based method to determine the properties of planar soft tissue was also described by Flynn et al. (1998). Unfortunately none of these models has been found sufficient to describe all properties of human and animal skin including the mechanical history before measurement and the time dependence during measurement. There is no comprehensive and unequivocally accepted model to describe completely the biorheology of skin.

Therefore, it is necessary to use several methods providing insight into clearly defined physical properties of skin.

STUDIES *IN VITRO* (*EX VIVO*)

Preparation of Samples

Subject animals, mostly rats, are sacrificed in anesthesia. The back skin is shaved and a flap of 5 × 5 cm removed. Subcutaneous fat is removed from the skin flaps and the sample placed between two pieces of plastic material with known thickness. In this way, skin thickness can be measured reliably with calipers to an accuracy of 0.1 mm. Perpendicular to the body axis two dumbbell-shaped specimens with a width of 4 mm in the middle of the sample are punched out (Vogel, 1969, 1970, 1989, 1993a). The samples are kept at room temperature on filter paper soaked with saline solution in petri dishes until testing. The specimens are fixed between the clamps of an INSTRON® instrument at a gauge length of 3 cm. All measurements are carried out within at least 1 hr. For long-lasting test procedures, such as relaxation or cyclic loading, the samples are wrapped with saline-soaked filter paper (Vogel, 1976a,b, 1989, 1993a,b).

Stress–Strain Curves *in Vitro*

After fixation of the specimens between the clamps of an INSTRON® instrument allowing a gauge length of 30 mm, stress–strain curves are registered at an extension rate of 5 cm/min; the curves show a characteristic shape (Figure 2.1). During low strain values, with a gradual increase of load, the curve has a concave section. The stress–strain curve ascends according to an exponential function (Vogel and Hilgner, 1977). Afterward an almost straight section is reached indicating dependence on Hook's law. At this section, the *ultimate modulus of elasticity* (Young's modulus) can be calculated (increase of load divided by the cross-sectional area). Then, the curve yields somewhat, ending in a sudden break of the specimen. At this point, *ultimate strain* and *ultimate load* can be measured. From ultimate load divided by the cross-sectional area (specimen width times original skin thickness measured at the beginning of the experiment), *tensile strength* can be calculated.

The parameters skin thickness, ultimate strain, ultimate load, tensile strength, and modulus of elasticity are influenced by many factors.

Several studies on *age dependence* of mechanical parameters in rat skin were performed at age 2 weeks and 1, 4, 12, and 24 months. The period up to 4 months can be regarded as maturation, and the period between 12 and 24 months as senescence. Skin thickness increases to a maximum at 12 months of age and decreases thereafter. The ultimate strain values follow another shape. A small increase is found during puberty, a maximum after 4 months, and a decrease thereafter. Ultimate load, tensile strength, and modulus of elasticity show a very sharp increase during puberty, a maximum at 12 months, and a slight decrease thereafter (Vogel et al., 1970; Vogel, 1976b, 1978a, 1983a, 1988a, 1989, 1993a). Similar changes were found in human skin (Holzmann et al., 1971; Vogel, 1987a,b).

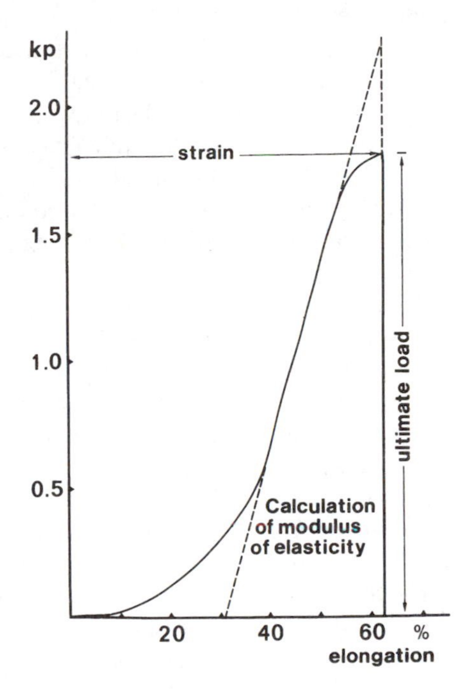

FIGURE 2.1 Original registration of a stress–strain experiment.

These parameters are also influenced by hormones and *desmotropic drugs*. The most-pronounced changes are seen after *glucocorticosteroids*, following both systemic and local application. Clinical experience unequivocally shows a decrease of skin thickness after long-term systemic or local treatment with corticosteroids. In rats, repeated administration of cortisol acetate in doses between 10 and 100 mg/kg cortisol acetate or 2 and 20 mg/kg prednisolone acetate induces a sharp dose-dependent increase of ultimate load and tensile strength up to 5 days. A continuous decrease of skin thickness and a decrease of ultimate load is noted up to 2 months. Ultimate load falls below controls, but tensile strength does not (Vogel, 1969, 1970a, 1974). From the increase of tensile strength after short-term administration, dose–response curves can be established for various corticosteroids vs. standard cortisol, allowing the calculation of potency ratios with confidence limits. The test can be used for evaluation of topical corticosteroids (Schröder et al., 1974; Alpermann et al., 1982; Vogel and Petri, 1985). Similarly, Töpert et al. (1990) measured skin atrophy and tensile strength of skin after 30 days of local administration of corticosteroids in rats. Van den Hoven et al. (1991) and Woodbury and Kligman (1992) recommended the hairless mouse model for assaying the atrophogenicity of topical corticosteroids. Oxlund and Manthorpe (1982) found after long-term glucocorticoid treatment of rats an increase of strength and a decrease of extensibility of skin strips. In agreement with studies by Vogel (1974b, 1978a), these results were explained by a change in collagen cross-linking pattern.

Jørgensen et al. (1989) found a dose-dependent increase of mechanical strength in intact rat skin after treatment with biosynthetic human growth hormone. Andreassen et al. (1981) studied the biomechanical properties of skin in rats with streptozotocin-induced diabetes. An increased stiffness and strength were found: maximal stiffness was increased by 20% and the strain rate at maximum stress was decreased by 10%. Oxlund et al. (1980) found that the stiffness of rat skin was increased in the early postpartum period. This increase was also found in adrenalectomized animals. *Other hormones*, such as ACTH, thyroid hormones, androgens, estrogens, and gestagens, influence the mechanical parameters of rat skin to a lesser degree.

Among desmotropic drugs, *lathyrogenic* compounds, such as amino-acetonitrile, and D-penicillamine decrease ultimate load and tensile strength without major influence on skin thickness (Vogel, 1971a, 1972a, 1974a). A decrease of the strength of skin strips in rats after treatment with D-penicillamine was also found by Oxlund et al. (1984). A dose-dependent increase of tensile strength in skin after treatment was found with *nonsteroidal anti-inflammatory drugs* (Vogel, 1977a). Strain rate influences the values of ultimate load, tensile strength, and modulus of elasticity, but not the effects of age and corticosteroids (Vogel, 1972b). Changes in tensile strength are correlated with the content of insoluble collagen (Vogel, 1974b).

The mechanical parameters skin thickness, ultimate strain, ultimate load, tensile strength, and modulus of elasticity were used to evaluate the effects of *ultraviolet* (UV-*irradiation*) and the prevention of skin damage in hairless mice (Alpermann and Vogel, 1978; Vogel et al., 1981).

Fry et al. (1964) prepared skin rings from the lower part of the leg in rats and studied the age dependence of the mechanical properties. Nimni et al. (1966) measured tensile strength of excised skin samples in rabbits during aging. Pan et al. (1998) studied ultrasound and viscoelastic and mechanical properties in rabbit skin, including stress relaxation, creep, and Young's modulus as a function of strain. Lofstrom et al. (1973) described circadian variations of tensile strength in the skin of two inbred strains of mice. The effect of radiation therapy on mechanical properties of skin was studied in mice by Hutton et al. (1977) and by Spittle et al. (1980).

Schneider et al. (1988) measured tensiometric properties in guinea pig skin from flaps of normal dorsal skin and after implantation of an ovoid *tissue expander* filled for 4 days with saline. Belkoff et al. (1995) studied the mechanical properties of skin in pigs after subcutaneous implantation and inflation of silicone tissue expanders. Mustoe et al. (1989) compared the effects of a conventional tissue expansion regimen of 6 weeks with the effects of an accelerated regimen of 2 weeks in a model in dogs; the study measured skin thickness, elasticity, creep, and stress relaxation.

Anisotropy of Skin

As is the case with human skin indicated by Langer's lines (Langer, 1861; Gibson et al., 1969; Wright, 1971; Stark et al., 1977; Daly, 1982), the skin of rats exhibits *directional differences* (Hussein, 1972, 1973; Vogel and Hilgner, 1979a,b; Vogel, 1981a, 1983a, 1985a,b, 1988b; Belkoff and Haut, 1991). Stress–strain curves of rat skin displayed a different shape when excised perpendicularly vs. longitudinally to the body axis. At low loads, extension is higher in the longitudinal than in the perpendicular direction. The situation is reversed at medium and high stress values. Ultimate extension shows remarkable differences between longitudinal and perpendicular samples. Specimens obtained perpendicular to the body axis showed an increase during maturation, a maximum at 4 months of age, and a decrease during further aging. The behavior of samples obtained longitudinal to the body axis was quite different. After an initial rise, a maximum was found at 3 weeks. Afterwards a slight decrease was noted. Between 1 and 4 weeks all values of ultimate extension were significantly higher in samples punched out longitudinally to the body axis; between 4 and 12 months they were considerably lower (Vogel, 1981a). Analysis of the low sections of the stress–strain curve revealed a *"step"-phenomenon* (Vogel and Hilgner, 1977, 1979a,b; Vogel, 1988b). If samples obtained perpendicular to the body axis are extended, a gradual increase of load is observed at low degrees of extension, which is suddenly interrupted by a decrease of the registered curve. Then the curve increases again, and is interrupted by a second or third step. The phenomenon can be explained by the different orientation of collagen fibers in the dermis and by the presence of a muscular layer in rat skin. The muscular fibers are in a longitudinal direction to the body axis. If samples are obtained perpendicular to the body axis, the muscle bundles are cut transversally. In further studies the muscle layer was removed in one specimen and compared with a control with the muscle layer intact. Investigation of the directional variation showed that the step phenomenon is mainly due to the muscular layer oriented longitudinal to the body axis and the connective tissue between the muscle bundles, whereas the anisotropic behavior of extensibility and ultimate strain is caused by the directional variation of the collagenous bundles in the dermis.

Directional variations of the stress–strain curves were also described in the skin of tight-skin mutant mice (Menton et al., 1978).

Hysteresis Experiments

In hysteresis experiments not only the elastic but also the viscous properties of skin are measured (Vogel, 1978b, 1983b). Using the INSTRON instrument, the samples are stretched to a given extension degree (e.g., 20%) with an extension rate of 20 mm/min. When the given extension is achieved, the crosshead is immediately returned to the starting position with the same velocity. From the upward curve the stress and the modulus of elasticity at the end of the loading phase at the given strain can be measured. When the sample is unstretched, the unloading curve shows a different pattern, reaching the baseline much earlier than the curve left it during the upward phase. From this point, the *residual extension* can be measured. Planimetry of the area below the upward curve, and of the area between the hysteresis loop, allows the calculation of the *energy input* and the *energy dissipation*, respectively, as well as the *ratio between energy dissipation and energy input* at each hysteresis cycle (Figure 2.2). Immediately after the first hysteresis cycle, the experiment is repeated to an extension degree of 30%, then to 40 and 50%, and finally to 60%. In rat skin, a maximum of the ratio between energy dissipation and energy input was found at 30 to 40% extension. This ratio is influenced by age. At low extension degrees, there is an increase with age, whereas at high extension degrees an age-dependent decrease is noted. The ratio of dissipation to input was slightly decreased by prednisolone, but definitively increased by D-penicillamine (Vogel, 1993b).

Relaxation Experiments

In the relaxation experiment, viscous properties of skin are measured (Vogel, 1973, 1976a, 1983b, 1985a, 1993b). Skin strips were fastened between the clamps of an INSTRON instrument and extended with the high strain rate of 1000 mm/min to 20% extension. This extension was kept constant for 5 min. The chart speed was initially 1000 mm/min, then 10 mm/min. In this way, the initial tension and the stress values at 0.001, 0.01, 0.1, 1, and 5 min could be measured. As a result of relaxation, the stress values drop roughly with the logarithm of time. For each sample, the relaxation was calculated according to the formula:

$$\sigma(t) = A_1 + A_2 \times \log t$$

resulting in two constants (A_1 and A_2) for each sample, where A_1 is the stress at $t = 0$ and A_2 the slope of the relaxation curve.

The ratio between the constants A_1 and A_2 must be considered the most characteristic parameter of the relaxation experiment. Furthermore, the *residual stress* after the 5-min relaxation period is measured and calculated as a percentage of the original stress. After 5 min the sample is returned to 90% of the original strain, for example, from 20 to 18%. The stress following such unloading is recorded and again calculated as percentage of the original stress. Immediately after unloading, the measured stress values rise again spontaneously, which is called *mechanical recovery* (Figure 2.3). Mechanical recovery is calculated as a percentage of initial tension and as a percentage of stress after unloading. The relaxation experiment is repeated with increasing degrees of extension of 40, 60, 80%, and eventually 100% until the specimen breaks.

In studies of age dependence in rat skin, a definitive decrease of the ratio of A_2 to A_1 is found at 40 and 60% extension degrees, indicating a decrease of plasticity with age. Mechanical recovery, as an indicator of secondary elasticity, at medium extension degrees is better in old animals than in young individuals. Stress relaxation is decreased after corticosteroids and increased after thyroid hormones and D-penicillamine (Vogel, 1973, 1993a,b).

Purslow et al. (1998) suggested that relaxation processes within the collagen fibers or at the fiber–matrix interface may be responsible for the viscoelastic behavior of skin.

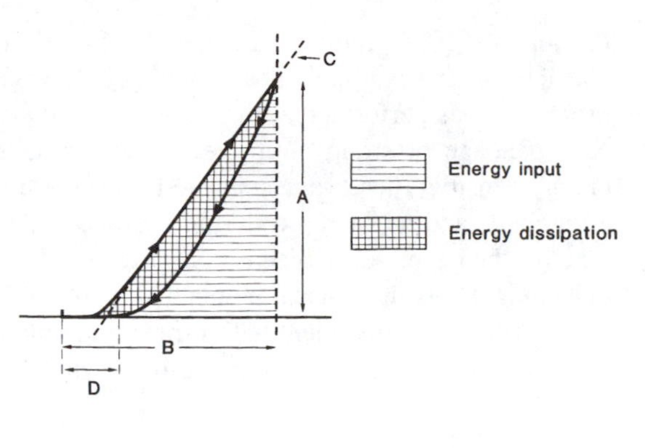

FIGURE 2.2 Original curve of the hysteresis experiment. A, Stress at the end of the loading phase; B, strain (%); C, line for calculation of the modulus of elasticity; D, residual extension.

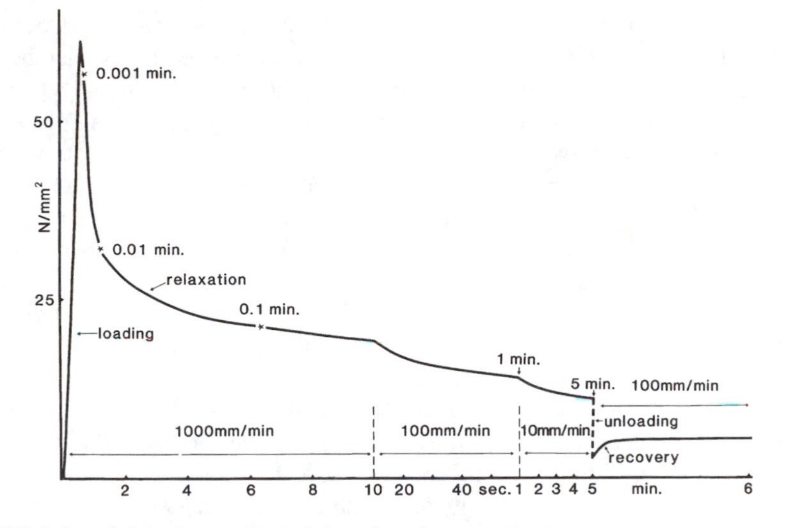

FIGURE 2.3 Original recording of the relaxation experiment.

Isorheological Behavior

Buss et al. (1976) demonstrated that the isorheological point is a valuable parameter for the mechanics of connective tissue. This method has been modified and elaborated for the skin strips of rats (Vogel, 1984, 1985b, 1987c). After fastening between the clamps of an INSTRON instrument, the specimen was expanded rapidly to 2 N and the corresponding strain measured. Keeping this strain constant, the load decay (relaxation) was measured for 5 min. Then the sample was again loaded to 2 N and a second relaxation period of 5 min was evaluated. In the third cycle, the sample was unloaded to 50% of the load observed after the 5-min relaxation period in the second cycle. The phenomenon of mechanical recovery was observed. With the crosshead driven up and down, the point was sought where neither immediate relaxation nor mechanical recovery could be observed. The load and strain at this point define the *isorheological point*, which is the point, under isometric conditions, that the measured load is constant for several minutes. Increasing and decreasing the load by 10% produced a sawtooth-shaped curve from which the modulus of elasticity at the isorheological point could be calculated (Figure 2.4). The same procedure was performed at higher initial loads such as 10 and 50 N. The product of percentage of strain multiplied by stress at the isorheological point indicates energy density. These values showed a decrease during maturation, a minimum at 12 months, and an increase during senescence. In studies with desmotropic compounds, the decreased viscosity after treatment with prednisolone acetate was more evident at the isorheological points than at the ultimate values. Similarly, the higher extensibility after treatment with D-penicillamine was indicated more clearly by the isorheological points than by the ultimate strain.

REPEATED STRAIN

The method using repeated strain also measures mainly the viscous properties of skin (Vogel and Hilgner, 1978; Vogel, 1987b). Skin specimens are fastened between the clamps of an INSTRON instrument, extended with a strain rate of 100 mm/min up to 20% extension, and immediately unloaded; this procedure is followed by further cycles with the same strain rate and extension degree. The peak of the second cycle is considerably lower than the first, followed by a further decrease in the next cycles. The number of cycles are counted until the stress value is only one half that of the first cycle. Immediately afterward, the degree of extension is increased to 30%. Again, the number of cycles is counted until the stress value is only one half that of the first cycle. In this way, the *number of cycles indicating the half-life of tension* due to relaxation is counted at each step of 20, 30, 40, 50, 60, 70, 80, 90%, and eventually 100% extension. The number of cycles decreases from the first step (20%) to the third step (40%) and increases continuously until the last step. This increase is almost an exponential function of the number of steps. The parameters measured with this method are influenced by several factors. An increase is noted from an age of 1 month to an age of 24 months. The values are decreased by D-penicillamine and increased by prednisolone treatment. Again, this method showed that plasticity of skin is decreased by age and by corticosteroids and increased by D-penicillamine.

CREEP EXPERIMENTS

In creep experiments, viscous behavior of skin is studied under constant load (Vogel, 1977b, 1987b). In a special apparatus, skin specimens are suddenly loaded with 100, 200, or 500 g and the extension degree measured. An immediate extension occurs, which is followed by a slow and almost continuous creep (Figure 2.5) that is measured as the *ultimate extension rate*. In addition, the extension achieved after 1 h is registered. An age-dependent decrease of these parameters was found in rat and human specimens indicating a decrease of viscosity or plasticity with maturation and age. Ultimate extension rate was decreased by prednisolone and increased by D-penicillamine.

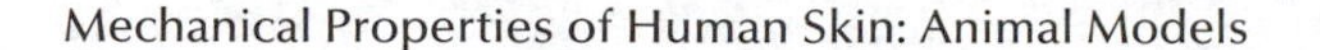

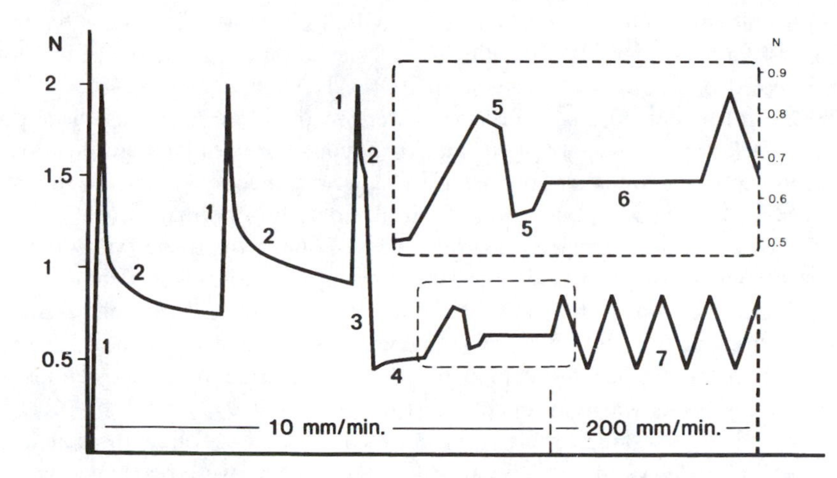

FIGURE 2.4 Repeated relaxation and determination of the isorheological point. 1, Loading to 2 N; 2, relaxation; 3, unloading; 4, mechanical recovery; 5, determination of the isorheological point; 7, sawtooth curve.

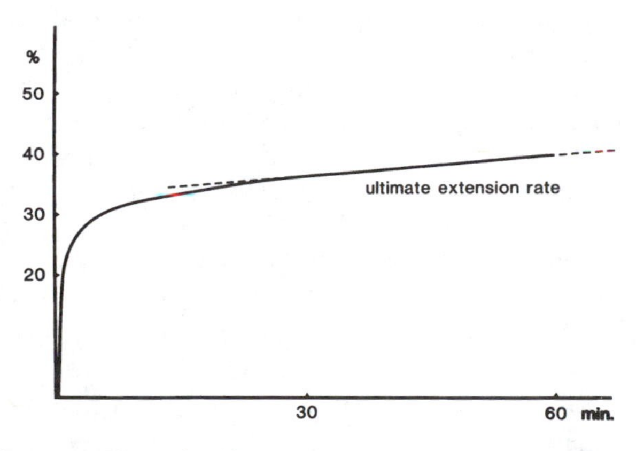

FIGURE 2.5 Creep experiment in skin.

Studies in Skin Wounds

The healing of skin wounds is a multiphasic process. The effect of drugs on the healing process was studied by measuring the mechanical strength at various time intervals after incision of the skin (Vogel, 1970b). Male Sprague–Dawley rats weighing 120 ± 5 g were anesthetized with ether. After shaving of the dorsal skin, an approximately 3-cm-long incision was made down to the fascia in a craniocaudal direction in the dorsolumbar region. Immediately afterward, the wound was closed with wound clips. The rats were treated subcutaneously with test drugs beginning with the day of surgery. The clips were removed the day before the tensile strength was tested or at the latest at day 10 postoperation.

For the measurement of wound tensile strength, the rats were sacrificed on days 3, 6, 9, or 12 after surgery. Wound clips were fastened on each side of the incision and connected by means of threads with the load cell and the crosshead of an INSTRON instrument. Stress–strain curves were recorded at an extension rate of 5 cm/min. *Dehiscence of the wound* resulted in a sudden drop of the registered load. For experiments of 3-week duration, *skin strips* were punched and tensile strength was tested as described for evaluation of tensile strength in normal skin.

Changes of tensile strength of skin wounds following treatment with drugs is expressed as a percentage of vehicle-treated controls. The dose-dependent decrease found after immediate postoperative treatment and the dose-dependent increase found after treatment in prolonged experiments with corticosteroids serve as comparative parameters for new test compounds. A biphasic effect of corticosteroids on wound healing in rats was also found by Oxlund et al. (1979).

Many authors measured tensile strength of skin wound to follow the course of wound healing under various conditions. Most experiments were performed in rats (Holm-Pedersen and Zederfeldt, 1971; Holm-Pedersen and Viidik, 1972a,b; Andreassen et al., 1977; Greenwald et al., 1993; Seyer-Hansen et al., 1993; Jyung et al., 1994; Adamson et al., 1996; Quirinia and Viidik, 1998; Canturk et al., 1999), some of them also using the INSTRON instrument (Phillips et al., 1993; Maxwell et al., 1998; Jimenez and Rampy, 1999; Kim and Pomeranz, 1999). In addition, mice (Butler et al., 1991; Celebi et al., 1994; Kashyap et al., 1995; Vegesna et al., 1995; Gonul et al., 1998; Matsuda et al., 1998), guinea pigs (Bernstein et al., 1991; Drucker et al., 1998; Silverstein and Landsman, 1999), rabbits (Sandblom, 1957; Wu and Mustoe, 1995; Pandit et al., 1999), dogs (Howes et al., 1929; Scardino et al., 1999), or pigs (Langrana et al., 1982; Higashiyama et al., 1992; Chang et al., 1998; Fung et al., 1999) were used. In some studies polyvinyl alcohol sponges were simultaneously implanted to determine collagen accumulation (Langrana et al., 1983; Albina et al., 1993; Schaffer et al., 1996; Koshizuka et al., 1997; Bitar, 1998; DaCosta et al., 1998; Witte et al., 1998).

Thermocontraction

If collagenous material such as skin or tendon is heated in a water bath, thermocontraction occurs. This phenomenon was described decades ago by Wöhlisch (Wöhlisch and du Mesnil de Rochemont, 1927; Wöhlisch, 1932). Verzár and other authors have used the phenomenon of thermocontraction of tendons and skin strips extensively to study the aging process (Verzár ,1955, 1956, 1957, 1964; Lerch, 1951; Rasmussen et al., 1964; Boros-Farkas and Everitt, 1967; Viidik 1969, 1973b, 1975, 1978, 1979a,b; Vogel, 1969). With increasing temperature, a sudden increase of isometric force is found, which is followed by a decrease. To evaluate this phenomenon, either the *shrinkage temperature* or the stress at and above the shrinkage temperature can be measured. The decrease of sample strength can also be measured.

Joseph and Bose (1962) tested the shrinkage temperature of skin in newborn rats, at 10 months and 2 years and found an increase from 51.2 to 61.1°C. Elden (1970) found that the melting temperature in samples of rat tail skin did not consistently change with age. However, the zero stress length of melted collagen fibers at 60°C relative to zero length at 50°C was greatly reduced with maturation and age. When isometric elevation of force at 60°C relative to that at 50°C was plotted against body weight, an increase was found. A decrease of percentage contraction of skin collagen melted at 60°C relative to that melted at 50°C was found, depending on age. Rupture stress of melted collagen fibers increased continuously with age.

Allain et al. (1977) tested pieces of dorsal skin of rats. The temperature of maximum tension was decreased from birth to 1 month, and then very slowly increased with age. A rapid relaxation was observed in young rats and in nonsenescent adult rats. Rundgren (1976) found changes of thermal contractility in skin of young and old female rats, as well as changes due to repeated pregnancies. Blackett and Hall (1980) found an increase of thermal shrinkage temperature in two strains of mice during the aging period.

Danielsen (1981) determined thermal stability measured as area shrinkage without tension during heating for membranes of collagen fibrils reconstituted from solutions of highly purified rat skin collagen. Le Lous et al. (1982a) and Flandin et al. (1984) applied the technique of differential scanning calorimetry to evaluate the denaturation process of collagen in rat skin.

Allain et al. (1980) built a device that measures not only hydrothermal shrinking but also swelling in rat skin. Le Lous et al. (1982b, 1983) studied hydrothermal isometric tension curves from rat and human skin and from different connective tissues. The curves obtained during a linear rise in temperature from 37 to 100°C at a rate of 1.15°C/min were classified into three major families, A, B, and C, depending on whether these curves show an early maximum, two shoulders, or a late maximum.

STUDIES *IN VIVO*

STRESS–STRAIN CURVES *IN VIVO*

In contrast to studies in humans, animal experiments allow measurement of mechanical properties both *in vivo* in anesthesia and later on *in vitro (ex vivo)* at the same site. This purpose required the development of special methods (Barbanel and Payne, 1981; Vogel, 1981b, 1982; Vogel and Denkel, 1982, 1985a; Denkel, 1983). The rats were anesthetized with 60 mg/kg Nembutal® i.p. The back skin was shaved mechanically. Skin thickness was measured by use of calipers on an elevated skin fold. Four small (14 × 14 mm) metal plates bearing a hook were used as tabs and were glued on the skin in both the longitudinal and perpendicular directions at a distance of 25 mm with a cyanoacrylate preparation. An operation table was mounted on the crosshead of an INSTRON instrument and a triangle was attached to the load cell. At the end of the triangle, threads were fastened that were conducted by reels and hooked to the tabs. The crosshead was moved down manually until the threads were stretched; however, no tension was yet measured. Then the crosshead was driven downward with a rate of 50 cm/min, which means that the actual extension rate was 100 cm/min. The load was measured only to limited values to prevent damage to the skin. In each rat the stress–strain curve was recorded in both directions, where the order (first longitudinal or first perpendicular) was changed from animal to animal. In this way, the influence of the first extension on the results of the second extension should be eliminated.

In spite of similar conditions, differences between the stress–strain curves from *in vitro* and *in vivo* experiments were found. These findings confirmed the statements of other authors (Barbanel and Payne, 1981; Wijn, 1980) regarding the importance of tab distance and tab geometry (Vogel, 1981c). Analysis of the data showed that the higher the ratio of distance between the tabs to the area below the tabs, the higher were the stress values. These findings can be explained: although skin consists of several layers, only the upper layer (epidermis) is fastened to the tabs in the *in vivo* experiment. The forces transmitted to the lower layers are transmitted by a larger area if the area under the tabs is larger. The lower layer can slide over a larger area and is therefore less extended, resulting in lower stress values. No sliding is possible if the sample is fastened from both sides as it is in the *in vitro* experiments. By taking all experimental conditions including the strain rate into account, *the* in vivo *results are comparable with those obtained* in vitro. This holds true for the age dependence as well as for the influence of desmotropic compounds (Vogel and Denkel, 1985a).

Cook et al. (1977) compared tension/extension ratio curves *in vivo* and *in vitro* in rats using the suction-cup method. Baker et al. (1988) described an apparatus for testing mechanical properties of normal and irradiated pig skin *in vivo.* In their system, the pads were attached to the skin of the pig rump with double-sided tape. They were moved apart at a predetermined rate using a motorized unit. Force was assessed using an S-shaped, center-point, double-beam load cell mounted on a movable crosshead. Displacement of the pads was measured using a floating-core linear variable displacement transducer. Using this system, Baker et al. (1989) studied the effect of single doses of X rays on the mechanical properties of pig skin *in vivo*.

Repeated Strain *in Vivo*

A special method has been developed to study the *mechanical properties of rat skin after repeated strain* in vivo and the course of recovery during different time intervals (Denkel, 1983; Vogel and Denkel, 1985; Vogel, 1988c). Tabs were fastened on the shaved back skin of anesthetized rats with a distance of 25 mm either perpendicular or longitudinal to the body axis. With an extension rate of 100 mm/min, the skin was extended 30 times for a 50% strain under anesthesia. Load was recorded and stress calculated by dividing of load by skin thickness measured from a skin fold obtained with calipers. Cycle 1, 5, 10, 20, 25, and 30 were recorded with more rapid paper speed to facilitate the evaluation of modulus of elasticity and stress values. The modulus of elasticity was calculated from the upper section of the stress–strain curve. Stress values decreased after repeated loading approximately with the logarithm of the number of cycles. The area under the curve of 30 cycles was evaluated by computerized calculation according to Simpson's formula (*Hütte*, 1915; Denkel, 1983; Vogel, 1988c). Stress values and the area under the curve were higher perpendicular than longitudinal to the body axis, as found in other *in vivo* experiments (Vogel and Denkel, 1985). The age dependence of the stress values was different for the direction vs. body axis. In longitudinal samples the stress values fell from 1 to 12 months and increased until 24 months. In perpendicular samples a decrease of stress values was found followed by an increase at 12 months and a decrease until 24 months. During the experiment, the modulus of elasticity increased in both directions from the first to the fifth cycle. This may be explained by the so-called conditioning of connective tissue (Nemetscheck, 1980). From cycle 5 to 30, a decay of the modulus of elasticity approximately with the logarithm of the number of cycles was noted. Age dependence of the modulus of elasticity was similar to that of stress values. In both directions a decrease during maturation was found. The behavior between 12 and 24 months was different: in longitudinal samples an increase was noted and in perpendicular samples a decrease was noted. The area under the curve calculated from stress values closely resembled the pattern of initial stress, indicating that this value dominates for the area under the curve and that the decay is only of secondary importance. In the *in vivo* experiments the values of initial load, initial stress, the modulus of elasticity, and the area under the curve show the same tendency during senescence: a decrease in the perpendicular and an increase in the longitudinal direction. This indicates that the aging of the skin in rats is different in the longitudinal vs. the perpendicular direction.

In Vivo Recovery after Repeated Strain

A method was developed to study the *in vivo recovery of mechanical properties of rat skin after repeated strain* (Denkel, 1983; Vogel and Denkel, 1985b; Vogel, 1988c, 1993a, 1993c). Full recovery, i.e., *restitutio ad integrum*, can be observed only by using *in vivo* experiments, not by using *in vitro* experiments. As in the experiment of repeated strain, tabs were fastened on the shaved back skin of anesthetized rats with a distance of 25 mm either perpendicular or longitudinal to the body axis. With an extension rate of 100 mm/min, the skin was extended 30 times for a 50% strain under anesthesia. Load was recorded and stress calculated by dividing load by skin thickness measured from a skin fold obtained with calipers. After the first run, the animals were returned to their cages with the tabs still in position. A second run of repeated strain was applied after different time intervals between 15 min and 16 h. The stress values in the second run were definitively lower than in the first run. When calculated as percentage of the first run, the differences diminished with extended time intervals. With this *in vivo* method not only the mechanical recovery, which also can be observed *in vitro*, but also the biological recovery, i.e., the *restitutio ad integrum*, could be measured. Almost full recovery was found after 16 h.

The age dependence as well as the influence of systemic treatment with desmotropic drugs on biological recovery was studied for both directions vs. body axis (Vogel, 1988c, 1993c). The stress values in the first cycle of the second run were calculated as a percentage of the first run. In both the longitudinal and perpendicular directions, the values were less than 50% after 15 min and 1 h, but rose continuously until 16 h, reaching or even exceeding 100% at this time. Apparently, the restoration process starts immediately and is detectable after 1 h. In both directions, the lowest values were for the youngest animals and the highest for 12-month-old rats. When cycle 30 was evaluated in the same manner, expressing the values of the second run as a percentage of the first run, smaller differences were observed. The values of the second run were between 70 and 80% of the first run after 15 min. These values dropped further, reaching a minimum after 3 h, and rose thereafter to around 100%. This may indicate that restoration of the skin has at least two components, one of which appears immediately and is apparent after 1 h and a second in which the damage reaches a maximum later and restoration starts later.

When the modulus of elasticity at the first cycle of the second run was expressed as percentage of the corresponding value of the first run, the percentage values were much higher than those for tension, indicating that the modulus of elasticity was less affected. The changes were the highest in young animals and the lowest in 12-month-old rats in both directions. A minimum after 1 or 3 h was observed, followed by an increase to almost normal values after 16 h. When the area under the curve was compared, expressing the data of the second run as a percentage of the first run, the values were between 60 and 70%. A decline until 3 h was observed, followed by a steady increase. The time course for the modulus of elasticity and the area under the curve indicated that several processes of damage and restoration are involved.

Surprisingly, *restitutio ad integrum* was the most rapid in old animals. The ability of the dermis to reconstitute the fibrous structure is apparently not negatively influenced by age.

In another experimental series, the mechanical behavior of rat skin during repeated strain *in vivo* and the restoration process thereafter were studied following 10-day treatment with desmotropic drugs (300 mg/kg p.o. D-penicillamine or 10 mg/kg s.c. prednisolone acetate). Repeated (30 times) elongations to 40% were performed under anesthesia both longitudinally and perpendicularly to the body axis. Evaluation of the first run of 30 cycles showed that stress values as well as moduli of elasticity decreased after D-penicillamine, whereas they were greatly increased after prednisolone treatment. The decay of stress values during repeated strain revealed similar effects of treatment. The restoration process was evaluated by a second run of 30 cycles after 1, 6, and 16 h. In the late phase of recovery, an even better *restitutio ad integrum* was found in treated animals. In contrast to other biomechanical parameters, the restoration process was found to be only slightly influenced by treatment with desmotropic compounds.

CONCLUSIONS

Studies on mechanical properties in animal skin are more advantageous than studies in human skin in several respects:

- They can be performed simultaneously *in vitro* and *in vivo*. Taking into consideration the different experimental and anatomical conditions, the results are comparable.
- Only in animal experiments can the biomechanical properties of the dermis at higher extension degrees be studied *ex vivo* or *in vivo* under anesthesia; studies in humans are limited by pain threshold or to tests in cadaver skin.

- Studies on age dependence show a similar pattern for skin thickness, ultimate load, tensile strength, ultimate extension, modulus of elasticity, hysteresis, relaxation, creep behavior, and biochemical data both in animal and human studies. Therefore, extrapolations from animal studies to behavior of human skin are justified.
- With the methods reported in this chapter, primarily the mechanical properties of the dermis are measured. Because of anatomical conditions (haired skin in animals), the biomechanics of the epidermis can be better studied in human rather than in animal experiments.

REFERENCES

Adamson B, Schwarz D, Klugston P, Gilmont R, Perry L, Fisher J, Lindblad W, and Rees R (1996) Delayed repair: the role of glutathione in a rat incisional wound model, *J. Surg. Res.*, 62: 159–164.

Albina JE, Gladden P, and Walsh WR (1993) Detrimental effects of an ω_3 fatty acid-enriched diet on wound healing, *J. Parenter. Enteral. Nutr.*, 17: 519–521.

Allain JC, Bazin S, Le Lous M, and Delaunay A (1977) Pathologie expérimentale — variations de la contraction hydrothermique de la peau du rat en fonction de l'âge des animaux, *C. R. Acad. Sci. Paris*, 284: 1131–1134.

Allain JC, Le Lous M, Cohen-Solal L, Bazin S, and Maroteaux P (1980) Isometric tensions developed during the hydrothermal swelling of rat skin, *Connective Tissue Res.*, 7: 127–133.

Alpermann H and Vogel HG (1978) Effect of repeated ultraviolet irradiation on skin of hairless mice, *Arch. Dermatol. Res.*, 262, 1: 15–25.

Alpermann HG, Sandow J, and Vogel HG (1982) Tierexperimentelle Untersuchungen zur topischen und systemischen Wirksamkeit von Prednisolon-17-ethylcarbonat-21-propionat, *Arzneim. Forsch./Drug Res.*, 32: 633–638.

Andreassen TT, Fogdestam I, and Rundgren Å (1977) A biomechanical study of healing of skin incisions in rats during pregnancy, *Surg. Gynecol. Obstet.*, 145: 175–178.

Andreassen TT, Seyer-Hansen K, and Oxlund H (1981) Biomechanical changes in connective tissue induced by experimental diabetes, *Acta Endocrinol.* (Copenhagen), 98: 432–436.

Baker MR, Bader DL, and Hopewell JW (1988) An apparatus for testing of mechanical properties of skin *in vivo*: its application to the assessment of normal and irradiated pig skin, *Bioeng. Skin*, 4: 87–103.

Baker MR, Bader DL, and Hopewell JW (1989) The effect of single doses of X rays on the mechanical properties of pig skin *in vivo*, *Br. J. Radiol.*, 62: 830–837.

Barbanel JC and Evans JH (1977) The time-dependent mechanical properties of skin, *J. Invest. Dermatol.*, 69: 318–320.

Barbanel JC and Payne PA (1981) *In vivo* mechanical testing of dermal properties, *Bioeng. Skin*, 3: 8–38.

Barbanel JC, Evans JH, and Jordan MM (1978) Tissue mechanics, *Eng. Med.*, 7: 5–9.

Belkoff SM and Haut RC (1991) A structural model used to evaluate the changing microstructure of maturing rat skin, *J. Biomech.*, 24: 711–720.

Belkoff SM, Naylor EC, Walshaw R, Lanigan E, Colony L, and Haut RC (1995) Effect of subcutaneous expansion on the mechanical properties of porcine skin, *J. Surg. Res.*, 58: 117–123.

Bernstein EF, Harisiadis L, Solomon G, Norton J, Sollberg S, Uitto J, Glatstein E, Glass J, Talbot T, Russo A, and Mitchel JB (1991) Transforming growth factor-β improved healing of radiation-impaired wounds, *J. Invest. Dermatol.*, 97: 430–434.

Bitar MS (1998) Glucocorticoid dynamics and impaired wound healing in diabetes mellitus, *Am. J. Pathol.*, 152: 547–554.

Blackett AD and Hall DA (1980) The action of vitamin E on the ageing of connective tissues in the mouse, *Mech. Ageing Dev.*, 14: 305-316.

Boros-Farkas M and Everitt AV (1967) Comparative studies of age tests on collagen fibers, *Gerontologia*, 13: 37–49.

Burlin TE (1980) Towards a standard for in vivo testing of the skin subject to uniaxial extension, *Bioeng. Skin*, 2: 37–40.

Burlin TE (1981) Towards a standard for in vivo testing of the skin subject to biaxial extension, *Bioeng. Skin*, 3: 47–49.

Buss V, Lippert H, Zech M, and Arnold G (1976) Zur Biomechanik menschlicher Sehnen: Zusammenhänge von Relaxation und Spannungsrückgewinn, *Arch. Orthop. Unfall-Chir.*, 86: 169–182.

Butler PEM, Barry-Walsh C, Curren B, Grace PA, Leader M, and Bouchier Hayes D (1991) Improved wound healing with a modified electrosurgical electrode, *Br. J. Plast. Surg.*, 44: 495–499.

Canturk NZ, Vural B, Esen N, Canturk Z, Oktay G, Kirkali G, and Solakoglu S (1999) Effects of granulocyte-macrophage colony-stimulating factor on incisional wound healing in an experimental diabetic rat model, *Endocr. Res.*, 25: 105–116.

Celebi N, Erden N, Gonul B, and Koz M (1994) Effects of epidermal growth factor dosage forms on dermal wound strength in mice, *J. Pharm. Pharmacol.*, 46: 386–387.

Chang HS, Hom DB, Agarwal RP, Pernell K, Manivel JC, and Song C (1998) Effect of basic fibroblast growth factor on irradiated porcine skin flaps, *Arch. Otolaryngol. Head Neck Surg.*, 124: 307–312.

Cook T, Alexander H, and Cohen M (1977) Experimental method for determining the 2-dimensional mechanical properties of living human skin, *Med. Biol. Eng. Comput.*, 15: 381–390.

DaCosta ML, Regan MC, Al Sader M, Leader M, and Bouchier-Hayes D (1998) Diphenylhydantoin sodium promotes early and marked angiogenesis and results in increased collagen deposition and tensile strength in healing wounds, *Surgery*, 123: 287–293.

Daly CH (1982) Biomechanical properties of dermis, *J. Invest. Dermatol.*, 79: 17s–20s.

Danielsen CC (1981) Thermal stability of reconstituted collagen fibrils. Shrinkage characteristics upon *in vitro* maturation, *Mech. Ageing Dev.*, 15: 269–278.

Danielson DA (1973) Human skin as an elastic membrane, *J. Biomech.*, 6: 539–546.

Denkel K (1983) *Vergleich rheologischer Parameter in vivo und in vitro an der Rückenhaut der Ratte*, Ingenieurarbeit, Frankfurt-Hoechst.

Drucker M, Cardenas E, Arizti P, Valenzuela A, Gamboa A, and Kerstein MD (1998) Experimental studies on the effect of lidocaine on wound healing, *World J. Surg.*, 22: 394–398.

Elden HR (1970) Biophysical properties of aging skin. In Montagna W (ed) *Advances in Biology of Skin*, Vol 10, pp. 231–252, Pergamon Press, New York.

Ferry JD (1970) *Viscoelastic Properties of Polymers*. 2nd ed. John Wiley & Sons, New York.

Flandin F, Buffevant C, and Herbage D (1984) A differential scanning calorimetry analysis of the age-related changes in the thermal stability of rat skin collagen, *Biochim. Biophys. Acta*, 791: 205–211.

Flynn DM, Peura GD, Grigg P, and Hoffman AH (1998) A finite element based method to determine the properties of planar soft tissue, *J. Biomech. Eng.*, 120: 202–210.

Frisén M, Mägi M, Sonnerup M, and Viidik A (1969a) Rheological analysis of soft collagenous tissue. Part I: Theoretical considerations, *J. Biomech.*, 2: 13–20.
Frisén M, Mägi M, Sonnerup M, and Viidik A (1969b)) Rheological analysis of soft collagenous tissue. Part II: Experimental evaluation and verification, *J. Biomech.*, 2: 21–28.
Fry P, Harkness MLR, and Harkness RD (1964) Mechanical properties of the collagenous framework of skin in rats of different ages, *Am. J. Physiol.*, 206: 1425–1429.
Fung LC, Mingin GC, Massicotte M, Felsen D, and Poppas DP (1999) Effects of temperature on thermal injury and wound strength after photothermal wound closure, *Laser Surg. Med.*, 25: 290–295.
Fung YC (1981) *Biomechanics. Mechanical Properties of Living Tissue*. Springer, New York.
Gibson T, Stark H, and Evans JH (1969) Directional variation in extensibility of human skin *in vivo*, *J. Biomech.*, 2: 201–204.
Gonul B, Soylemezoglu T, Babul A, and Celebi N (1998) Effect of epidermal growth factor dosage forms on mice full-thickness skin wound zinc levels and relation to wound strength, *J. Pharm. Phamacol.*, 50: 641–644.
Greenwald DP, Shumway S, Zachary LS, LaBarbera M, Albear P, Temaner M, and Gottlieb LJ (1993) Endogenous versus toxin-induced diabetes in rats: a mechanical comparison of two skin wound-healing models, *Plast. Reconstr. Surg.*, 91: 1087–1093.
Harkness RD (1968) Mechanical properties of collagenous tissues. In Gould BS (Ed) *Treatise on Collagen*. Vol 2, Biology of Collagen, Part A. Academic Press, New York, pp. 247–310.
Harkness RD (1971) Mechanical properties of skin in relation to its biological function and its chemical components. In Elden HR (ed) *Biophysical Properties of the Skin*. Wiley-Interscience, New York, pp. 393–436.
Higashiyama M, Hashimoto K, Takada A, Fujita K, Kido K, and Yoshikawa K (1992) The role of growth factor in wound healing, *J. Dermatol.*, 19: 676–679.
Hirsch C and Sonnerup L (1968) Macroscopic rheology in collagen material, *J. Biomech.*, 1: 13–18.
Holm-Pedersen P and Viidik A (1972a) Maturation of collagen in healing wounds in young and old rats, *Scand. J. Plast. Reconstr. Surg.*, 6: 16–23.
Holm-Pedersen P and Viidik A (1972b) Tensile properties and morphology of healing wounds in young and old rats, *Scand. J. Plast. Reconstr. Surg.*, 6: 24–35.
Holm-Pedersen P and Zederfeldt B (1971) Strength development of skin incisions in young and old rats, *Scand. J. Plast. Reconstr. Surg.*, 5: 7–12.
Holzmann H, Korting GW, Kobelt D, and Vogel HG (1971) Prüfung der mechanischen Eigenschaften von menschlicher Haut in Abhängigkeit von Alter und Geschlecht, *Arch. Klin. Exp. Dermatol.*, 239: 355–397.
Howes EL, Sooy JW, and Harvey SC (1929) The healing of wounds as determined by their tensile strength, *J. Am. Med. Assoc.*, 92: 42–45.
Hütte Des Ingenieurs Taschenbuch (1915) Verlag von Wilhelm Ernst & Sohn, Berlin, Vol II, pp. 630–631.
Hussein MAF (1972) The orientation of connective tissue fibres in rat skin, *Acta Anat.*, 82: 549–564.
Hussein MAF (1973) Skin cleavage lines in the rat, *Eur. Surg. Res.*, 5: 73–79.
Hutton WC, Burlin TE, and Ranu HS (1977) The effect of split dose radiations on the mechanical properties of the skin, *Phys. Med. Biol.*, 22: 411–421.
Jamison CE, Marangoni RC, and Glaser AA (1968) Viscoelastic properties of soft tissue by discrete model characterization, *J. Biomech.*, 1: 33–46.
Jenkins RB and Little RW (1974) A constitutive equation for parallel-fibered elastic tissue, *J. Biomech.*, 7: 397–402.

Jiminez PA and Rampy MA (1999) Keratinocyte growth factor-2 accelerates wound healing in incisional wounds, *J. Surg. Res.*, 81: 238–242.

Jørgensen PH, Andreassen TT, and Jørgensen KD (1989) Growth hormone influences collagen deposition and mechanical strength in intact rat skin. A dose-response study, *Acta Endocrinol.* (Copenhagen), 120: 767–772.

Joseph KT and Bose SM (1962) Influence of biological ageing on the stability of skin collagen in albino rats. In Ramanathan N (ed) *Collagen*. Interscience, New York, pp. 371–393.

Jyung RW, Mustoe TA, Busby WH, and Clemmons DR (1994) Increased wound-breaking strength induced by insulin-like growth factor I in combination with insulin-like growth factor binding protein-1, *Surgery*, 115: 233–239.

Kashyap A, Beezhold D, Wiseman J, and Beck WC (1995) Effect of povidone iodine dermatologic ointment on wound healing, *Am. Surg.*, 61: 486–491.

Kim LR and Pomeranz B (1999) The sympathomimetic agent, 6-hydroxydopamine, accelerates cutaneous wound healing, *Eur. J. Pharmacol.*, 376: 257–264.

Koshizuka S, Kanazawa K, Kobayashi N, Takazawa I, Waki Y, Shibusawa H, and Shumiya S (1997) Beneficial effects of recombinant human insulin-like growth factor-I (IGF-I) on wound healing in severely wounded senescent mice, *Surg. Today*, 27: 946–952.

Langer K (1861) Zur Anatomie und Physiologie der Haut. I. Über die Spaltbarkeit der Cutis, *Sitzungsber. Akad. Wien*, 44: 19–46.

Lagrana NA, Alexander H, Strauchler I, Metha H, Ricci J, (1983) Effect of mechanical load in wound healing, *Ann. Plast. Surg.*, 10: 200–208.

Lanir Y (1979) A structural theory for the homogeneous biaxial stress-strain relationships in flat collagenous tissue, *J. Biomech.*, 12: 423–436.

Lanir Y and Fung YC (1974) Two-dimensional properties of rabbit skin. II. Experimental results, *J. Biomech.*, 7: 171–182.

Larrabee WF, Jr (1986) A finite element model of skin deformation. I. Biomechanics of skin and soft tissue: a review, *Laryngoscope*, 96: 399–405.

Larrabee WF, Jr and Galt JA (1986) A finite element model of skin deformation. III. The finite element model, *Laryngoscope*, 96: 413–419.

Larrabee WF, Jr and Sutton D (1986) A finite element model of skin deformation. II. An experimental model of skin deformation, *Laryngoscope*, 96: 406–412.

Le Lous M, Flandin F, Herbage D, and Allain JC (1982a) Influence of collagen denaturation on the chemorheological properties of skin, assessed by differential scanning calorimetry and hydrothermal isometric tension measurement, *Biochim. Biophys. Acta*, 717: 295.

Le Lous M, Allain JC, Cohen-Solal L, and Maroteaux P (1982b) The rate of collagen maturation in rat and human skin, *Connective Tissue Res.*, 9: 253–262.

Le Lous M, Allain JC, Cohen-Solal L, and Maroteaux P (1983) Hydrothermal isometric tension curves from different connective tissues. Role of collagen genetic types and noncollagenous components, *Connective Tissue Res.*, 11: 199–206.

Lerch H (1951) Über Wärmeschrumpfungen des Kollagengewebes, *Gegenbaur's Morphol. Jahrb.*, 90: 206–220.

Lofstrom DE, Felts WJL, and Halberg F (1973) Circadian variation in skin tensile strength of two inbred strains of mice, *Int. J. Chronobiol.*, 1: 259–267.

Matsuda H, Koyama H, Sato H, Sawada J, Itakura A, Tanaka A, Matsumoto M, Konno K, Ushio H, and Matsuda K (1998) Role of nerve growth factor in cutaneous wound healing: accelerating effects in normal and healing-impaired diabetic mice, *J. Exp. Med.*, 187: 297–306.

Maxwell GL, Soisson AP, Brittain PC, Harris R, and Scully T (1998) Tissue glue as an adjunct to wound healing in the porcine model, *J. Gynecol. Tech.*

Menton DN, Hess RA, Lichtenstein JR, and Eisen AZ (1978) The structure and tensile properties of the skin of tight-skin (Tsk) mutant mice, *J. Invest. Dermatol.*, 70: 4–10.

Mustoe TA, Bartell TH, and Garner WL (1989) Physical, biochemical, histologic, and biochemical effects of rapid versus conventional expansion, *Plast. Reconstr. Surg.*, 83: 787–691.

Nemetscheck T, Riedl H, Jonak R, Nemetscheck-Gansler H, Bordas J, Koch MHJ, and Schilling V (1980) Die Viskoelastizität parallelsträhnigen Bindegewebes und ihre Bedeutung für die Funktion, *Virchow's Arch.*, 386: 125–151.

Nimni ME, de Guia E, and Bavetta LA (1966) Collagen, hexosamine and tensile strength of rabbit skin during aging, *J. Invest. Dermatol.*, 47: 156–158.

Oxlund H and Manthorpe R (1982) The biochemical properties of tendon and skin as influenced by long term glucocorticoid treatment and food restriction, *Biorheology*, 19: 631–641.

Oxlund H, Fogdestam I, and Viidik A (1979) The influence of cortisol on wound healing of the skin and distant connective tissue response, *Surg. Gynecol. Obstet.*, 148: 867–880.

Oxlund H, Rundgren A, and Viidik A (1980) The influence of adrenalectomy on the biomechanical properties of collagenous structures of rats in the post-partum phase, *Acta Obstet. Gynecol. Scand.*, 59: 453–458.

Oxlund H, Adreassen TT, Junker P, Jensen PA, and Lorenzen I (1984) Effect of D-penicillamine on the mechanical properties of aorta, muscle tendon and skin in rats, *Atherosclerosis*, 52: 243–252.

Pan L, Zan L, and Foster FS (1998) Ultrasonic and viscoelastic properties of skin under transverse mechanical stress in vitro, *Ultrasound Med. Biol.*, 24: 995–1007.

Pandit A, Ashar R, and Feldman D (1999) The effect of TGF-β delivered through a collagen scaffold on wound healing, *J. Invest. Surg.*, 12: 89–100.

Phillips LG, Abdullah KM, Geldner PD, Dobbins S, Ko F, Linares HA, Broemeling LD, and Robson MC (1993) Application of basic fibroblast growth factor may reverse diabetic wound healing impairment, *Ann. Plast. Surg.*, 31: 331–334.

Potts RO and Breuer MM (1983) The low-strain, viscoelastic properties of skin, *Bioeng. Skin*, 4: 105–114.

Purslow PP, Wess TJ, and Hukins DW (1998) Collagen orientation and molecular spacing during creep and stress-relaxation in soft connective tissue, *J. Exp. Biol.*, 201: 135–142.

Quirinia A and Viidik A (1998) The effect of recombinant basic fibroblast growth factor (bFGF) in fibrin adhesive vehicle on the healing of ischemic and normal incision skin wounds, *Scand. J. Plast. Reconstr. Surg. Hand Surg.*, 32: 9–18.

Rasmussen DM, Wakim KG, and Winkelmann RK (1964) Effect of aging on human dermis: studies of thermal shrinkage and tension. In Montagna W (Ed) *Advances in Biology of Skin*. Vol 6, 151–162.

Ridge MD and Wright V (1964) The description of skin stiffness, *Biorheology*, 2: 67–74.

Ridge MD and Wright V (1965) The rheology of skin. A bio-engineering study of the mechanical properties of human skin in relation to its structure, *Br. J. Dermatol.*, 77: 639–649.

Ridge MD and Wright V (1966) Rheological analysis of connective tissue. A bio-engineering analysis of skin, *Ann. Rheum. Dis.*, 25: 509–515.

Riedl H and Nemetscheck Th (1977) *Molekularstruktur und mechanisches Verhalten von Kollagen*. Sitzungsberichte der Heidelberger Akademie der Wissenschaften. Mathematisch-naturwissenschaftliche Klasse. pp. 216–248.

Rundgren A (1976) Age changes of connective tissue in the rat as influenced by repeated pregnancies, *Akt. Gerontol.*, 6: 15–18.

Sandblom P (1957) Wundheilungsprobleme, mit Reißfestigkeitsmethoden untersucht, *Langenbeck's Arch. Dtsch. Z. Chir.*, 287: 469–480.
Sanjeevi R (1982) A viscoelastic model for the mechanical properties of biological materials, *J. Biomech.*, 15: 107–109.
Scardino MS, Swaim SF, Morse BS, Sartin MA, Wright JC, and Hoffman CE (1999) Evaluation of fibrin sealants in cutaneous wound closure, *J. Biomed. Mater. Res.*, 48: 315–321.
Schaffer MR, Tantry U, Gross SS, Wasserkrug HL, Barbul A (1996) Nitric oxide regulates wound healing, *J. Surg. Res.*, 63: 237–240.
Schneider MS, Borkow JE, Cruz IT, Marangoni RD, Shaffer J, and Grove D (1988) The tensiometric properties of expanded guinea pig skin, *Plast. Reconstr. Surg.*, 81: 398–403.
Schröder HG, Babej M, and Vogel HG (1974 Tierexperimentelle Untersuchungen mit dem lokal wirksamen 9-Fluor-16-methyl-17-desoxy-prednisolon, *Arzneim. Forsch.*, 24, 3–5.
Seyer-Hansen M, Andreassen TT, Jørgensen PH, and Oxlund H (1993) Influence of biosynthetic human growth hormone on the biomechanical strength development in skin incisional wounds of diabetic rats, *Eur. Surg. Res.*, 25: 162–168.
Silverstein RJ and Landsman AS (1999) The effects of a moderate and a high dose of vitamin C on wound healing in a controlled guinea pig model, *J. Foot Ankle Surg.*, 38: 333–338.
Soong TT and Huang WN (1973) A stochastic model for biological tissue elasticity in simple elongation, *J. Biomech.*, 6: 451–485.
Spittle RF, Ranu HS, Hutton WC, Challoner AV, and Burlin TE (1980) A comparison of different treatment regimens on the visual appearance and mechanical properties of mouse skin, *Br. J. Radiol.*, 53: 697–702.
Stark HL, Strath PD, Eng C, Mech MI, and Aust MIE (1977) Directional variations in the extensibility of human skin, *Br. J. Plast. Surg.*, 30: 195–114.
Töpert M, Olivar A, and Opitz D (1990) New developments in corticosteroid research, *J. Dermatol. Treatment*, 1, Suppl. 3: S5–S9.
Van den Hoven WE, van den Berg TP, and Korstanje C (1991) The hairless mouse as a model for study of local and systemic atrophogenic effects following topical application of corticosteroids, *Acta Derm. Venereol.* (Stockholm), 71: 29–31.
Vegesna V, McBride WH, Taylor JGM, and Withers HR (1995) The effect of interleukin 1β or transforming growth factor β on radiation impaired murine skin wound healing, *J. Surg. Res.*, 59: 600–704.
Veronda DR and Westman RA (1970) Mechanical characterization of skin — finite deformations, *J. Biomech.*, 3: 111–124.
Verzár F (1955) Veränderungen der thermoelastischen Kontraktion von Haut und Nerv bei alternden Tieren, *Experientia* (Basel), 11: 230–231.
Verzár F (1956) Das Altern des Kollagens, *Helv. Physiol. Acta*, 14: 207–221.
Verzár F (1957) The ageing of connective tissue, *Gerontologica*, 1: 363–378.
Verzár F (1964) Factors which influence the age-reaction of collagen in the skin, *Gerontologica*, 9: 209–221.
Viidik A (1968) A rheological model for uncalcified parallel-fibred collagenous tissue, *J. Biomech.*, 1: 3–11.
Viidik A (1969) The aging of collagen as reflected in its physical properties. In Engel A and Larsson T (eds) *Aging of Connective and Skeletal Tissue*. Thule International Symposia, Nordiska Bokhandels Förlag, Stockholm, pp. 125–152.

Viidik A (1973a) Rheology of skin with special reference to age-related parameters and their possible correlation to function, *Front Matrix Biol.*, 1: 157–189.
Viidik A (1973b) Functional properties of collagenous tissues. In Hall DA and Jackson DS (eds) *International Review of Connective Tissue Research.* Vol 6, Academic Press, New York, pp. 127–215.
Viidik A (1975) The "dynamic" connective tissue, *Ann. Estonian. Med. Assoc.*, pp. 155-169.
Viidik A (1977) Thermal contraction — relaxation and dissolution of rat tail tendon collagen in different ages, *Akt. Gerontol.*, 7: 493–498.
Viidik A (1978) On the correlation between structure and mechanical function of soft connective tissue, *Verh. Anat. Ges.*, 72: 75–89.
Viidik A (1979a) Biomechanical behavior of soft connective tissues. In Akkaˌ N (ed) *Progress in Biomechanics.* Sijthoff & Noordhoff, Alphen aan den Rijn, the Netherlands, pp. 75–113.
Viidik A (1979b) Connective tissues — possible implications of the temporal changes for the aging process, *Mech. Ageing Dev.*, 9: 267–285.
Vogel HG (1969) Zur Wirkung von Hormonen auf physikalische und chemische Eigenschaften des Binde- und Stützgewebes, *Arzneim. Forsch./Drug Res.*, 19: 1495–1503, 1732–1742, 1790–1801, 1981–1996.
Vogel HG (1970a) Beeinflussung der mechanischen Eigenschaften der Haut von Ratten durch Hormone, *Arzneim. Forsch./Drug Res.*, 20: 1849–1857.
Vogel HG (1970b) Tensile strength of skin wounds in rats after treatment with corticosteroids, *Acta Endocrinol.*, 64: 295–303.
Vogel HG (1971) Antagonistic effect of aminoacetonitrile and prednisolone on mechanical properties of rat skin, *Biochim. Biophys. Acta*, 252(3): 580–585.
Vogel HG (1972a) Effects of D-penicillamine and prednisolone on connective tissue in rats, *Connective Tissue Res.*, 1: 283–289.
Vogel HG (1972b) Influence of age, treatment with corticosteroids and strain rate on mechanical properties of rat skin, *Biochim. Biophys. Acta*, 286: 79–83.
Vogel HG (1973) Stress relaxation in rat skin after treatment with hormones, *J. Med.*, 4: 19–27.
Vogel HG (1974a) Organ specificity of the effects of D-penicillamine and of lathyrogen (aminoacetonitrile) on mechanical properties of connective and supporting tissue, *Arzneim. Forsch.*, 24: 157–163.
Vogel HG (1974b) Correlation between tensile strength and collagen content in rat skin. Effect of age and cortisol treatment, *Connective Tissue Res.*, 2: 177–182.
Vogel HG (1975) Collagen and mechanical strength in various organs of rats treated with D-penicillamine or amino-acetonitrile, *Connective Tissue Res.*, 3: 237–244.
Vogel HG (1976a) Measurement of some viscoelastic properties of rat skin following repeated load, *Connective Tissue Res.*, 4: 163–168.
Vogel HG (1976b) Tensile strength, relaxation and mechanical recovery in rat skin as influenced by maturation and age, *J. Med.*, 2: 177–188.
Vogel HG (1977a) Mechanical and chemical properties of various connective tissue organs in rats as influenced by non-steroidal antirheumatic drugs, *Connective Tissue Res.*, 5: 91–95.
Vogel HG (1977b) Strain of rat skin at constant load (creep experiments). Influence of age and desmotropic agents, *Gerontology*, 23: 77–86.
Vogel HG (1978a) Influence of maturation and age on mechanical and biochemical parameters of connective tissue of various organs in the rat, *Connective Tissue Res.*, 6: 161–166.
Vogel HG (1978b) Age dependence of mechanical parameters in rat skin following repeated strain, *Akt. Gerontol.*, 8: 601–618.

Vogel HG (1981a) Directional variations of mechanical parameters in rat skin depending on maturation and age, *J. Invest. Dermatol.*, 76: 493–497.

Vogel HG (1981b) Attempts to compare "in vivo" and "in vitro" measurement of mechanical properties in rat skin, *Bioeng. Skin*, 3: 39–46.

Vogel HG (1981c) Comments on the paper by Barbenel and Payne "In vivo" mechanical testing of dermal properties, *Bioeng. Skin*, 3: 53–56.

Vogel HG (1982) Mechanical properties of rat skin as compared by in vivo and in vitro measurement, *Bioeng. Skin*, 3: 198–209.

Vogel HG (1983a) Effects of age on the biomechanical and biochemical properties of rat and human skin, *J. Soc. Cosmet. Chem.*, 34: 453–463.

Vogel HG (1983b) Age dependence of viscoelastic properties in rat skin. Directional variations in stress-strain and hysteresis experiments, *Bioeng. Skin*, 4: 136–155.

Vogel HG (1984) Messung der Relaxation und des isorheologischen Punkts an Hautstreifen von Ratten, *Z. Rheumatol.*, 43, Suppl. 1: 46–47.

Vogel HG (1985a) Age dependence of viscoelastic properties in rat skin; directional variations in relaxation experiments, *Bioeng. Skin*, 1: 157–174.

Vogel HG (1985b) Repeated relaxation and determination of the isorheological point in skin strips of rats as influenced by maturation and ageing, *Bioeng. Skin*, 1: 321–335.

Vogel HG (1986) In vitro test systems for evaluation of the physical properties of skin. In Marks R, Plewig G (eds) *Skin Models*. Springer Verlag, Berlin, pp. 412–419.

Vogel HG (1987a) Age dependence of mechanical and biochemical properties of human skin. Part I: Stress-strain experiments, skin thickness and biochemical analysis, *Bioeng. Skin*, 3: 67–91.

Vogel HG (1987b) Age dependence of mechanical and biochemical properties of human skin. Part II: Hysteresis, relaxation, creep and repeated strain experiments, *Bioeng. Skin*, 3: 141–176.

Vogel GH (1987c) Repeated loading followed by relaxation and isorheological behaviour of rat skin after treatment with desmotropic drugs, *Bioeng. Skin*, 3(3): 255–269.

Vogel HG (1988a) Age-dependent mechanical and biomechanical changes in the skin, *Bioeng. Skin*, 4: 75–81.

Vogel HG (1988b) Further studies on directional variations and the "step-phenomenon" in rat skin depending on age, *Bioeng. Skin*, 4: 297–309.

Vogel HG (1988c) Restitution of mechanical properties of rat skin after repeated strain. Influence of maturation and ageing, *Bioeng. Skin*, 4: 343–359.

Vogel HG (1989) Mechanical properties of rat skin with aging. In Balin AK and Kligman AM (eds) *Aging and the Skin*. Raven Press, New York, pp. 227–275.

Vogel HG (1993a) Mechanical measurements in assessing aging. In Frosch PJ and Kligman AM (eds) *Noninvasive Methods for the Quantification of Skin Functions. An Update on Methodology and Clinical Applications*. Springer-Verlag, Berlin, pp. 145–180.

Vogel HG (1993b) Strength and viscoelastic properties of anisotropic rat skin after treatment with desmotropic drugs, *Skin Pharmacol.*, 6: 92–102.

Vogel HG (1993c) In vivo recovery of repeatedly strained rat skin after systemic treatment with desmotropic drugs, *Skin Pharmacol.*, 6: 103–110.

Vogel HG and Denkel K (1982) Mechanical properties of rat skin as compared by in vivo and in vitro measurement, *Bioeng. Skin*, 3: 198–209.

Vogel HG and Denkel K (1985a) Influence of maturation and age, and of desmotropic compounds on the mechanical properties of rat skin in vivo, *Bioeng. Skin*, 1: 35–54.

Vogel HG and Denkel K (1985b) In vivo recovery of mechanical properties in rat skin after repeated strain, *Arch. Dermatol. Res.*, 277: 484–488.

Vogel HG and Hilgner W (1977) Analysis of the low part of stress-strain curves in rat skin. Influence of age and desmotropic drugs, *Arch. Derm. Res.*, 258: 141–150.

Vogel HG and Hilgner W (1978) Viscoelastic behaviour of rat skin after repeated and stepwise increased strain, *Bioeng. Skin*, 1: 22–33.

Vogel HG and Hilgner W (1979a) The "step phenomenon" as observed in animal skin, *J. Biomech.*, 12: 75–81.

Vogel HG and Hilgner W (1979b) Influence of age and of desmotropic drugs on the step phenomenon observed in rat skin, *Arch. Dermatol. Res.*, 264: 225–241.

Vogel HG and Petri W (1985) Comparison of various pharmaceutical preparations of prednicarbate after repeated topical administration to the skin of rats, *Arzneim. Forsch./Drug Res.*, 35: 939–946.

Vogel HG, Kobelt D, Korting GW, and Holzmann H (1970) Prüfung der Festikeitseigenschaften von Rattenhaut in Anbhängigkeit von Alter und Geschlecht, *Arch. Klin. Exp. Dermatol.*, 239: 296–305.

Vogel HG, Alpermann HG, and Futterer E (1981) Prevention of changes after UV-irradiation by sunscreen products in skin of hairless mice, *Arch. Dermatol. Res.*, 270: 421–428.

Wijn PFF (1980) The Alinear Viscoelastic Properties of Human Skin in Vivo for Small Deformations, doctoral dissertation, Katholieke Universiteit, Nijmegen.

Wilkes GL, Brown IA, and Wildnauer RH (1973) The biomechanical properties of skin. In Fleming D (ed) *Critical Reviews in Bioengineering*. CRC Press, Boca Raton, FL.

Witte MB, Thornton FJ, Kiyama T, Efron DT, Schulz GS, Moldawer LL, Barbul A, and Hunt TK (1998) Metalloproteinase inhibitors and wound healing: a novel enhancer of wound strength, *Surgery*, 124: 464–470.

Wöhlisch E (1932) Die thermischen Eigenschaften der faserig strukturierten Gebilde des tierischen Bewegungsapparates, *Ergeb. Physiol.*, 34: 406–493.

Wöhlisch E and du Mesnil de Rochemont (1927) Die Thermodynamik der Wärmeumwandlung des Kollagens, *Z. Biol.*, 85: 406.

Woodbury R and Kligman AM (1992) The hairless mouse model for assaying the atrophogenicity of topical corticosteroids, *Acta Derm. Venereol.* (Stockholm), 72: 403–408.

Wright V (1971) Elasticity and deformation of skin. In Elden HR (ed) *Biophysical Properties of Skin*. Wiley Interscience, New York, pp. 437–449.

Wu L and Mustoe TA (1995) Effect of ischemia on growth factor enhancement of incisional wound healing, *Surgery*, 117: 570–576.

3 Mechanical Properties of Human Skin: Elasticity Parameters and Their Relevance

Jørgen Serup

CONTENTS

Introduction 41
Essential Elasticity Parameters 42
Measuring Skin Elasticity 43
Main Function of Skin as an Elastic Integument 44
Usefulness and Relevance of Skin Elasticity Parameters 46
References 47

INTRODUCTION

Biologically, the elasticity of the skin is a simple bimodal parameter with only two major components: (1) the ability under the influence of stress to undergo distension or strain and (2) the ability to subsequently retract or recover to the original state (Figure 3.1).

Skin is composite and consists of layers with entirely different structures, functions, and mechanical properties. The low-humidity stratum corneum with densely packed keratin is thought to be quite resistant; the cellular stratum Malpighii and the papillar dermis with high proliferation and high perfusion are thought to be soft and pliable (albeit with a denser basement membrane zone interposed); and the reticular dermis packed with collagen bundles and some elastic fibers is thought to be highly mechanically resistant. Skin is mechanically a viscoelastic system, and it shows major histological and microanatomical variation in different anatomical sites, among individuals, and between men and women. Simple theories, developed by mechanical engineers from studies of nonliving materials, cannot be applied uncritically to a complex biological system such as full-thickness human skin.

0-8493-7521-5/02/$0.00+$1.50

Elasticity parameters are manifold and can be measured and expressed in various ways *in vivo* and *ex vivo* depending on aims and usefulness. This chapter mainly addresses the background and aims related to usefulness in dermatology and cosmetology in an attempt to simplify and outline the essentials. Elasticity is considered from many different angles, and readers are referred to the chapters and the references given in the *Handbook of Noninvasive Methods and the Skin.*[2]

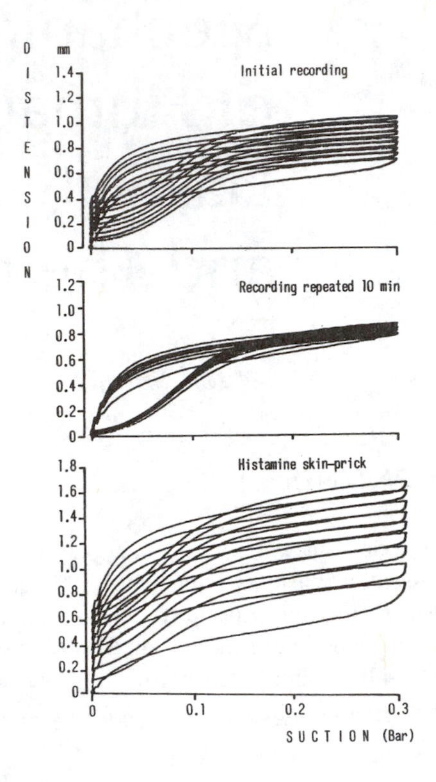

FIGURE 3.1 Examples of skin mechanics illustrating basic stress–strain functions, recorded with the suction chamber device Dermaflex A (Cortex Technology, Hadsund, Denmark). At initial recording, repeated suction results in further distension, so-called creeping or hysteresis. Re-recording after 10 min at the same skin site results in higher initial distension compared with initial recording but with less creeping. Recordings illustrate the functional mechanical memory. Histamine wheal and acute accumulation of water in the dermis result in markedly increased creeping. (From Serup, J. and Northeved, A., *J. Dermatol.* (Tokyo), 12, 52–62, 1985. With permission.)

ESSENTIAL ELASTICITY PARAMETERS

The essential parameters are, as mentioned, *distensibility* (or stress–strain) and *recovery* (or elastic retraction). These are both dependent on *time*, in particular with respect to recovery following a stress, which is "remembered" for quite a period (Figure 3.2).[3] Repeated stress results in further distension, a phenomenon known as *creeping* (or hysteresis); see Figure 3.1. This functional memory is probably a manifestation of the interstitial fluid and fiber network interactions in live skin, the expression of viscoelasticity of a highly specialized biological network.

In the past it was often postulated, but never experimentally supported, that distensibility is related to collagen fibers and that retraction is related to elastic fibers. It is, of course, not possible in a complex system such as the skin to study isolated elements without intruding into the system; and the beauty of simplicity of opinion is no justification in itself in view of the present understanding of the complex nature of skin elasticity.

Experimental stress of skin has been induced as monoaxial elongation, as biaxial elongation (push, pull, and suction), by torsional maneuvers, and by elaborate testing such as applying different vibrating forces.

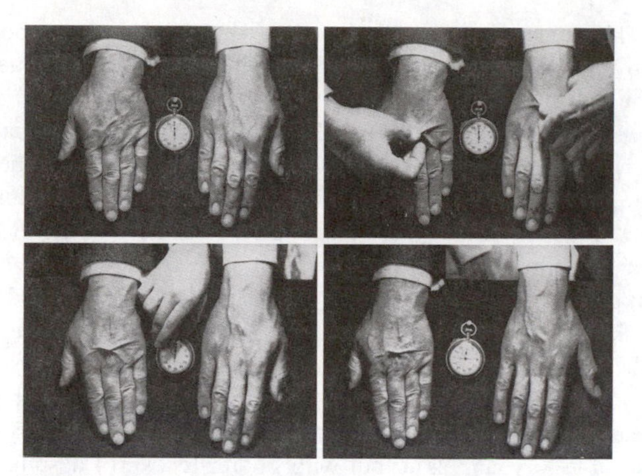

FIGURE 3.2 Elasticity of the skin in a man 68 years old (person on the left) contrasted to that of a man 26 years old (person on the right). At rest (upper left), stretched (upper right), immediately after release (lower left), and after 15 s (lower right). (From Lockhard, R.D., Ed., *Living Anatomy, a Photographic Atlas of Muscles in Action and Surface Contours*, 2nd ed., Faber & Faber, London, 1962. With permission.)

The value of a measuring system can be related to its usefulness in biological applications such as dermatology or cosmetology, and to its role in experimental research. The biaxial and monoaxial elongation systems have a fundamental rationale in biological situations and clinical states. Statements in the following sections mainly relate to this group of methods.

MEASURING SKIN ELASTICITY

There is no general reference or standard for measuring the elasticity of live human skin, and the state of the art is — despite engineering sophistication — chaotic. There are no standard operating procedures regulating typical sources of variation and error during measurement, such as joint position, skin hydration state (epidermal and dermal), dermal perfusion and fluidity, intradermal texture, Langer line orientation, underlying adipose panniculus, skin temperature, effect of drinking/sweating, effect of stressful physical activity, etc. The state of the art is reflected by the range of elastometers available in the market, used in variable settings and by operators with variable educational backgrounds. Much activity is market-driven.

Most equipment operates on a suction-cup principle, where skin under a defined circular area is pulled in under the influence of negative pressure applied and released over a specific time (available systems typically have resistance in the pressure-regulating system with unknown pressure-to-time conditions in the suction chamber, and no tank with a negative-pressure reservoir). The circular area can vary from less than 1 to about 10 mm in diameter. With a small opening, the deformation of the skin is disproportionate; and only the epidermis and outer dermis can fold and undergo strain. With a larger diameter, the full-thickness skin can elongate.

Another type of equipment, torsional devices, induce only a few degrees of rotation and thereby typically only twist the very outer skin with a disproportionate elongation. This is also the case with the gas-bearing electrodynamometer, which exerts repetitive monoaxial vibrations parallel to the skin. Thus, some devices operate with a suction pull exerted to a larger area and *proportionate elongation of the full-thickness skin,* whereas others operate with a small area and a tangential deformation with a *disproportionate elongation especially of the outer skin.* The devices and principles are not directly comparable. The former type of recording is directed toward dermatology and biology, the latter toward cosmetology.

Comparing healthy skin with abnormal skin involves a *reference/baseline problem*, which is difficult to solve. Aged skin is already sacked, elongated, and different from young skin before the device is applied. Measurements resulting from applied suction to these alterations are not directly comparable with measurements performed in a young person, who does not have the same structural abnormality and baseline. Measurement of aged vs. young skin typically demonstrates old skin to be equally distensible to young skin (but with poorer recovery) despite a sacked and obviously distended state on clinical inspection. The measured distensibility is technically correct, but experimentally not meaningful.

The *structure/elasticity relationship of skin* is complicated and seemingly very dependent on the structure and fluid relationship in the skin as exemplified by findings in scleroderma and histamine wheals. Interstitial fluid acts as oil between the fibers (see Figure 3.1). The more water, the softer and more distensible the skin and the easier the recovery following a strain. Skin *turgor* (from the Latin *turgor*, swelling) is the water tension of skin. In medicine, turgor is known to reflect the intradermal and general hydration state, as well as the fiber systems of the skin, particularly age-related changes (see Figure 3.2). Intradermal water and structural texture interactions are measured by ultrasound image analysis as low echogenic pixels with high precision. Intradermal water accumulates in most conditions in the outer dermis under the epidermis, rather than evenly throughout the skin. The skin is, among other tissues, a water reservoir that is mobilized when the body water balance is negative.

It is surprising that skin elasticity (distensibility) is variably or poorly correlated to thickness and that elasticity, even in normal skin (female vs. male, acral skin vs. truncal), is inversely correlated to thickness. The thinner and more echodense skin of extremities is less distensible compared with the thicker skin of the trunk. Females with thinner and more echodense skin show decreased distensibility compared with males. In addition, the thin skin of Pasini–Pierini atrophic morphoea is less distensible. There is no expected gross difference in intradermal water content of thin skin compared with thick skin in different sites and in different genders; and the mentioned differences are likely to be related to the dermal structures and the way the fiber network is woven, rather than to total skin thickness.

MAIN FUNCTION OF SKIN AS AN ELASTIC INTEGUMENT

The important mechanical functions of the skin can be outlined as follows:

- To provide a protective and supporting cover of the body
- To remain tense but allow free motion of joints
- To counteract gravitation

- To be soft and pliable to allow effective contact with physical surfaces as a basis for simple and complex sensory perceptions (touch including stereological perception of objects, pain, heat, cold, and others)

The *protective function* is quite obvious. The skin must resist trauma and wounds must be avoided. The skin must support the adnexa, vessels, etc. Reduction of resistance related to connective tissue deficiency is seen in dermatosparaxis in cattle, in skin fragility during penicillamine treatment, in corticostroid atrophy, and in aged skin as traction wounds or tangential wounds in the outer, less-resistant dermis, leaving pseudoscars.

The *function in relation to joint motions* (Figure 3.3) is vital in evolution and is directly reflected in the Langer lines, which are oriented in directions that clearly respect free motion of rotator joints and hinge joints. Langer systematically punctured cadaver skin with a round trocar and noted the direction of the resulting ellipsoid wounds. The lines represent the predominant direction of collagen and other fibers of the dermis. The Langer experiment at the same time demonstrated the variable pretension of the skin in different sites and directions, a pretension well appreciated in surgery and taken into account when incisions are made. Lichenification furrows in atopic dermatitis also follow the orientation of hinge joints. Fissures in chronic dermatitis do the same. Joint motion in acroscleroderma may be hampered by dermal sclerosis with excessive dermal collagen; conditions with loose connective tissue such as Ehlers–Danlos syndrome and pseudoxanthoma elasticum are associated with increased joint and skin motility. Increased skin distensibility may occur without altered joint motility.

Counteraction of gravitation is a function not well appreciated despite its fundamental significance. It has been known for decades that the giraffe, because of the tight connective tissue anatomy of its legs, avoids the development of orthostatic fluid accumulation despite its height. Valves of the veins have long been known to function to carry the hydrostatic force, and defective valves may result in swollen legs. In humans, it was clearly demonstrated in a study of 22 different body regions that there is a vertical vector in skin distensibility with increased spontaneous

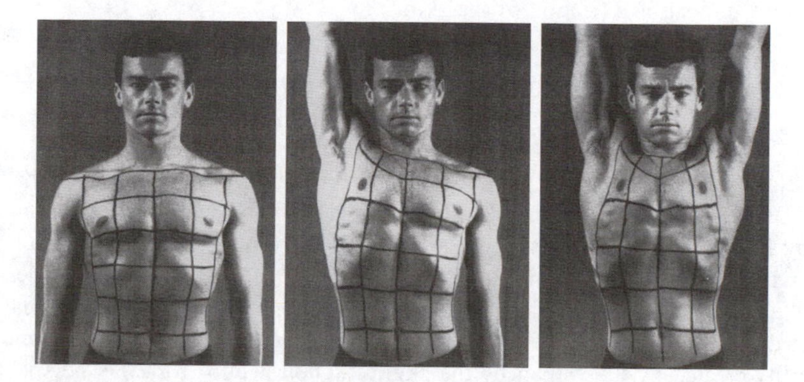

FIGURE 3.3 Elasticity of the skin relative to shoulder joint positions. Reproduced from Lockhard, R.D., Ed., *Living Anatomy, a Photographic Atlas of Muscles in Action and Surface Contours*, 2nd ed., Faber & Faber, London.

pretension of the skin from the head toward the foot.[2] This was confirmed in a study by Gniadecka, who also found a deficient vertical vector in sufferers of venous leg ulcers. Ultrasound image analysis of elderly persons shows an increase in low echogenic pixels in the morning 2 h after standing up, indicating orthostatic fluid accumulation in the legs. Recently, it was demonstrated that, in contrast, persons tilted in a negative-gravity bed develop fluid accumulation in the head, explaining the known phenomenon that astronauts develop significant head edema when they are no longer under the influence of earth's gravity. Russian astronauts applied cuffs around their thighs to avoid this problem. Thus, the vertical vector of skin elasticity in humans representing a systematic regional difference in the mechanical resistance of skin is, together with tense muscle fascia and venous valves, oriented to direct blood centrally, counteracting orthostatic pressure and gravity. It is a fundamental trait arising from evolution that is needed for physical performance and survival without swollen legs and leg ulcers at a young age.

A soft and pliable skin that can adjust to the surface of objects is a prerequisite in *sensory perception* including the different sensory modalities but particularly important for touch, which is sensed by Ruffini and Meissner end organs laid down in the skin texture. Touch is a three-dimensional modality that can become acutely sensitive as occurs in blindness, when touch replaces vision and may allow reading of complicated texts formed with letters made of small raised dots. Further, touch and the pressure exerted by pulses against an object are used as a "method" to assess the elasticity of objects. Resultant stress–strain of the skin replicates stress–strain of the object touched. Finally, touch combined with a tangential maneuver following the surface of the object serves to assess the three-dimensional surface contour and friction of objects, a highly developed function with central nervous system and even psychological elements. Tangential touch (and surface friction) plays an important role for feeling pleasure and for sexuality including sexual acts, well understood but little studied in spite of its huge importance in human interrelations from birth to death. Thus, surface, softness, pliability, and sensibility of the skin act together with intellect and psyche in different neurosensory functions important for daily life. In general, there is only a vague distinction between appreciation of stress–strain mechanics and surface friction of the skin.

USEFULNESS AND RELEVANCE OF SKIN ELASTICITY PARAMETERS

Skin elasticity has no obvious, general, and direct user relevance except perhaps in some selected clinical situations. Elasticity is not commonly appreciated in normal life by healthy people or by patients. Aged people are more concerned about how their wrinkled and sacked skin looks than about some altered stress–strain measure when a piece of skin is pulled or twisted. A patient with scleroderma may be primarily concerned about the hampered joint motion, the fingertip ulcers, and the disability created by these than by the skin affliction itself. From this perspective, there is often limited relevance of measuring skin elasticity with very high precision except perhaps in selected medical conditions to monitor disease course and the effect of medication.

Elasticity measurement in humans has the added limitation that a measurement is not very likely to be representative since normally only a small piece of skin is evaluated, typically a few square centimeters, and the whole skin is some 10,000 times larger and shows gross local site variations. Thus, the role of measuring skin elasticity mainly lies in experimental research when selected areas are treated or handled in a way that can illustrate general phenomena, effect of treatments, etc. that are due to a wider design perspective. However, extrapolation from observation in a small piece of skin to some broad field must be done with great reservation.

Elasticity methods have been somewhat overused to document efficacy of cosmetic products. This field is even more problematic because of the overlap in the appreciation of skin surface friction and surficial skin mechanics as depicted with a tangential movement by consumers spreading a cream or touching their skin (normally involves no twisting or vertical pull or push). The Young's modulus, valid for isotropic materials and widely used in engineering, does not appear meaningful in the biomechanics of anisotrophic skin simply because the relation to thickness is variable and thus follows no generally applicable equation for skin as outlined in engineering texts.

There is a strong need for standardization in the field of skin elasticity measurement. It is a difficult variable to measure, and there are obviously many limitations and controversies related to instruments, experimental designs, preconditioning, data handling, and data interpretation. Skin elasticity measurement suffers a low status in the medical community and the techniques are generally not accepted by legal bodies. To some degree methods have been miscredited following their misuse. There are, nevertheless, a number of situations when elasticity measurements and proper experiments handled with excellence should receive the recognition they deserve. After decades of engineering and pragmatic use, it is time for bioengineers to review the field critically and to agree upon a set of realistic standards. Application, interpretation, and relevance must certainly be considered.

REFERENCES

1. Serup, J. and Northeved, A., Skin elasticity in localized scleroderma (morphoea), introduction of a biaxial *in vivo* method for measurement of tensile distensibility, hysteresis and resilient distension of diseased and normal skin, *J. Dermatol.* (Tokyo), 12, 52–62, 1985.
2. Serup, J. and Jemec, G.B.E., Eds., *Handbook of Noninvasive Methods and the Skin*, CRC Press, Boca Raton, FL, 1995.
3. Lockhard, R.D., Ed., *Living Anatomy, a Photographic Atlas of Muscles in Action and Surface Contours*, 2nd ed., Faber & Faber, London, 1962.

4 Mechanical Properties of the Skin During Friction Assessment

H. Zahouani, J. Asserin, and Phillippe Humbert

CONTENTS

Introduction 49
Friction Apparatus 50
Procedure 50
Measurements of the Friction Force 51
Friction Force Measurements of Calibrated Steel Roughness Plate 51
Friction Force of a Silicone 52
In Vivo Measurements of Skin Friction Force 52
Moisturization Effect in Mechanical and Frictional Properties of the Skin 52
Effect in Mechanical Parameters 53
Effect in Static and Dynamic Frictional Properties of the Skin 54
The Skin Friction Model 55
Deformation Component of Friction 56
Conclusions 57
References 57

INTRODUCTION

Previous studies on the tribology of human skin have attempted to demonstrate a correlation between certain tactile sensations and the friction between the skin surface and variety of probes.[1-10] In addition, friction measurements have been used to provide *in vivo* information about the effects of age, hydration, dermatitis, and cosmetic products[11–15] on both the interfacial and bulk properties of skin. Most previous studies on the friction of human skin, using sliding probes, have involved torque measurements on disks or flat cones rotating at constant velocity. Typical instruments have employed rotating wheels, oscillating disks, or adapted viscometers.

The physical nature of the forces involved during friction and the significance of the change of friction properties after cosmetic application are not clearly established.

0-8493-7521-5/02/$0.00+$1.50

The current techniques that simultaneously measure the normal load F_z between the contacting surfaces and the friction force F_x can be used to determine the skin friction coefficient defined as $\mu = F_x/F_z$. The survey of the published literature on skin friction shows a wide range of measured values of μ. These differences emphasize that the assessment of the friction coefficient of skin is a highly complex problem. It involves skin elasticity (Young's modulus $E = 10^5$ to 10^7 Pa for the forearm),[16–18] the anisotropy and the range of skin furrows, the physicochemical nature of skin (free energy of the skin $\gamma_s = 38$ mJ m^{-2}),[19–21] the hydrophobia of the sensor used, the normal load, speed, and the mean frequency deformation during friction.

This chapter describes measurements of the static and dynamic friction force of a spherical indentor on an inner human forearm. Through the results obtained by means of original *in vivo* tribometer,[22] which enables measurement of the components of friction force in a wide range of pressure and sliding, this study demonstrates the significance of the moisturization effect in terms of mechanical properties and changes in friction forces.

FRICTION APPARATUS

PROCEDURE

The equipment used in this study employs a more conventional tribological configuration based on a linear sliding action[22] on an inner human forearm (Figure 4.1). This system allows the control of the applied normal load, penetration depth, and sliding velocity, coupled with an accurate measurement of the frictional response. A smooth spherical indentor was selected as a probe to facilitate theoretical modeling, and to aid comparison with extensive earlier results on friction and lubrication of ball–elastomer contacts. During the friction test, the volar forearm axis was perpendicular to the sliding direction. The measured friction force corresponds to the friction of the probe on the skin, generating flexion of the two copper sheets. This flexion was transduced in an electrical signal by the strain gauges. The output signal was converted to digital data and collected by the computer, which displayed strain gauge deformations in related forces (mN).

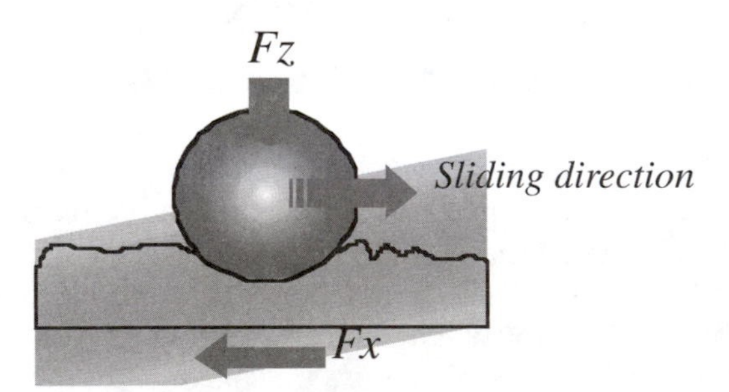

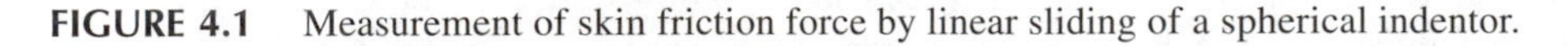
FIGURE 4.1 Measurement of skin friction force by linear sliding of a spherical indentor.

Measurements of the Friction Force

Friction Force Measurements of Calibrated Steel Roughness Plate

The ability of the equipment used is studied by measuring the friction force of rough steel, with different mean arithmetic roughness Ra = 1.6 to 50 μm. Tests were performed using a normal load of 20 g. All curves represent the friction force vs. displacement. The measured profiles indicated irregularities of the surface obtained by manufacturing which produced changes in the frictional force measurements (Figure 4.2).

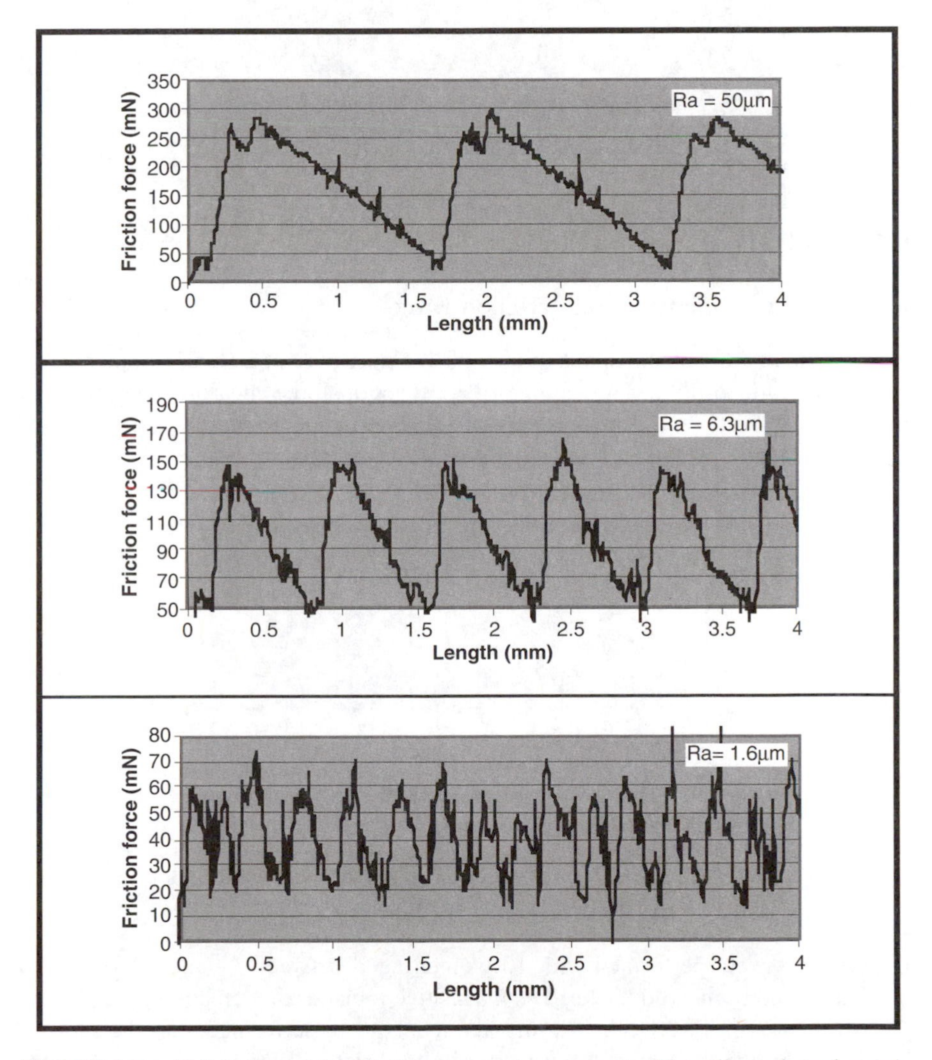

FIGURE 4.2 Validation of the friction force measurement with calibrated steel roughness plate.

Friction Force of a Silicone

The friction forces vs. the normal load were measured on the polymeric material at a low velocity (v = 0.27 mm s^{-1}) (Figure 4.3). The static force, defined as the critical force required to allow friction, was higher than friction force with an observed relaxation.

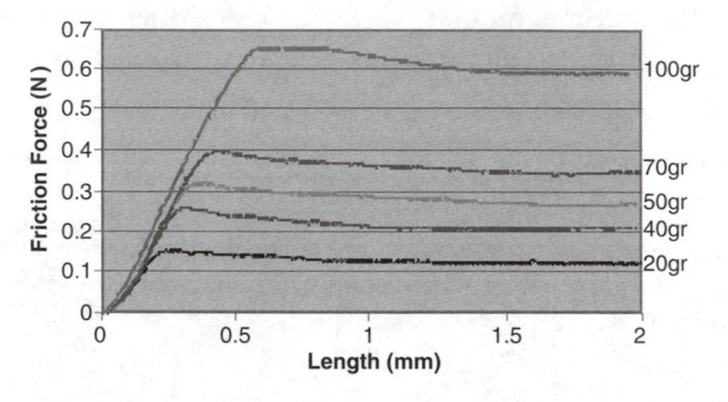

FIGURE 4.3 Friction force of silicone.

In Vivo Measurements of Skin Friction Force

Measurements were obtained in a room with a temperature of 20 to 25°C and relative humidity of 40 to 50%. An example of the friction force measured on the volar forearm of a 45-year-old woman is presented in Figure 4.4. The tests were performed at normal load of 5 g and a low velocity of 0.27 mm s^{-1}.

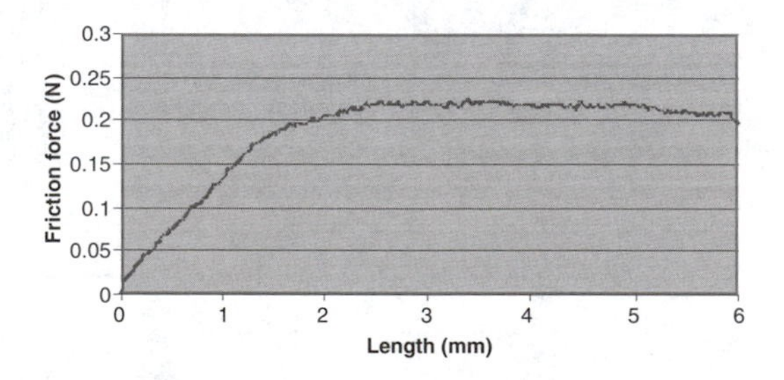

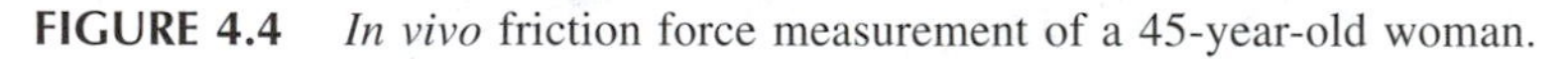

FIGURE 4.4 *In vivo* friction force measurement of a 45-year-old woman.

MOISTURIZATION EFFECT IN MECHANICAL AND FRICTIONAL PROPERTIES OF THE SKIN

Experiments were performed with 10 women aged between 40 and 49 years. All participants were in good health. The insensitive moisturizing cream was applied to the volar forearm for 2 weeks. All measurements were performed after the subjects had been physically inactive for at least 15 min. Friction measurements were made on the volar aspect of their forearms before and after the application of the cream. The normal applied load used for mechanical and frictional properties was fixed at 50 mN. The contact was moved repeatedly along the arm over a length of about

6 mm, with imposed sliding velocity of 0.27 mm s^{-1}. The mechanical and frictional parameters were assessed with ten tests.

Effect in Mechanical Parameters

During the first contact between the spherical indentor and skin surface under a normal load F_z, it was assumed that the skin deforms in an elastic manner. If the indentor has radius R, the radius of the circle of contact formed under a load is given by[23]

$$a^3 = \frac{3}{4} F_z R \frac{1 - \nu^2}{E} \tag{4.1}$$

where ν is Poisson's coefficient and E is Young's modulus of the skin. The Hertzian theory[24] gives the pressure p at any point distant r from the center as

$$p = \frac{3}{2} \frac{F_z}{\pi a^2} \left(1 - \frac{r^2}{a^2}\right) \tag{4.2}$$

The indentation depth, δ, as a function of load, indentor radius, and skin parameters, is given by[24] (Figure 4.5)

$$\delta = \frac{a^2}{R} = \left(\frac{9}{16R}\right)^{1/3} \left(\frac{1 - \nu^2}{E}\right)^{2/3} F_z^{2/3} \tag{4.3}$$

In this first stage of contact, an approximate modulus can be evaluated by introducing into Equation 4.3 the measured value of the indentation depth δ, the radius of curvature of the indentor R, and the skin Poisson's coefficient ($\nu \approx 0.45$).

The effect of moisturization in mechanical parameters is presented in Figure 4.6. The results show an increase of the mean contact radius and the mean indentation depth vs. the age group. On the basis of Equations 4.1 and 4.3, this increase in contact parameters was the result of Young's modulus reduction (Figure 4.7). The reduction in the elastic modulus of the skin as the moisture content is increased has

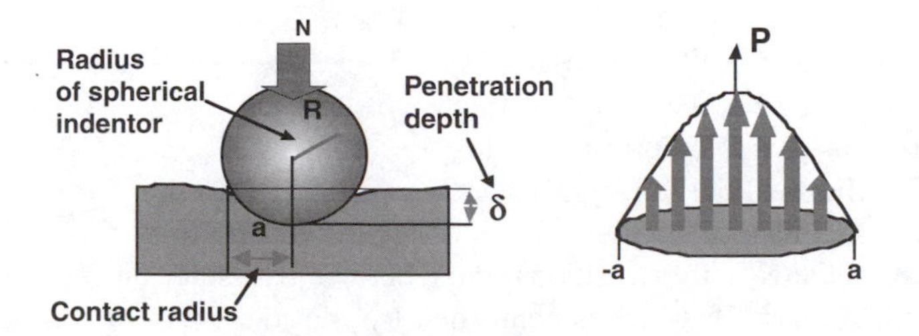

FIGURE 4.5 Contact parameters during a static contact between the spherical indentor and skin surface.

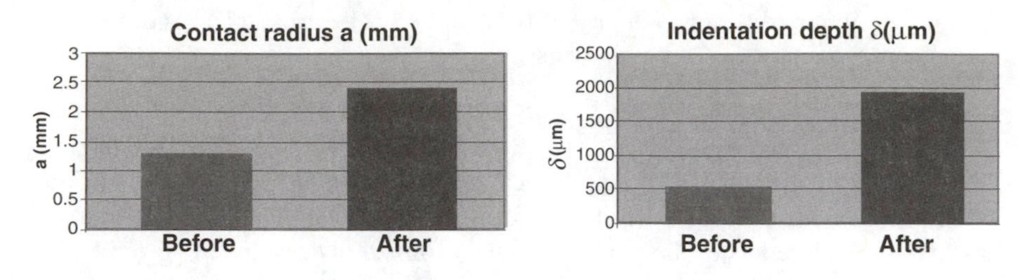

FIGURE 4.6 Effect of moisturization in contact mechanics.

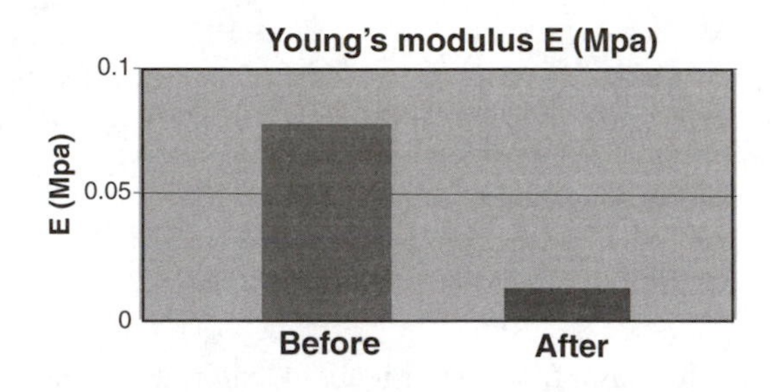

FIGURE 4.7 Effect of moisturization in elasticity of the skin surface.

been observed both *in vitro*[25] and *in vivo*. According to the literature,[26] water reduces the elastic modulus of skin in the range 2 to 2000. This study shows that after 2 weeks of application of an insensitive moisturizing cream, the factor of reduction of Young's modulus varies in the range of 4 to 7, depending on the age group.

Effect in Static and Dynamic Frictional Properties of the Skin

When the contact is moved along the arm over a length of about 6 mm with a sliding velocity of about 0.25 mm/s, the friction force F_x of the skin surface can be characterized in three distinct phases (Figure 4.8).

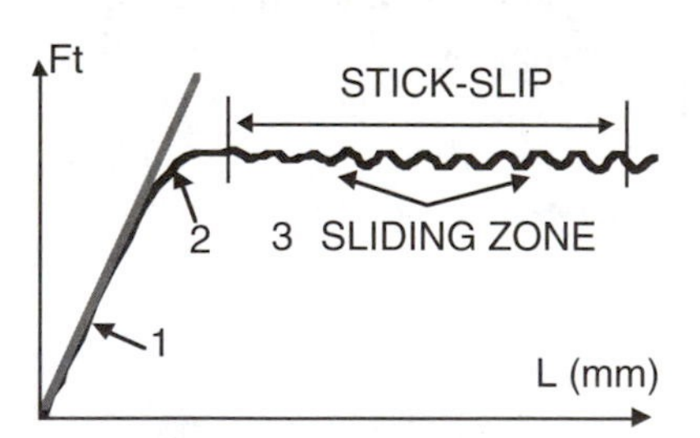

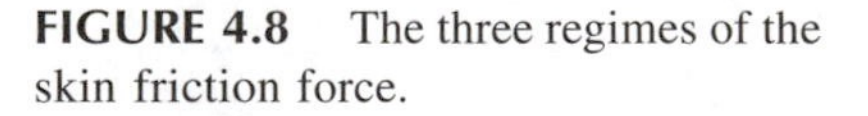
FIGURE 4.8 The three regimes of the skin friction force.

In phase I the lateral force increases linearly to reach a maximum that corresponds to the static limit friction force. During this sliding resistance phase, the slope of the linear curve corresponds to the lateral stiffness of the skin K_x and the shear elastic modulus G.

The slope change in the transition regime between the static and dynamic friction corresponds to phase II, which is controlled by the maximum of the lateral stress and the creep of the skin relief during the contact.

During the dynamic friction phase (phase III), an intermittent "stick–slip" motion can be observed depending on the conditions. The slopes of the "stick" phases of the motion provide information about the local stiffness of the skin. Figure 4.9 shows the result of the friction force before and after hydration over the course of 3 weeks. The increase of the friction force was observed for all participants, as shown in Figure 4.10. It is reasonable to conclude that the increase in friction force is due to the increase in the real area of contact and the reduction of the elastic modulus.

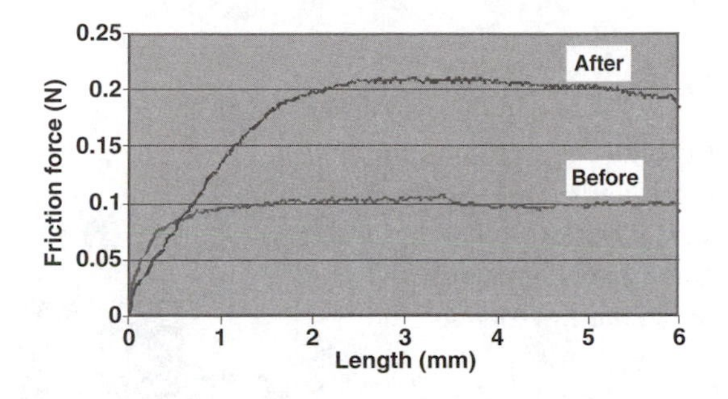

FIGURE 4.9 Effect of moisturization in the total friction force.

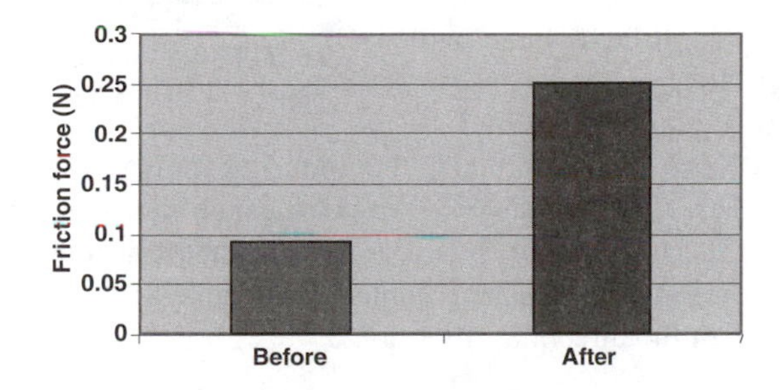

FIGURE 4.10 The increase of the friction force of moisturized skin surface.

THE SKIN FRICTION MODEL

The total friction force, F_{tot}, may be given by the sum of two noninteracting terms — a deformation term, F_{def}, and an interfacial adhesion term, F_{int}; thus,

$$F_{tot} = F_{def} + F_{int} \tag{4.4}$$

In general, the friction deformation term, F_{def}, of viscoelastic material is explained as a fraction of the frictional work done per unit sliding distance:

$$F_{def} = \alpha W \tag{4.5}$$

The parameter W describes the deformation work done; for example, per unit sliding distance, and α the fraction dissipated.

The interfacial adhesion term, F_{int}, is the interface frictional work described by

$$F_{int} = SA \tag{4.6}$$

where A is the real contact area and S is a parameter termed the interface shear stress. This is the adhesion model of friction and it is supposed that the work is transmitted to the surface layer by the action of adhesive forces operating at the area of contact.

Deformation Component of Friction

Friction may be usefully regarded as the energy dissipated per unit sliding distance. For a deformable substrate, which is expended at both the interface and in the bulk of the material during friction, the deformation component arises from the work dissipated by the rigid indentor in deforming the material. For viscoelastic materials, such as skin, the deformation–recovery cycle is associated with viscous energy losses. According to the experimental work examined by Greenwood and Tabor,[27] the deformation component over the front half of the circle of contact (Figure 4.11) can be integrated over the annulus at constant point distance r from the center. Then the pressure p will be the same at every point of each annulus. The sum of the mean forces in the contact intervalue (0 to a) gives the total deformation force in the direction of motion as

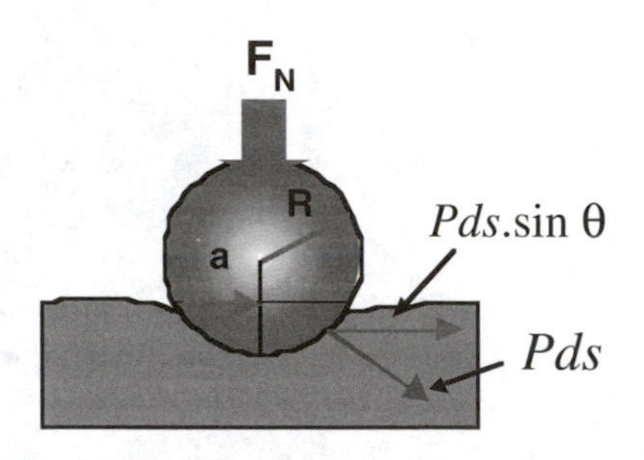

FIGURE 4.11 The deformation component of the friction force.

$$F_{def} = \frac{3F_z a}{16R} \tag{4.7}$$

by replacing Expression 4.1 in Equation 4.5, the total deformation force is given by

$$F_{def} = \left(\frac{9}{128R}\right)^{2/3} \left(\frac{1-\nu^2}{E}\right)^{1/3} F_z^{4/3} \tag{4.8}$$

If the skin were ideally elastic, the skin behind the sphere would yield an identical amount and no energy would be lost. If it is now assumed that, in fact, a constant fraction α of the input elastic energy is lost as a result of elastic hysteresis in the skin, the deformation force is given by

$$F_{def} = \alpha\left(\frac{9}{128R}\right)^{2/3} \left(\frac{1-\nu^2}{E}\right)^{1/3} F_z^{4/3} \tag{4.9}$$

where α is the viscoelastic hysteresis loss fraction. For sinusoidal-solicitation systems, α is equivalent to $\pi \tan \gamma$, where γ is the loss tangent of the viscoelastic substrate and is given by the ratio of the loss and storage modulus.

Some cyclic loading–unloading measurements similar to the normal indentation test have shown that for rates of deformation the first cycle involved an energy loss due to hysteresis, which varied in the range of α = 20 to 35% of the total elastic energy deformation. On the basis of Equation 4.7, the deformation component was determined for each subject before and after hydration. The result shows an increase of the deformation component as a result of the change of Young's modulus and contact parameters after hydration (Figure 4.12).

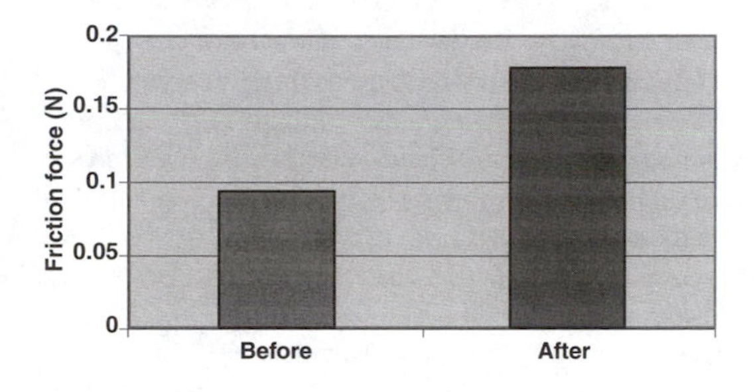

FIGURE 4.12 Effect of moisturization on skin deformation.

CONCLUSIONS

The present study describes the physical significance of the friction force in the static and dynamic regimes by computing the functional parameters connected with the intrinsic response of the skin surface. The equipment used employed a more conventional tribological configuration based on a linear sliding of a rigid spherical probe under a small load, on an inner human forearm. The large increase in friction of moisturized skin is a result of moisture-dependent mechanical properties of the stratum corneum. A better interpretation of the frictional property changes can be achieved by taking into account the physicochemical properties of the skin area tested.

REFERENCES

1. Naylor, P.F.D., The skin surface and friction, *Br. J. Dermatol.*, 67, 239–248, 1955.
2. Comaish, S. and Bottoms, E., The skin and friction: deviations from Amonton's laws, and the effects of hydration and lubrication, *Br. J. Dermatol.*, 84, 37–43, 1971.
3. Cosmaish, S., Glutaraldehyde lowers skin friction and enhances skin resistance to acute friction injury, *Acta Derm. Venereol.* (Stockholm), 53, 455–460, 1973.
4. Cosmaish, J.S., Harborow, P.R.H., and Hofman, D.A., A hand-held friction meter, *Br. J. Dermatol.*, 89, 33–35, 1973.
5. El-Shimi, A.F., *In vivo* skin friction measurements, in *Journal of Consumers Studies and Home Economics*, Rees, A.M., Ed., Blackwell Scientific Publications, London, 1976, 37–51.
6. Highley, D.R., Coomey, M., Denbeste, M., and Wolfram, L.J., Frictional properties of skin, *J. Invest. Dermatol.*, 69, 303–305, 1977.

7. Hills, R.J., Unsworth, A., and Ive, F.A., A comparative study of the frictional properties of emollient bath additives using porcine skin, *Br. J. Dermatol.*, 130, 37–41, 1994.
8. Prall, J.K., Instrumental evaluation on the effects of cosmetic products on skin surfaces with particular reference to smoothness, *J. Soc. Cosmet. Chem.*, 24, 693–707, 1973.
9. Henricsson, V., Svensson, A., Olsson, H., and Axell, T., Evaluation of a new device for measuring oral mucosal surface friction, *Scand. J. Dent. Res.*, 98, 529–536, 1990.
10. Elsner, P., Wilhelm, K-P., and Maibach, H.I., Frictional properties of human forearm and vulvar skin: influence of age and correlation with transepidermal water loss and capacitance, *Dermatologica*, 181, 88–91, 1990.
11. Lodén, M., Olsson, H., Skare, L., and Axell, T., *J. Soc. Cosmet. Chem.*, 43, 13, 1992.
12. Cua, A.B., Wilhelm, K-P., and Maibach, H.I., Frictional properties of human skin: relation to age, sex and anatomical region, stratum corneum hydration and transepidermal water loss, *Br. J. Dermatol.*, 123, 473–479, 1990.
13. Cua, A.B., Wilhelm, K-P., and Maibach, H.I., Frictional properties of human skin: relation to age, sex and anatomical region, stratum corneum hydration and transepidermal water loss, *Br. J. Dermatol.*, 123, 473–479, 1990.
14. Cua, A.B., Wilhelm, K-P., and Maibach, H.I., Skin surface lipid and skin friction: relation to age, sex and anatomical region, *Skin Pharmacol.*, 8, 246–251, 1995.
15. Manuskiatti, W., Schwindt, D.A., and Maibach, H.I., Influence of age, anatomic site and race on skin roughness and scaliness, *Dermatology*, 196, 401–407, 1998.
16. Wijn, P.F.F., The Alinear Viscoelastic Properties of Human Skin *in Vivo* for Small Deformations, Ph.D. thesis, Catholic University of Nijmengen, the Netherlands, 1980.
17. Parot, S. and Bourliere, F., A new technique for measuring compressibility of skin and subcutaneous tissue. Influence of sex, age and body area, *Gerontology*, 13, 95–110, 1967.
18. Piérard, G.E., Mechanical properties of aged skin: indentation and elevation experiments, in *Aging Skin*, Leveque, J.L. and Agache, G.P., Eds., Marcel Dekker, New York, 1993, 49–55.
19. Zahouani, H., Lee, S.H., Vargiolu, R., Asserin, J., and Humbert, P., Mathematical microscopy analysis of the skin relief, Presented at International Symposium of Bioengineering and the Skin, Boston, June 25–27, 1998.
20. Mavon, A., Zahouani, H., Redoules, D., Agache, P., Gall, Y., and Humbert, P., Sebum and stratum corneum lipids increase human skin surface free energy as determined from contact angle measurements: a study on two anatomical sites, *Colloids Surfaces B Biointerfaces*, 8, 147–155, 1997.
21. Corcuff, P., de Rigal, J., and Leveque, J.L., Skin relief and aging, *J. Soc. Cosmet. Chem.*, 34, 177–190, 1983.
22. Asserin, J., Zahouani, H., Humbert, P., Couturaud, V., and Mougin, D., Measurements of the friction coefficient of the human skin *in vivo*. Quantification of the cutaneous smoothness, *Colloids Surfaces B Biointerfaces*, 19, 1–12, 2000.
23. Timoshinko, S.P. and Goodier, J.N., *Theory of Elasticity,* McGraw-Hill, New York, 1970.
24. Hertz, H., On the elastic contact of elastic solids, *J. Reine Angew. Math.*, 92, 156–171, 1881.
25. Park, A.C. and Baddiel, C.B., *J. Soc. Cosmet. Chem.*, 23, 3, 1972.
26. Takahashi, M., Yamad, M., Machida, Y., and Tsuda, Y., *J. Soc. Cosmet. Chem.*, 36, 335, 1985.
27. Greenwood, J.A. and Tabor, D., The friction of hard sliders on lubricated rubber: the importance of deformation losses, *Proc. Phys. Soc.*, 71, 989–1001, 1957.

Section II

Elasticity and Viscoelasticity

Part 1

General Aspects

5 Hardware and Basic Principles of the Dermal Torque Meter

Jean de Rigal

CONTENTS

Introduction 63
Basic Principle of the Hardware 65
The Probe 65
The Control Unit 66
Computers, Printers, and Plotters 66
Basic Mechanical Principle of the Twistometer and DTM 66
Interpreting the Results 68
The Various Rheological Models 68
The Data Processing Modes Used by the DTM 70
The Basic Model Tool 71
The PEXP Curve-Fit Tool 71
Precautions and Remarks 71
Standard Precautions 72
Influence of Probe Position 72
Influence of Preconditioning or Serial Stresses 74
Conclusions 75
References 75

INTRODUCTION

Complete determination of the mechanical properties of a material requires exposing it to a stress that is as clearly defined as possible over a perfectly determined geometry. The simplest example is in determination of the mechanical characteristics of an isotropic material that can be simply deformed by elongation in a single direction. The use of geometrically well-defined test assays then enables determination of all the intrinsic characteristics.

If the material is not homogeneous, anisotropic, and, like the skin, is subject to natural stresses, the problem becomes particularly complex. It is necessary to conduct a large number of tests under different configurations to stress the test specimen in

0-8493-7521-5/02/$0.00+$1.50

all directions. Failing that, determination of the intrinsic parameters can only be approximate. A structure consisting of a pile of layers, as is the case with the skin, cannot be isotropic. At best, if the layers are isotropic, the structure will exhibit transverse isotropism, necessitating determination of five elastic constants to characterize it completely. A certain number of finite-element modeling procedures[1] have shown that that degree of asymmetry is not sufficient to describe the behavior of the skin, which is at least orthotropic (8 elastic constants) or even completely anisotropic (21 elastic constants). It should also be noted that the skin is not elastic, but viscoelastic, and so the viscous characteristics must also be determined to describe completely the mechanical behavior of the skin. That approach, already highly complex *in vitro*, on test assays well-defined, becomes even more difficult *in vivo* when the material is attached to deep lying layers such as the adipose tissue. In addition, measuring skin properties *in vivo* is further complicated because only one surface of the sample is accessible.

It is thus clear that it is impossible to use simple means to characterize the skin fully. The objective will thus be to determine its properties using a rheological model that is as pertinent as possible, while clearly stating the set of measurement conditions (geometry, stresses, stress mode, etc.).

Numerous methods have been proposed to resolve the problem of *in vivo* determination of the mechanical characteristics of the skin. Several modes of applying stress have been proposed, such as uniaxial extensometry; in the plane[2] of the skin, or perpendicular to that plane; as in ballistometry,[3] the propagation of mechanical or sound waves,[4] and torque deformations.[5]

Concerning the most widely used methods, the general principles of determination are of two types: (1) constant-force stress and measurement of the deformation (i.e., creep mode) or (2) deformation of imposed amplitude and measurement of the stresses (i.e., relaxation mode). The stress or deformation may be imposed abruptly or slowly. The rate of imposition will condition the accounting for, and hence the determination of, the viscosity phenomena and the associated time constants.

The Twistometer and the Dermal Torque Meter (DTM), the latter a copy of the former, thus attempt to resolve certain of the above problems. The deformations are applied in the plane of the skin to minimize the contribution of the deep layers, as Wlasbloom[5] has shown. The stresses are applied in rotation, hence with an axis of symmetry. This solution does not eliminate the contribution of natural tensions, but averages them and makes measurement independent of the orientation of the imposed stress relative to the natural tension lines (Langer's lines), in contrast to uniaxial methods.[6] The general principle of the DTM is of the creep type, with an abrupt rise in stress.

Torsion methods have been available for some time.[7–9] The development of miniaturized electromechanical components enabled design of the Twistometer, then the DTM. Those instruments are now portable and can be used on all body areas of sufficient flatness and width (about 5 cm). The applications are in the fields of cosmetics (determination of hydration of the *stratum corneum*,[10] effects of sun,[11,12] effects of aging,[13] and effects of photoaging[14]) and dermatology (follow-up of disease course[15] and treatment effect monitoring).[16]

BASIC PRINCIPLE OF THE HARDWARE

THE PROBE

The probe consists of a torque motor and an angle sensor (Figure 5.1). The torque motor is rigidly locked to a flat disk at the probe head, which is referred to as the "torque disk." A ring, referred to as the "guard ring" or "torque ring," surrounds the disk. Both are glued onto the skin by means of two-sided adhesive tape. The torque disk imposes the stress on the skin and the immobile guard ring limits the propagation of the deformation.

The gap between the torque disk and the guard ring is the "operating gap" (OG). Three torque rings are available; depending on the question asked and the problem to be solved, a 1-, 3-, or 5-mm ring may be selected. For assessment of the mechanical properties of the skin surface, a 1-mm OG is required. For determination of deeper mechanical properties, a 3- or 5-mm OG is required. It should be borne in mind that the reproducibility of the measurements decreases as the OG width increases.

When the torque is applied for a given duration the torque disk moves to a degree dependent on the suppleness and viscosity of the tissue, and the resulting angular deformation is recorded. When the torque is switched off, the remaining tensile strength of the skin moves the torque disk back toward the initial position, and this phase of the response is recorded. During this phase, the skin is the only source of the forces acting in opposition to the mechanical friction of the probe. The deformation angle recording stops when the equilibrium between the two opposing forces is reached, so the remaining deformation is not simply a pure characteristic of the skin.

A two-way switch on the body of the probe enables one to zero the torque motor before measurement, and the second position of the switch applies the torque. The

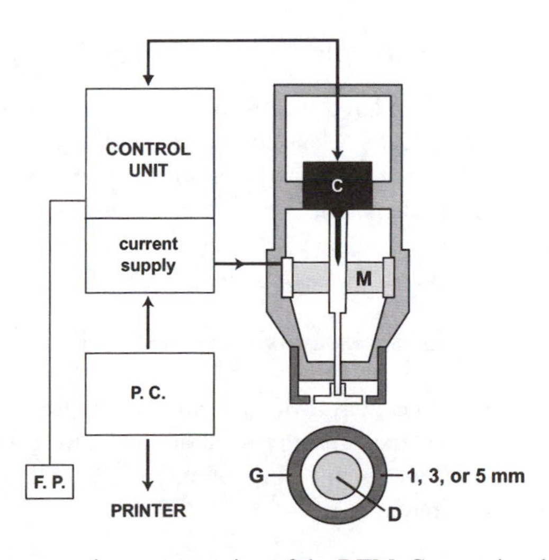

FIGURE 5.1 Diagrammatic representation of the DTM; C = rotational sensor, M = torque motor, D = torque disk, G = torque ring, D = operating gap.

period when the torque is applied is referred as the "torque-on" phase, and when the torque is removed is referred to as the "torque-off" phase. A complete measurement consists of these two phases.

The Control Unit

The control unit (see Figure 5.1) contains all the electronic circuits necessary to power the torque motor, the amplifier, the rotating sensor, and for relaying data to and from the computer. A foot pedal connected to the control unit enables the operator to initiate the measurements. The probe cable is also connected to the control unit. Thus, all the instructions and characteristics of the measurement or the set of determinations are controlled by computer via the control unit.

Computers, Printers, and Plotters

The DTM is connected to the serial port of a computer (see Figure 5.1), which cannot be used in stand-alone mode. Rather, a PC with a printer or a plotter to print the results or curves is required. The setup enables a standard computer to be used with no change in hardware. The user-friendly program contains all the functions commonly used to collect, store, and analyze the response curve of the skin submitted to the torque. The program is window-based and consists of various menus for the main functions.

All the measurement parameters (torque value, duration of the torque-on phase, duration of the torque-off phase) are defined and activated by the computer. The response curve is divided into two sections; torque-on and torque-off, which are analyzed separately on the basis of a rheological model. The essential parameters are extracted from each of the curves and stored.

Basic Mechanical Principle of the Twistometer and DTM

The two systems may be considered electromechanical oscillators connected to a spring exhibiting a variable degree of viscous recoil in the skin. The oscillator has its own inertia, friction, and viscosity characteristics, which may be added to those of the skin or may oppose them (friction in the torque-off phase). The dynamic response of the mobile equipment may be described by a differential equation of the following type (harmonic oscillator-type response):

$$Jd^2\theta/dt^2 + \eta d\theta/dt + K_o\theta = C_o$$

where J is the moment of the inertia of the mobile component, C_o the imposed torque, K_o the stiffness of the recoil spring (the skin), and η the coefficient of viscous friction of the system skin plus mobile component.

The solution of this differential equation is given by

$$\theta(t) = C_o/K_o + A\exp(-at)\sin(bt + c)$$

where $a = \eta/2J$ and $b = {}^1/_2J(4K_oJ - \eta2)^{1/2}$.

In the creep direction (torque-on), there is an inertial system with a recoil torque (skin). In the torque-off direction, there is an inertial system with no recoil torque but with friction.

For very short time periods, the profile of the response thus depends on the ratio between the values of the two parameters, J and η. If damping is marked, relative to the inertial moment, the response will have a continuous profile (Figure 5.2). If the damping is very small, an oscillating response will occur with duration depending on the ratio of the inertia to the damping. In a purely elastic material, the limit value for a long time period is C_o/K_o. These considerations show that analysis of the signal recorded by the DTM or Twistometer over very short time periods are conditioned by the ratio between the inertia and damping. If damping is marked, the elastic response, U_e, can never be achieved; the response would pass continuously from the elastic phase to the viscous one. For short time intervals, the determination is related more to the apparatus and the viscous response of the skin and less to the elastic component.

An illustration of the profile of the response recorded on skin using the Twistometer is given in Figure 5.3. The recording was conducted at two different rates to clearly image the dynamic section (176 ms) and the quasi-static section (5 s). The recording was obtained at "critical" damping, yielding a single oscillation or incipient oscillation for an amplitude of rotation, $U_e = 10.32°$. The dynamic phase is even more clearly imaged in the torque-off section since the resisting forces are weak. In the case of the skin, which is not a pure elastic material, the viscous deformation takes over and is added to the elastic deformation, as can be observed in the curve in Figure 5.3.

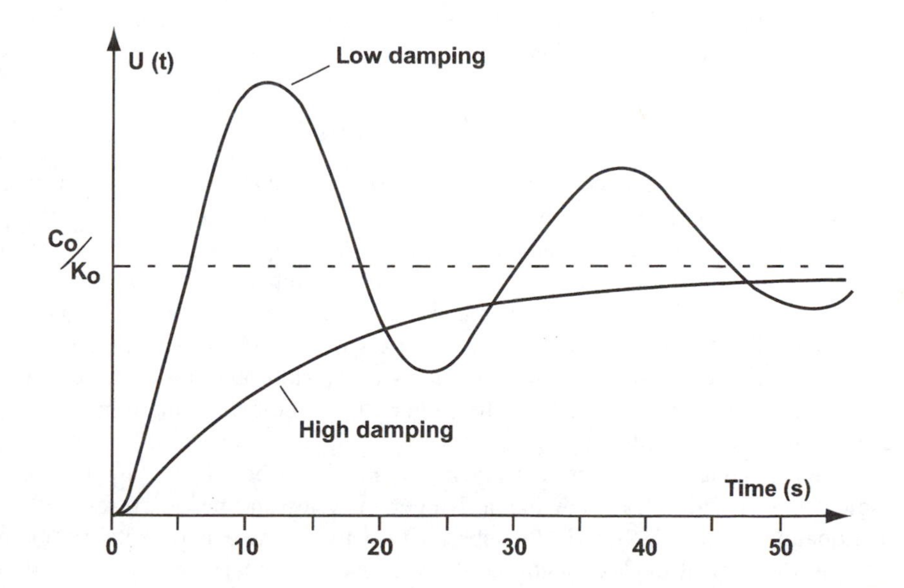

FIGURE 5.2 Response profile for a harmonic oscillator with a force scale.

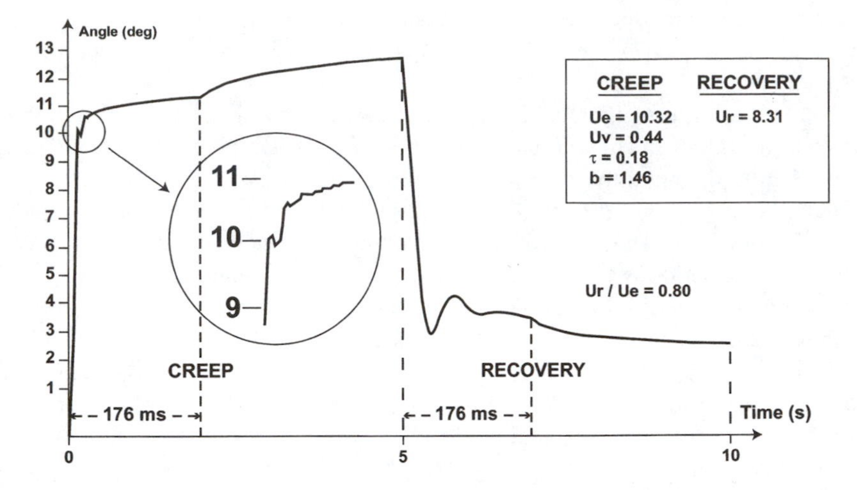

FIGURE 5.3 Skin response profile recorded using the Twistometer (critical damping mode) and parameters calculated using the rheological model of Equation 5.3.

INTERPRETING THE RESULTS

The Various Rheological Models

Numerous analytical methods for the response of skin subjected to stress have been proposed. Their aim is to extract the extrinsic parameters that characterize the behavior of the tissue under study, based on the simplest rheological model possible, but a model that describes all the properties of the skin. The oldest rheological model consists in the following equation:

$$U(t) = U_e + U_v(1 - \exp(-t/\tau)) \tag{5.1}$$

The correspondences between the main parameters and their physical significance are outlined in Figure 5.5b below.

This model, consisting of an immediate elastic response, U_e, and a delayed viscoelastic deformation, is not totally satisfactory. Comparison of the curves generated experimentally with the mathematical fit (Equation 5.1) shows that a bias is present and the deviations between the two curves are not randomly distributed. The model systematically overevaluates the values at the start and end of deformation and underevaluates those in the middle (Figure 5.4). The bias is more marked the greater the OG and/or the torque.

Another equation, proposed by Pichon et al.,[17] and based on Burger's rheological model with a non-Newtonian damper in series and hence corresponding to Equation 5.2, yields a markedly superior fit between the experimental response and the rheological model. Pichon et al.[17] studied the response curve of the skin using the finite-element method proposed by de Lacombe in 1939.[18] The method has the advantage of independence of immediate extensibility. Pichon et al.[17] have shown that the value of 1/3, commonly known as Andrade's constant,[19] applies equally

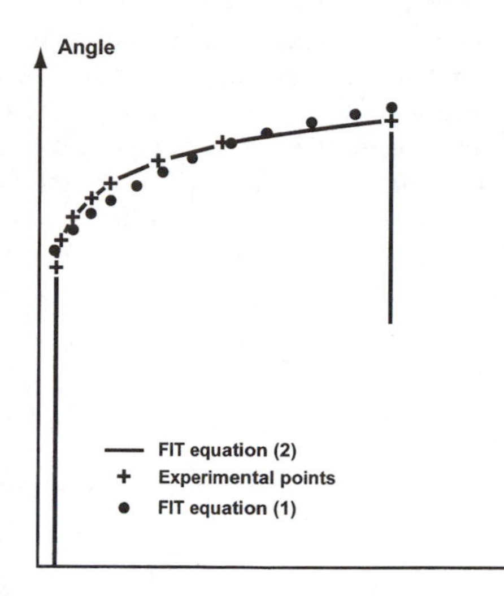

FIGURE 5.4 Comparison of the rheological models for Equations 5.1 and 5.2 with the experimental curve.

well to determinations of the mechanical properties of the dermis (OG = 3 or 5 mm) and epidermis (OG = 1 mm) and that the value is a mean valid for all subjects.

$$U(t) = U_e + U_v(1 - \exp(-t/\tau)) + bt^{1/3} \tag{5.2}$$

The term U_e is the elastic response, the term U_v is the fast or transient creep, and the term $bt^{1/3}$ is the stationary creep (Figure 5.5c).

A diagrammatic representation of Burger's model and the fit between the model and the calculated parameters are given in Figure 5.6.

Another value for the time dependence of the stationary creep term has been proposed by Sanders,[7] Wjin,[8] and Finlay,[9] who all assigned it the value of 1. This becomes a Newtonian Burger model:

$$U(t) = U_e + U_v(1 - \exp(-t/\tau)) + bt \tag{5.3}$$

The model can be enhanced by adding Voight's component to the Burger model.

Much more recently, another model was proposed by Salter et al.,[20] based on Weibull's theory. The equation is as follows:

$$U(t) = A(1 - \exp(Ct)^n) \tag{5.4}$$

In this equation, it is much more difficult to provide a physical interpretation for parameters A, C, and n. Wickett and Murray[21] have shown that the equation does not generate any additional information and is not totally satisfactory.

The publication presenting the model[20] based on Weibull's theory shows that it is impossible to image and measure the component of elastic deformation denoted U_e. The impossibility is due only to the markedly damped response of the mobile component (see Figure 5.2). Thus, the model is only applicable under those conditions and will not fit a system with little damping and with oscillations.

Fundamentally, the rheological models, as presented above (Equations 5.1 through 5.3), cannot take into account the dynamic part of the response and only apply to static or quasi-static deformations. The analysis of the dynamic response, skin plus mobile equipment, necessitates modeling the differential equation indicated above and separating the specific contributions of the apparatus and the skin. Currently, no one has advanced in that direction.

The Data Processing Modes Used by the DTM

Two processing modes are available on the DTM:

1. The basic model tool
2. The Pexp curve fit tool

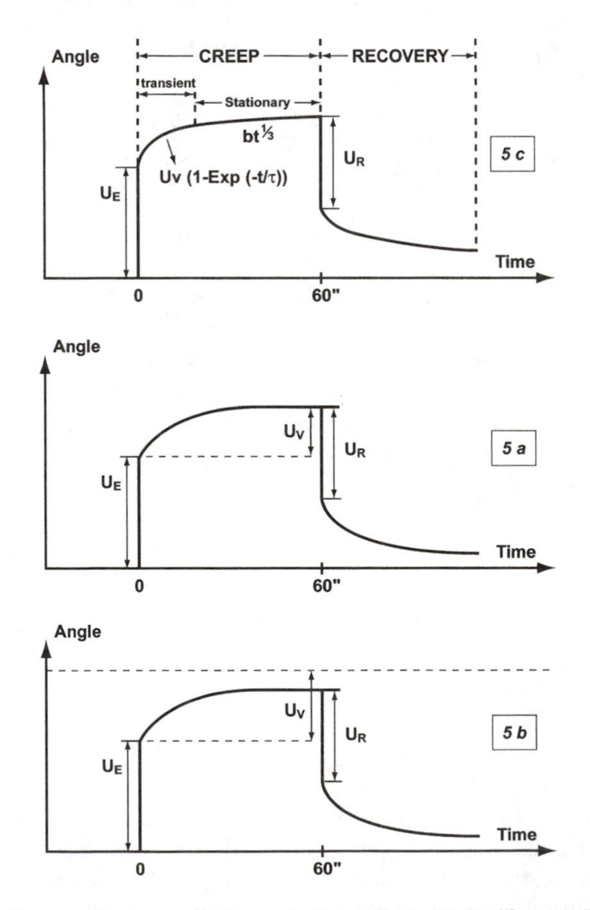

FIGURE 5.5 The various parameters and their physical significance for each of the rheological models covered.

Those data processing modes can be run online or subsequently. The program processes the torque-on and torque-off phases separately.

The Basic Model Tool

The determination of parameters U_e, U_v, U_r, and U_r/U_e is conducted directly on the curve recorded 50 ms after application or discontinuation of the torque (Figure 5.5a). Normally the measurement times have been defined to minimize the influence of the dynamic response.

The PEXP Curve-Fit Tool

The DTM uses the Equation 5.3 model (linear dependence over time for stationary creep). To ensure the condition for static or quasi-static deformation, processing does not take into account the first 100 ms in the torque-on or torque-off phases. It is clear that, in the example given in Figure 5.3, in the torque-off phase, the dynamic phase is not complete and the U_r determination will be subject to error.

Precautions and Remarks

It is important to note that the parameters determined, while they bear the same name, do not all have the same significance. In fact, the basic model parameters, U_e and U_r, may include some viscosity (Figure 5.5a). Parameters U_v and U_f are deformation amplitude parameters determined for a given time period and are only meaningful if the duration of torque application is specified. As is the case with all mechanical properties, the results of the determinations depend on the geometric

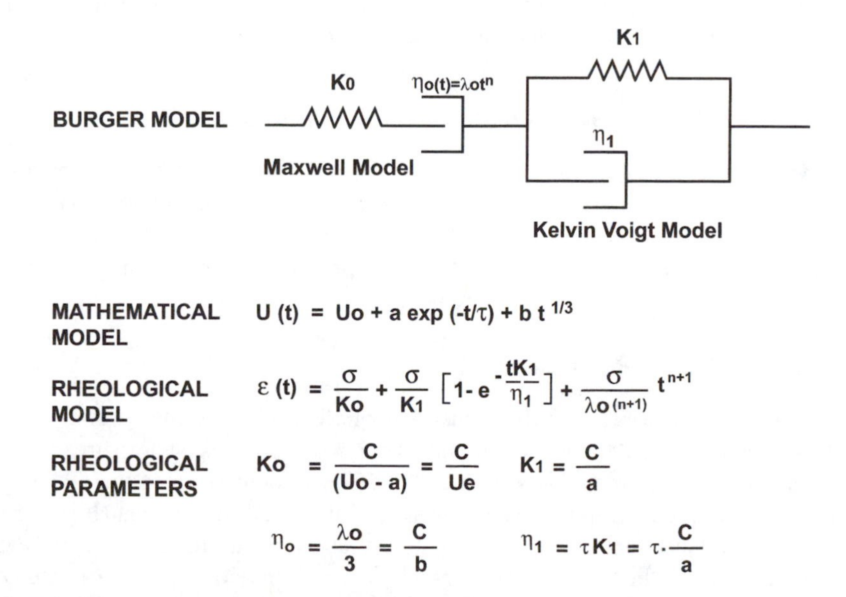

FIGURE 5.6 Burger's model and the rheological parameters calculated from the mathematical model, fitting the response curve.

characteristics of the test assay, in this case the thickness of the skin and the width of the OG. It is tempting to calculate the ratios between the various magnitudes determined to obtain independent parameters for the thickness and/or without dimensions. That is the case, for example, with U_r/U_e (elasticity).

In contrast, the parameters U_v/U_e, U_f/U_e, and U_f/U_v are not dimensionless magnitudes and depend on time and the duration of torque application, if generated using the basic model. Using those parameters, it is possible to calculate the intrinsic magnitudes, taking into account the thickness of the skin. The latter should be measured in all long-duration tests in which a variation in tissue thickness may be suspected and in all tests in which differences in thickness exist (solar effect, age, disease, etc.). The formulas enabling calculation of those intrinsic magnitudes (modulus of elasticity or Young's modulus and shear modulus) are obtained by considering the material to be isotropic (but that is an approximation).

For example, for elastic parameters and in the case of torque determination, the formula for the shear modulus is given by

$$G = (1/r^2 - 1/R^2)\ (C \sin \theta)\ /4\pi e$$

where e is the thickness of the tissue, C the torque applied, θ the angle of rotation, and r and R the internal and external radii of the crown of skin.

Other parameters concerning the viscous phase may be calculated.[22]

In the measurement probe, mechanical friction is inevitable. For the torque-on phase, friction normally remains negligible in view of the imposed torque. That does not apply, however, in the torque-off phase, at the end of which the equilibrium between the recoil force of the skin and mechanical friction restricts the return to the initial position. During the torque-off phase the values of the residual deformation and the viscous deformation phase are thus to be considered with caution.

STANDARD PRECAUTIONS

The stress applied to the skin is transmitted to it through the two-sided adhesive tape used; thus, the tape must be capable of transmitting the torque. It is further necessary to ensure that the adhesive forces are sufficient and that, in the case of cosmetic or other formulation efficacy tests, the presence of product traces does not interfere with or influence the measurement.

Influence of Probe Position

As in all determinations, the probe must not interfere with the measurements. The probe should be applied to the measurement zone with the weakest pressure possible, thus allowing the skin to retain its natural plane. The influence of pressure (Table 5.1) and an angle of inclination relative to the skin of the forearm are great (Figure 5.7). The effects induced on the measurement results by positioning errors are more marked, the greater the OG. An almost complete block may occur with a 1-mm OG.

The position of the measurement zone on the body must be clearly specified. Numerous property gradients exist, as is the case with the forearm (Figure 5.8). The same applies to the top and bottom of the back. Concerning determinations on the forearm of athletic subjects, a laterality effect exists (Figure 5.8). The skin is normally subject to natural tensions. Muscle contraction induces variations (Figure 5.9), which, as above, are a function of the OG.

TABLE 5.1
Influence of Probe Application Pressure on the Skin during Determination (relative to the no-pressure value)

	Operating Gap	
Applied Pressure, g/cm²	**5-mm**	**1-mm**
0	1	1
50	0.81	0.96
150	0.77	0.80
500	0.70	0.40

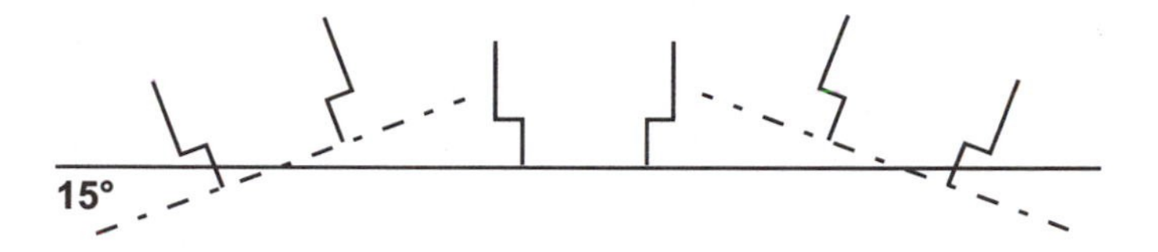

O.G.: 5 mm	1.07	1	1.45
O.G.: 1 mm	0.96	1	1.07

FIGURE 5.7 Influence of measurement probe angle of inclination (expressed relative to the normal value).

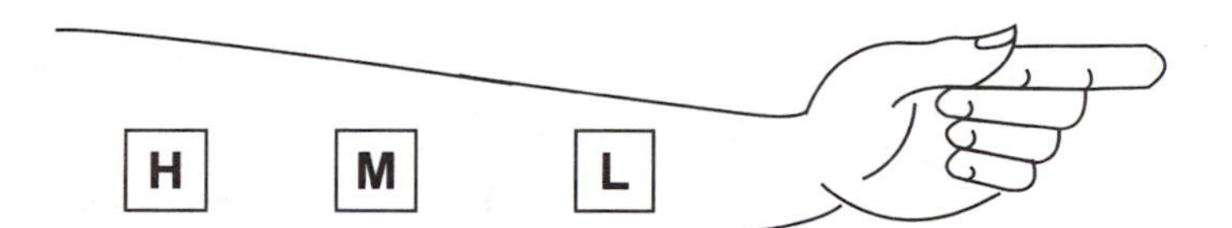

	H	M	L
Left	175 ± 15	156 ± 7	109 ± 11
Right	151 ± 15	135 ± 15	120 ± 15

FIGURE 5.8 Influence of forearm measurement zone and laterality effect (OG = 3 mm, mean ± SEM, nine models).

Influence of Preconditioning or Serial Stresses

The statement that the skin should be prestressed before recording and measuring the mechanical characteristics is sometimes encountered. Some call the procedure normalization. If the torque is applied serially several times, the profile of the curves recorded complies with that in Figure 5.10. This type of test is conducted in mechanics, when it is used as a fatigue test designed to determine the ability of a material to withstand serial stresses. Under those conditions, parameter U_e is no longer the immediate extensibility but includes an increasing viscosity component. It will be observed, using, for example, the model of Equation 5.2, that the sum, $U_e + U_v$, remains constant, but the U_e value increases at the expense of the U_v value.

A study (Table 5.2) was conducted on two groups of 12 models (mean age: 30 and 50 years, respectively). Torque duration was 10 s and each torque was separated from the next by 1 min, without removing the probe from the skin (OG = 3 mm). The rheological model used was that of Equation 5.3. It is clear that repeated stresses impair the mechanical properties, with the decrease in elasticity particularly marked at high torque values. The mechanisms for transfer of viscosity toward extensibility are different for low and high torque and are also a function of skin condition.

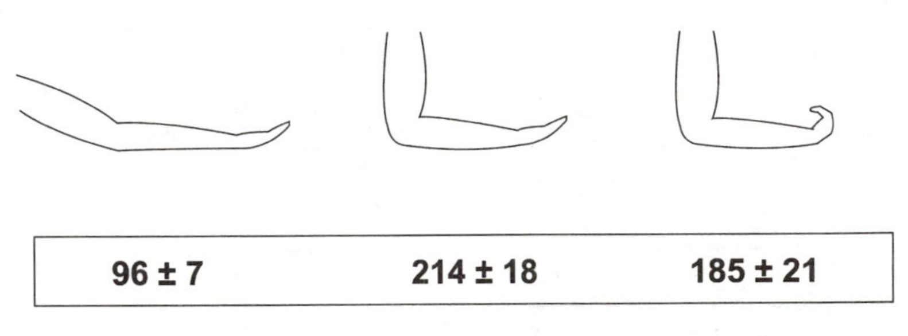

FIGURE 5.9 Influence of forearm position: relaxed extended arm, arm bent 90° and relaxed, arms and hands bent and muscles contracted (OG = 3 mm, mean ± SEM, nine models).

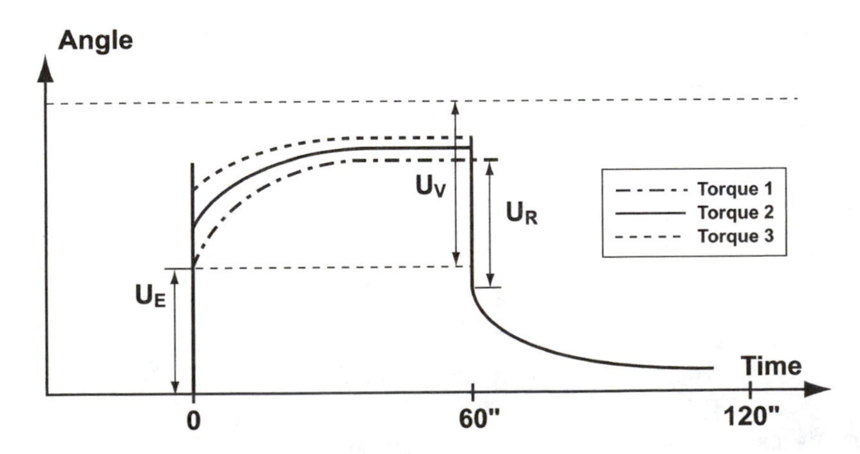

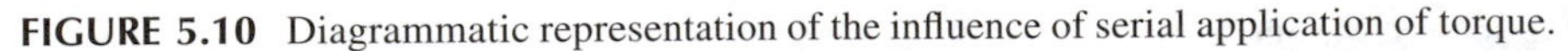

FIGURE 5.10 Diagrammatic representation of the influence of serial application of torque.

TABLE 5.2
Influence of Three Serial Applications of Torque (relative to the initial value, mean ± SEM)

Applied Torque	Parameter	"Young Population" Torque 2 / Torque 1	"Young Population" Torque 3 / Torque 1	"Old Population" Torque 2 / Torque 1	"Old Population" Torque 3 / Torque 1
Low torque, 60 cm·cN	U_e	1.11 ± 0.01	1.15 ± 0.02	1.12 ± 0.02	1.15 ± 0.03
	U_v	0.95 ± 0.05	0.57 ± 0.07	0.64 ± 0.2	0.43 ± 0.12
	b	0.82 ± 0.10	0.75 ± 0.05	0.86 ± 0.05	0.82 ± 0.09
	U_r/U_e	0.98 ± 0.01	0.97 ± 0.02	0.96 ± 0.02	0.93 ± 0.03
High torque, 90 cm·cN	U_e	1.17 ± 0.01	1.24 ± 0.01	1.16 ± 0.03	1.21 ± 0.03
	U_v	0.71 ± 0.04	0.75 ± 0.06	0.81 ± 0.07	0.75 ± 0.06
	b	0.81 ± 0.01	0.76 ± 0.02	0.79 ± 0.03	0.73 ± 0.03
	U_r/U_e	0.91 ± 0.01	0.87 ± 0.02	0.92 ± 0.01	0.90 ± 0.01

CONCLUSIONS

Although the effect that incorrect handling may induce must be taken into account, use of the DTM or Twistometer is relatively simple. The reproducibility of the determinations is good and enables study of numerous phenomena, particularly in the cosmetic field. Caution is nonetheless required with respect to comparing the results derived by different methods, since a full understanding of the meaning of the parameters and the stress modes is necessary. Interpretation of the results is not always easy, and the researcher must exercise both caution and a precise understanding of the phenomena and techniques involved.

REFERENCES

1. Ouisse, M., Etude des propriétés mécaniques de la peau, *Mémoire DEA*, ENS, Cachan, 1997.
2. Wjin, P., The Alinear Viscoelastic Properties of the Human Skin *in Vivo* for Small Deformations, Ph.D. thesis, Catholic University of Nijmegen, the Netherlands, 1980.
3. Tosti, A., Compagno, G., Fazzini, M.L., and Villardita, S., Ballistometry for the study of the plasto-elastic properties of the skin, *J. Invest. Dermatol.*, 69, 315–317, 1977.
4. Potts, R.O., Chrisman, D.A., and Buras, E.M., The dynamic mechanical properties of human skin *in vivo*, *J. Biomech.*, 16, 365–372, 1980.
5. Wlasbloom, D.C., Skin Elasticity, Ph.D. thesis, University of Utrecht, the Netherlands, 1967.
6. Gigbon, T., Stark, T., and Evan, J.H., Directional variation in extensibility of human skin *in vivo*, *J. Biomech.*, 2, 201–204, 1969.
7. Sanders, R., Torsional elasticity of human skin *in vivo*, *Pflugers Arch.*, 342, 255–260, 1973.
8. Wjin, P., The Alinear Viscoelastic Properties of the Human Skin *in Vivo* for Small Deformations, Ph.D. thesis, Catholic University of Nijmegen, the Netherlands, 1980.

9. Finlay, B., The torsional characteristics of the skin *in vivo*, *Biol. Med. Eng.*, 5, 567–573, 1971.
10. de Rigal, J. and Leveque, J.L., *In vivo* measurement of the stratum corneum elasticity, *Bioeng. Skin*, 1, 13–23, 1985.
11. Leveque, J.L., Porte, G., de Rigal, J., Corcuff, P., Francois, A.M., and Saint Leger, D., Influence of chronic exposure on some biophysical parameters on human skin, *J. Cutaneous Ageing Cosmet. Dermatol.*, 1, 265–275, 1988/1989.
12. Seite, S., Moyal, D., Richard, S., de Rigal, J., Leveque, J.L., Hourseau, C., and Fourtanier, A., Effects of repeated suberythemal doses of UVA in human skin, *Cutaneous Photobiol. Photoimmunol.*, 253–258, 199X,
13. Escoffier, C., de Rigal, J., Rochefort, A., Vasselet, R., Leveque, J.L., and Agache, P., Age related mechanical properties of human skin. An *in vivo* study, *J. Invest Dermatol.*, 93, 353–357, 1989.
14. Richard, S., de Rigal, J., De Lacharriere, O., Berardesca, E., and Leveque, J.L., Non invasive measurement of effect of lifetime exposure to the sun on the aged skin, *Photodermatol. Photoimmunol. Photomed.*, 10, 164–169, 1994.
15. Kalis, B., de Rigal, J., Leonard, F., Leveque, J.L., Riche, O., Le Corre, Y., and De Lacharriere, O., *In vivo* study of scleroderma by non-invasive techniques, *Br. J. Dermatol.*, 122, 785–791, 1990.
16. De Lacharriere, O., Escoffier, C., Gracia, A.M., Teillac, D., Saint Leger, D., Berrebi, C., Debure. A., Leveque, J.L., Kreis, H., and De Prost, Y., Reversal effects of topical acid on the skin of kidney transplant recipients under systemic corticotherapy, 95(5), 516–522, 1990.
17. Pichon, E., de Rigal, J., and Leveque, J.L., *In vivo* rheological study of the torsional characteristics of the skin, presented at Congress of Bioengineering and the Skin, Stresa, 1990.
18. de Lacombe, J., *Rev. Metall.*, 36, 178, 1939.
19. Andrade, *Proc. R. Soc. London*, A 84, 1, 1910; A 90, 329, 1914.
20. Salter, D.C., McArthur, H.C., Crosse, J.E., and Dickens, A.D., Skin mechanics measured *in vivo* using torsion, a new and accurate model more sensitive to age, sex, and moisturising treatment, *Int. J. Cosmet. Sci.*, 15, 200–218, 1993.
21. Wickett, R. and Murray, B.C., Comparison of cutometer and DTM for skin elasticity measurements, presented at 20th Anniversary Symposium, Int. Soc. for Bioengineering and the Skin, Miami, 15–17 February 1996.
22. Agache, P. and Varchon, D., Exploration fonctionnelle mécanique, in *Physiologie de la peau et explorations fonctionnelles cutanées*, P. Agache et al., Eds., E.M.I.,

6 *In Vivo* Tensile Tests on Human Skin: The Extensometers

Paul Vescovo, D. Varchon, and Phillippe Humbert

CONTENTS

Introduction 77
Mechanical Properties Measurements 78
In Vitro and *in Vivo* Tests 78
Measurement of Mechanical Properties of Skin with an Extensometer 79
Extensometers: Examples 79
The Portable Hand-Held Extensometer 80
Magnetic Extensometer 81
Extensometer of Baker, Bader, and Hopewell 83
The Cutech Extensometer 83
Use of an Extensometer 85
Bond with Skin 85
Movements and Interference Vibrations 86
Types of Test Conducted and Precision of Results 86
Rheological Equations 86
Effect of Hypoderma 88
Use of a Support Plate 88
Prospects 89
References 89

INTRODUCTION

Dermatologists conduct physical measurements on skin to diagnose and to follow the development of pathologies. They also need, as do cosmetologists, to measure the efficacy of the treatments they apply to skin. The measurement of skin mechanical properties can offer a wealth of information to this field but, at the present time, studies are too often carried out in a qualitative way, which leads to relatively subjective assessments. It is essential to quantify precisely the evolution of at least one mechanical parameter of skin representative of its state. When studies are

0-8493-7521-5/02/$0.00+$1.50

conducted in hospitals or in research laboratories, the means used can be sophisticated, whereas the methods must be rather simple when they are intended to be used by practitioners in their offices.

MECHANICAL PROPERTIES MEASUREMENTS

The mechanical behavior of a material is defined by rheological equations linking loadings, deformations, and such external or internal parameters as temperature, hygrometry, material constituents, etc. In the simplest case the behavior is considered linear elastic. In many applications and to simplify the equations, it is possible to assume that the state of stress is plane; then the behavior can be modeled as follows:

$$\begin{pmatrix} \sigma_1 \\ \sigma_2 \\ \sigma_6 \end{pmatrix} = \begin{bmatrix} C_{11} & C_{12} & 0 \\ C_{12} & C_{22} & 0 \\ 0 & 0 & C_{66} \end{bmatrix} \begin{pmatrix} e_1 \\ e_2 \\ e_6 \end{pmatrix} \quad \text{or} \quad \begin{pmatrix} e_1 \\ e_2 \\ e_6 \end{pmatrix} = \begin{bmatrix} S_{11} & S_{12} & 0 \\ S_{21} & S_{22} & 0 \\ 0 & 0 & S_{66} \end{bmatrix} \begin{pmatrix} \sigma_1 \\ \sigma_2 \\ \sigma_6 \end{pmatrix} \tag{6.1}$$

Terms of rigidity C_{ij} and suppleness S_{ij} are often written according to elastic constants that characterize material in an elastic range. An isotropic material is completely defined by two constants called Young's modulus E and Poisson's ratio ν, with

$$S_{11} = S_{22} = \frac{1}{E} \; ; \; S_{12} = S_{21} = -\frac{\nu}{E} \; ; \; S_{66} = \frac{1}{G} \; ; \; G = \frac{E}{2(1+\nu)} \tag{6.2}$$

In the case of an anisotropic material, the number of independent constants is higher, nine for an orthotropic material.[1] Experimental determination of the constants requires that the number of independent experiences equal the number of parameters to be determined. When the behavior is more complex (viscous, for example), the types of tests are identical but the loadings to implement are different (creep experiment, stress relaxation, etc.). Human skin is a material that can be considered isotropic or transverse isotropic. In its plane, two Young's moduli E_1, E_2, one Poisson's ratio ν_{12}, and one shear modulus G_{12} can be calculated.

IN VITRO AND *IN VIVO* TESTS

Skin studied by *in vitro* tests comes from cadaver or plastic surgery; the number of experiments, the variety of skins which can be tested, and the variety of their original localizations on the body are thus very limited. In contrast, with *in vivo* experiments it is easy to conduct a great number of tests on various kinds of skin anywhere on the body. From a mechanical point of view, experimental conditions are better during *in vitro* tests than during *in vivo*. Because of the specimens used, it is possible to have homogeneous stress fields on large surfaces, which allow accurate measurements of mechanical properties. However, human skin does not lend itself at all well to these kinds of experiences because it is very difficult to obtain samples of good quality. Indeed, skin used for experiment comes from cadavers or plastic surgery. In the first case, the mechanical behavior of the skin may have been modified by time; in the second, the skin is often greatly damaged by stretch marks and may exhibit behavior close to that of pathological skin. Several research teams have focused on that subject; Schneider et al.[2] even carried out biaxial tensile tests with a good instrumentation. Vescovo et al.[3] brought to the fore the difficulties involved in obtaining specimens that would allow proper measurements and interpretation of results, because the initial conditions of the skin specimens are different from those of skin *in vivo*. The main problems with *in vivo* tests are the poor experimental conditions, which make analysis very complex. Despite their disadvantages, *in vivo* tests can certainly bring a better understanding of skin mechanical behavior than *in vitro* experiments can provide.

Several kinds of apparatus have been designed in the past; few have been marketed. One of the most-developed devices is the Cutometer. It is a suction apparatus marketed by Courage and Khazaka (Cologne, Germany); results are analyzed by a developed software. However, it is unfortunate that results are not provided via intrinsic parameters like Young's modulus by means of a simple modeling as are those proposed by Vescovo et al.[4] or Diridolou et al.[5] Another interesting apparatus is the Twistometer, which has not been as important as the Cutometer. However, it is quite easy to measure the coulomb modulus in the skin plane with the Twistometer. These two devices have the same disadvantage: they cannot provide information about the anisotropy of skin. Therefore, it is necessary to use another type of apparatus to measure this important feature of skin.

MEASUREMENT OF MECHANICAL PROPERTIES OF SKIN WITH AN EXTENSOMETER

The principle of the test is simple: either two pads fixed on skin are moved with respect to each other and forces applied on them are measured, or the forces are applied on pads and the displacements between them are measured. The test seems easy, but reduction of the results is rather complicated. Indeed, the specimen boundary conditions are not well defined and the test is not a pure uniaxial tensile test. However, it is a directional test, which allows measurements in the plane of skin and thus a quantification of anisotropy. This feature makes the extensometer very different from the other devices, and is the reason this process is especially interesting and is analyzed more thoroughly later in this chapter.

EXTENSOMETERS: EXAMPLES

Extensometers are not used commonly at the present time; however, several research teams have taken an interest in this technique in the past. The technologies used did not allow a large enough optimization of the apparatus and the process was partially abandoned. A few examples of devices that seem to be more representative and accomplished are presented here.

The Portable Hand-Held Extensometer

The extensometer was built by Gunner et al.[6] in 1978; it is a development of a device reported previously.[7] It can apply a uniform extension rate to skin *in vivo*. Its functioning principle is relatively simple (Figure 6.1). Two arms, one fixed, the other movable by means of a motor-driven lead screw, have at their ends rectangular pads that are stuck to the skin surface by means of double-sided adhesive tape. Strain gauges attached at the reduced section on each arm and a linear variable differential transformer sensor (LVDT) measure the forces applied to the pads and their displacements. The force–displacement curve can be worked out easily.

The main advantage of that apparatus is its simplicity, which makes it very easy to use; its disavantages are its size, weight, and significant dissymmetry, which can disrupt the results because of interference forces. The main features of this apparatus are listed in Table 6.1.

The device can conduct tests with loadings at an imposed extension rate with subsequent relaxation. To analyze the recovery appearing after the end of the loading the same authors propose a device that can measure the return of skin to its initial state.[8] This apparatus consists of two pads; one is fixed to the device body, the other is free on a slide rail and has an LVDT sensor attached on it to measure the displacement (Figure 6.2). It can be used alone; the displacement of the mobile pad is then applied manually or with the extensometer described previously. It is then possible to record loading, relaxation, and recovery (Figure 6.3). The simplicity of this apparatus, its light weight (50 g), and the fact that it can be used without being held by the experimenter or placed on a stand are its main advantages. Friction is certainly a disadvantage for its proper functioning.

These two apparatuses allowed their designers to conduct many tests on various zones of the human body. Their experiences allowed the designers, for example, to propose a rheological model identified on normal and pathological skin and to measure the effect of treatments on the mechanical behavior of human skin.

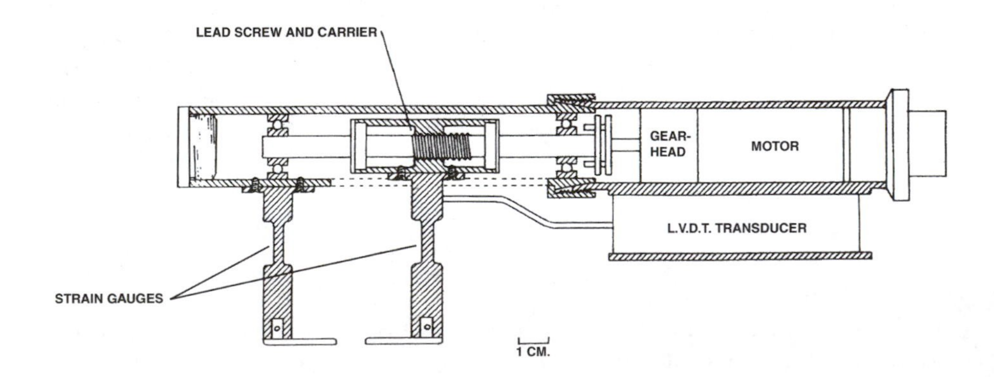

FIGURE 6.1 Schematic diagram of the extensometer designed by Gunner et al.[6] (From Gunner, C.W. et al., *Br. J. Dermatol.*, 100, 161, 1979. With permission of Blackwell Science.)

Magnetic Extensometer

This apparatus was described in 1979 by Wijn et al.[9] and was used in 1986 by Manschot et al.[10] Two pads are stuck to the skin surface by means of double-sided adhesive tape or cyanoacrilate adhesive; the force producing the displacement of the movable pad is obtained with a magnet situated inside a coil. The force applied is then proportional to the electric current, and the displacement is measured with an LVDT-type transducer (Figure 6.4).

The principal advantage of the device is its operating mode, which allows creep tests to be conducted because it works by imposing force. Its disadvantages are due to its large weight, which requires fixing the apparatus on a table. Thus, it is not possible to carry out tests on all locations on the body. Tests are mostly performed on arms and legs. Friction may disrupt the measurements.

TABLE 6.1
Features of the Various Presented Extensometers

Study (extensometer's name)	Gunner et al.,[6] 1979	Gunner et al.,[8] 1979 (Recoil apparatus)	Manschot et al.,[10] 1986	Baker et al.,[11] 1988	Vescovo,[13] 1998 (Cutech)
Pads anchorage on skin	Double-sided adhesive tape	Double-sided adhesive tape	Cyanoacrylate	Double-sided adhesive tape	Double-sided adhesive tape
Portable	Yes	Yes	No	Yes	Yes
Initial tab distance (mm)	10	10	3–20	4	6
Tab size (length × width) (mm)	25×?	?	10 × 10	10 × 20	17.8 × 10
Maximum strain (%)	40	30	Up to 50	50	40
Maximum force (N)	6	10	12	5	3
Loading	Uniform extension rate (0.35 mm/s)	Manually or with the extensometer used by Gunner et al.[6] in 1979	Sawtooth-shaped force loading	Uniform extension rate (0.05 or 0.33 mm/s)	Uniform extension rate (0.1 or 0.33 mm/s)

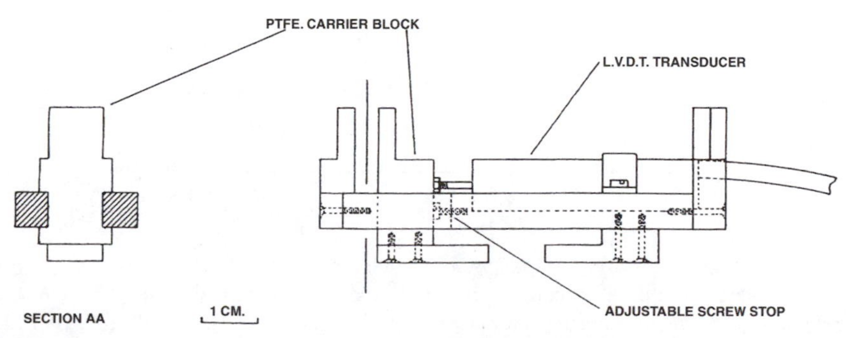

FIGURE 6.2 Schematic drawing of the *in vivo* recoil apparatus designed by Gunner et al.[8] (From Gunner, C.W. et al., *Med. Biol. Eng. Comput.*, 17, 142, 1979. With permission of IEE Publishing.)

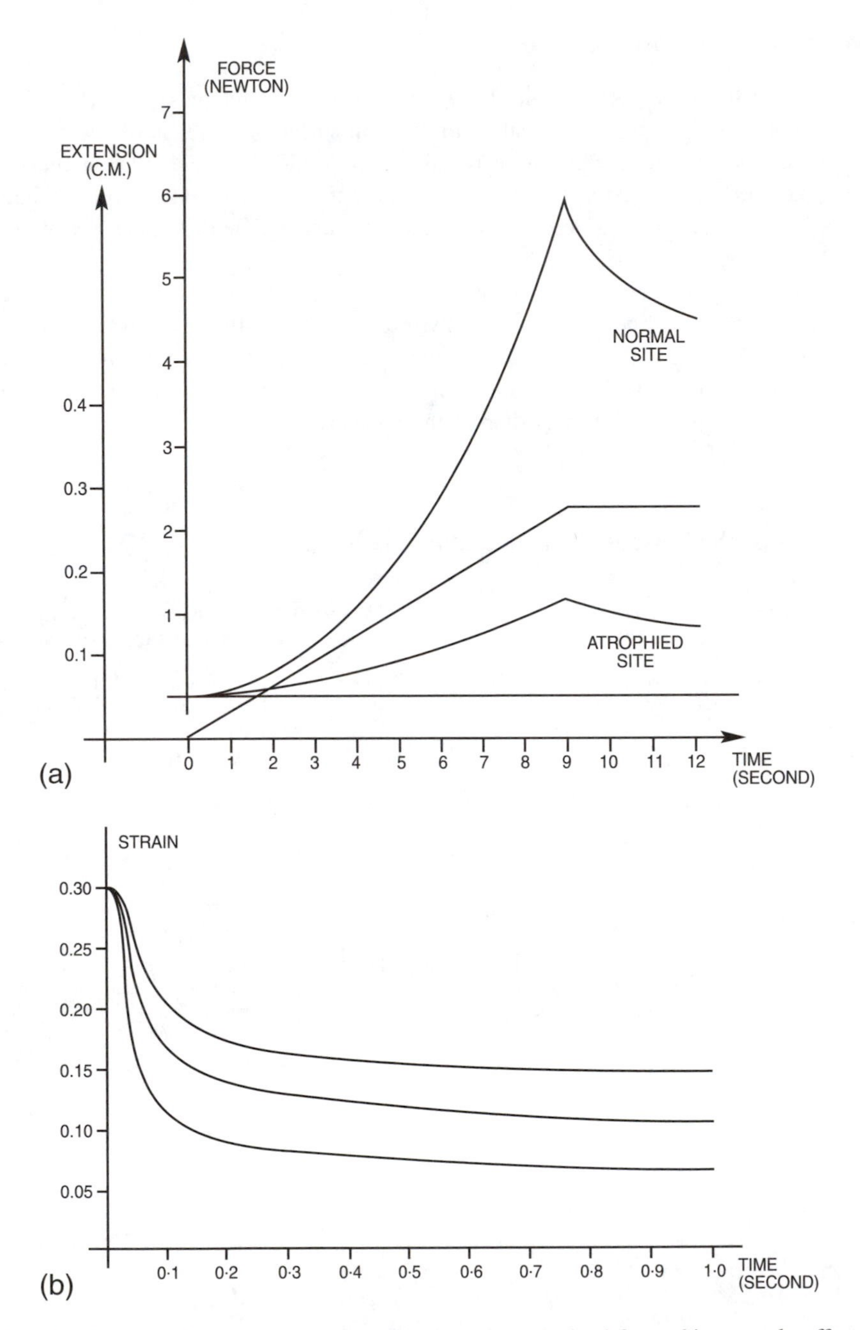

FIGURE 6.3 (a) Comparison of extensometer outputs obtained from skin severely affected by steroid atrophy and adjacent normal skin near axilla. Tests were carried out with the extensometer designed by Gunner et al.[6] (From Gunner, C.W. et al., *Br. J. Dermatol.*, 100, 161, 1979. With permission of Blackwell Science.) (b) Recoil characteristics obtained from adjacent sites on the forearm of a normal subject with the recoil apparatus designed by Gunner et al.[8] (From Gunner, C.W. et al., *Med. Biol. Eng. Comput.*, 17, 142, 1979. With permission of IEE Publishing.)

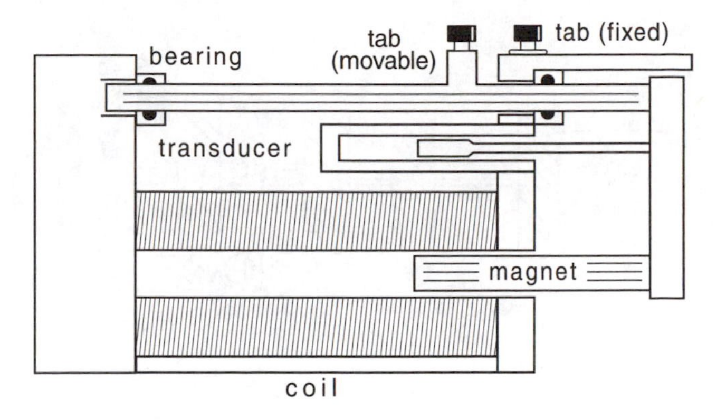

FIGURE 6.4 Schematic drawing of the uniaxial strain apparatus.[10] (From Manschot, J.F.M. and Brakkee, A.J.M., *J. Biomech.*, 19, 511, 1986. With permission of Elsevier Science.)

Extensometer of Baker, Bader, and Hopewell

This apparatus was built and used in 1988 by Baker et al.[11] Here, again, there is a screw and bolt system with an LVDT sensor. But force measurement is performed differently by means of a load cell placed between the movable pad and the apparatus body (Figure 6.5). This device has been used to carry out stiffness measurements in parallel with A scan ultrasonic tests to study irradiated pig skin. It is heavy and has not been optimized.

The Cutech Extensometer

Two 10 × 18 mm pads are adhered to skin 6 or 9 mm apart. The pads are moved closer together (tension test) or farther apart (compression test) at a constant but adjustable speed (from 3 to 10 mm/s). The arms supporting the pads have gauges attached to them that make it possible to measure the applied load (4 N max.) (Figures 6.6 and 6.7). The apparatus can be used with or without a support plate, which is a plate with a rectangular hole in it through which the pads are adhered to skin surface. This support plate is fixed to the device body and is also adhered to skin by means of double-sided adhesive tape. When the support plate is used, skin around the investigated zone is fixed to the apparatus. The apparatus can carry out loading followed by creep and recovery; with analog recording. This device was marketed by Cutech and the Stiefiel Laboratories about 10 years ago, but is no longer available. Its design is similar to that of the Gunner et al. apparatus, but with improved performance.

This apparatus was used by Asserin[12] to investigate the anisotropic properties of human skin. His study was carried out *in vivo* on arms, by means of a finite-element model whose meshing used the natural patterns of skin. He was able to prove that an isotropic model does not properly represent the elastic behavior of skin. Analysis of deformations in every direction allows optimization of skin mechanical constants when modeled as an isotropic material. Vescovo[13] completely validated this extensometer through tests carried out on materials whose mechanical

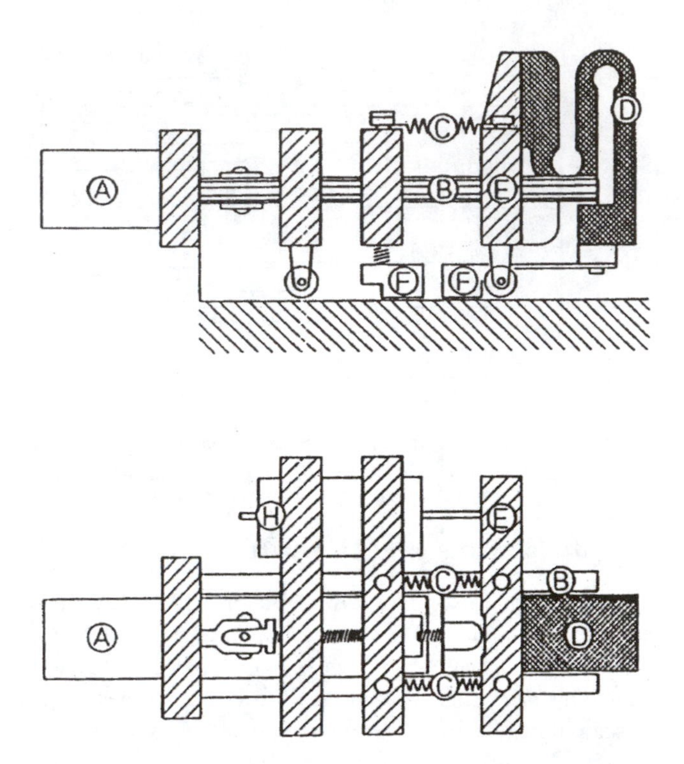

FIGURE 6.5 Diagrammatic representation of the transducer in plane and side view.[11] (From Baker, M.R. et al., *Bioeng. Skin*, 4, 87, 1988. With permission of Kluwer Academic Publishers.)

FIGURE 6.6 View of the Cutech extensometer.

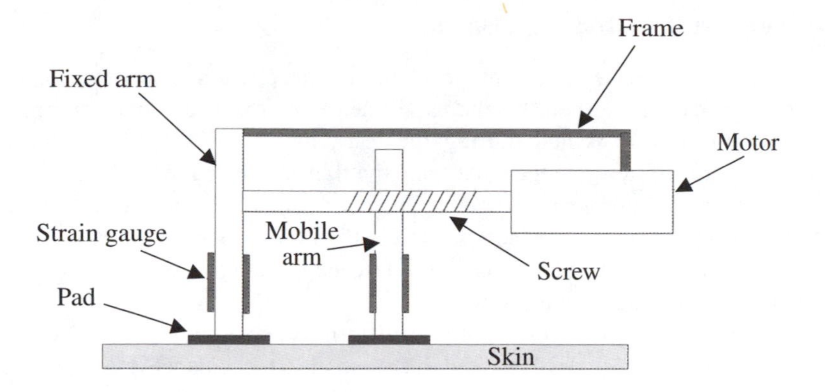

FIGURE 6.7 Schematic diagram of the Cutech extensometer.

behaviors are simpler and easier to work out by means of standard tensile tests. He showed the performance limits of this apparatus and highlighted operating instructions leading to quality results.

The main features of the extensometers presented here are summarized in Table 6.1. The devices and the research conducted with them are similar because designers used the same-generation technologies.

USE OF AN EXTENSOMETER

In the case of physical measurements, environmental influences cannot be completely suppressed. No matter the apparatus, the material, or the experimenter, environmental influences always disrupt the measurements. The best device is the one yielding results closest to reality, while being accessible to all practitioners. Whatever their differences, the devices presented here all have a very important mission. The main problems can be overcome or, at least, they can be taken into account in the results analysis.

BOND WITH SKIN

Measurements are correct only if the pad displacements represent those of the skin. This is valid for every type of device (including the Cutometer, Twistometer, and so forth) and is possible only if the pads are properly anchored on skin. Because tests are carried out *in vivo*, invasive techniques are not acceptable and most of the time double-sided adhesive tape is simply used. However, this process is not completely satisfactory if the forces applied are large. Especially during creep tests, the measured displacements are incorrect because of adhesive coat creep. So it is necessary to check the bonding quality before carrying out experiments. Other techniques are being studied at the present time but the results are not yet final; however, a rapid improvement in this fundamental issue can be anticipated.

Movements and Interference Vibrations

Except for the device designed by Gunner et al.[8] (recoil apparatus), the extensometers presented here have to be held by the experimenter or fixed to a stand during the tests. When the device is held during the tests, low-amplitude movements and vibrations are created between the device and the skin area tested, which significantly disrupt the force measurements. When the apparatus is fixed to a stand, if the patient is put in a good position (which is in practice difficult to achieve; indeed, it requires a system suitable for every body type), interference vibrations are very low but the body areas that can be tested are few.

The future lies either in sufficiently miniaturized and lightweight devices that do not have to be held or in devices fixed to a stand via a mechanical system allowing tests to be carried out on a very large number of body zones.

Types of Test Conducted and Precision of Results

Apparatus designed in the past allowed either loading with imposed forces or displacements, but no one device could do both. Moreover, for most extensometers, loading is not programmable (except, for example, the device used by Manschot et al.[10]), which significantly restricts the number of feasible tests (for example, the device used by Baker et al.[11] can have constant extension rates of only 0.05 or 0.33 mm/s). Conducting combined tests, such as creep and relaxation tests, is of great interest to identify the parameters of skin rheological equations. With the aid of current technological means it seems possible to design high-performance devices allowing every type of loading.

User aims are of two types:

1. Conducting simple tests that allow a qualitative assessment of skin mechanical behavior during pathology or medical or cosmological treatment. If experimental conditions are always identical, result comparison is possible and sufficient.
2. Working out intrinsic mechanical constants essential to identify models which can greatly help to understand biological phenomena, forecast pathology evolution, or optimize surgery. It is then necessary to carry out tests with high precision on the loading applied and the measurements performed. Because of the viscoelastic behavior of skin, the apparatus must be able to perform tests with a closed-loop servo on force or displacement; only those having these advanced technologies are suitable.

Rheological Equations

Because rheological equations are solved at the level of the representative volume element, it is necessary to know all stresses and deformations at work. These constants are tensorial-type constants; in the case of skin it is often possible, to simplify, to model the state of stress as a plane one. The equation between stresses and deformations in the elastic range is then written as Equation 6.1. It is, of course, necessary to know the three stresses and the three deformations to calculate the elastic constants.

In the case of an *in vitro* tensile test, because of the experimental conditions, the state of stress is sufficiently homogeneous to have tensile stress alone in the investigated zone. It is possible to work out stresses and strains applied to a volume element from the forces and displacements applied to the specimen. Under these conditions, it is possible to work out the engineering constants directly.

$$F = ES\,\frac{\Delta l}{l} \Rightarrow \sigma_1 = E\varepsilon_1 \tag{6.3}$$

where F = tensile force, E = Young's modulus, S = specimen section, Δl = elongation, l = initial length, σ = stress, ε = strain.

For a test carried out *in vivo* with an extensometer, the area of skin that is loaded depends on the test geometry and cannot be defined easily. The specimen is very different from that used in a standard tensile test, and movement of the pads creates a state of strain (stress) between them that stretches the whole skin area. The strain (stress) decreases when the distance from central zone increases and theoretically becomes zero to infinity. In practice, as was shown by Vescovo et al.,[4] the loaded zone can be modeled as finite,but even so it is not easy to work out the stress level from the force applied to the pads.

A skin sample was modeled with the software Ansys 5.5 (Figure 6.8). The skin was taken as a pure elastic material with a Young's modulus E of 1 MPa. In many studies, the authors consider that only the area of skin between pads is loaded, which means that they model the test as a standard tensile test (specimen width = pad width, specimen length = distance between pads). With this model there is only a longitudinal stress, which is constant in the specimen and is equal to 0.87 MPa. It can be seen with the finite-element modeling that the longitudinal stress is not constant and that its maximum level is about 0.5 MPa, which is far less than the value calculated previously with the simple model. Moreover, finite-element modeling shows that there are shear and cross stresses. With an optical method to measure the strain field and finite-element modeling, it is possible to work out mechanical constants correctly. Finite-element modeling of an *in vivo* extensometer test presents many problems that are due to the specific mechanical behavior of skin. For most standard materials, it is acceptable to consider that their elastic behavior is identical in tension and compression. This hypothesis is not true for skin. Indeed, when skin undergoes compression loading, it creases; its effective stiffness is very low. Skin has a mechanical behavior similar in a way to that of a tissue. Retel et al.[14] have taken that particular feature into account to model the state of stress around a scar with ablation. Few experimenters take into account the state of heterogeneous stress when calculating skin mechanical constants (Asserin[12]).

Rigorous study of the extensometer test shows that it can provide information about skin anisotropy, but it shows as well that its analysis is tricky.

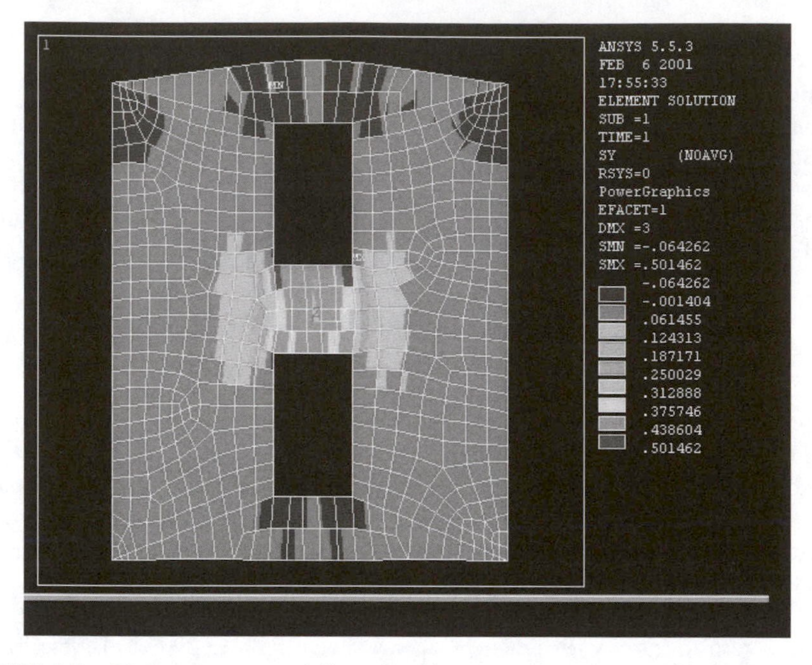

FIGURE 6.8 Finite-element modeling of a skin zone undergoing a test by extensometer. Each grayscale tone represents a longitudinal stress level.

Effect of Hypoderma

During a test with an extensometer *in vivo*, not only skin is loaded but also hypoderma. The problem is to work out skin mechanical constants from a test carried out on a structure composed of two materials, skin and hypoderma, whose mechanical behaviors are unknown. The hypothesis posed most of the time to solve that problem is that hypoderma, during an extensometer test, is much less loaded than skin and that its stiffness is much lower than that of skin.

Use of a Support Plate

A support plate can be used to define more precisely the skin area that is loaded. This process must be used with care because it can be a source of error. For some devices, the measurement of forces applied to the pads are calculated by means of gauges on the two arms. This technique is appropriate when the test is symmetrical (forces applied to the two pads are identical), which is not the case when only one arm moves and when a support plate is used. Another problem with a support plate that completely frames the tested area is that the compressed zone is situated behind the pads, which can alter the force calculation. To avoid that, the support plate can be open behind the pads.

PROSPECTS

At the present time, extensometer tests are not commonly used. However, these authors think that the extensometer can offer practitioners precious information, especially on the anisotropic nature of human skin. To expand, this technique must improve by offering easier-to-use devices with higher performance. An up-to-date apparatus should meet the following requirements:

- It should be miniaturized (present microtechnologies should allow great improvements).
- It should carry out force or displacement servo-controlled tests.
- It should fit to an optical device that can analyze in real time the tested zone in order to get the strain field.
- It should fit to an ultrasonic device.
- It should be sufficiently simple and easy to use to allow common practitioners to conduct comparative tests.
- It should be a high-performance apparatus to allow researchers to study and analyze precisely skin mechanical behavior.

REFERENCES

1. Jones, R.M., *Mechanics of Composite Materials*, Hemisphere Publishing Corporation, 1975.
2. Schneider, D.C., Davidson, T.M., and Nahum, A.M., *In vitro* biaxial stress–strain response of human skin, *Arch. Otolaringol.*, 110, 329, 1984.
3. Vescovo, P., Jacquet, E., Burtheret, A., Varchon, D., Coral, H.-P., and Humbert, P., Méthodologie expérimentale sur matériaux biologiques — application à la peau humaine, in *Mécano Transduction Matériaux et Structures des Sciences de l'Ingénieur et du Vivant,* Editions Tec & Doc, 2000, 317.
4. Vescovo, P., Jacquet, E., Varchon, D., Asserin, J., Humbert, P., and Agache, P., Analyse mécanique rigoureuse de trois techniques expérimentales de mesure des propriétés elastiques de la peau *in vivo,* in *Proc. 1er Colloque National d'Ingéniérie Cutanée,* 1999, 197.
5. Diridollou, S., Patat, F., Gens, F., Vaillant, L., Black, D., Lagarde, J.M., Gall, Y., and Berson, M., *In vivo* model of mechanical properties, *Skin Res. Technol.,* 6, 214, 2000.
6. Gunner, C.W., Hutton, W.C., and Burlin, T.E., The mechanical properties of skin *in vivo* — a portable hand-held extensometer, *Br. J. Dermatol.*, 100, 161, 1979.
7. Hutton, W.C., Ranu, H.S., and Burlin, T.E., An apparatus for measuring the effects of radiotherapy on the elastic properties of human skin *in vivo*, *Med. Biol. Eng.,* 13, 584, 1975.
8. Gunner, C.W., Hutton, W.C., and Burlin, T.E., An apparatus for measuring the recoil characteristics of human skin *in vivo*, *Med. Biol. Eng. Comput.*, 17, 142, 1979.
9. Wijn, P.F.F., Brakkee, A.J.M., Kuiper, J.P., and Vendrik, A.J.H., The alinear viscoelastic properties of human skin *in vivo* related to sex and age, in *Bioengineering and the Skin,* MTP Press, 1981.

10. Manschot, J.F.M. and Brakkee, A.J.M., The measurement and modelling of the mechanical properties of human skin *in vivo*, I. The measurement, *J. Biomech.*, 19, 511, 1986.
11. Baker, M.R., Bader, D.L., and Hopewell, J.W., An apparatus for testing of mechanical properties of skin *in vivo*: its application to the assessment of normal and irradiated pig skin, *Bioeng. Skin*, 4, 87, 1988.
12. Asserin, J., Etude par test d'extension des propriétés mécaniques de la peau humaine *in vivo* et étude d'un derme équivalent *in vitro*, Thesis, Université de Franche-Comté, 1996.
13. Vescovo, P., Validation de méthodes de mesure des modules d'élasticité de la peau humaine, Diplôme d'Etudes Approfondies, Université de Franche-Comté, 1998.
14. Retel, V., Vescovo, P., Jacquet, E., Trivauday, F., Varchon, D., and Burtheret, A., Nonlinear model of skin mechanical behaviour analysis with finite element method, *Skin Res. Technol.*, in press, 2001.

7 Hardware and Measuring Principle: The Cutometer®

Undine Berndt and Peter Elsner

CONTENTS

Introduction 91
Measuring Principle 91
Measuring Device 92
Measuring Modes 92
Measuring Results 94
Dermatological and Cosmetic Applications 94
References 96

INTRODUCTION

The biomechanical properties of human skin are a complex combination of elastic and viscous components. Elasticity correlates with the function of elastin fibers; viscosity is controlled by the collagen fibers and the surrounding intercellular ground substance, which consists primarily of water and proteoglycans.

The Cutometer allows the measurement of the viscoelastic properties of the skin *in vivo*, which provides valuable information on physiological and pathological changes of human dermis as well as on the efficacy of topical treatments. It is recognized as a standard tool in dermatological and cosmetic research.

MEASURING PRINCIPLE

The measuring principle of the Cutometer is based on suction. A defined negative air pressure is created and applied on the skin surface through the opening of a probe drawing the skin into its aperture. The resulting vertical deformation of the skin surface is measured by determining the depth of skin penetration into the probe. This is achieved by a noncontact optical system that consists of a light transmitter

0-8493-7521-5/02/$0.00+$1.50

and a light recipient. Two glass prisms project the light from transmitter to recipient, where the diminution of the infrared light beam depending on the penetration depth of the skin is measured (Figure 7.1).[1]

MEASURING DEVICE

The Cutometer consists of a hand-held probe that is connected to the main unit through an air tube and an electric cable (Figure 7.2). The hand-held measuring probe is a 3 × 10 cm cylinder with a weight of approximately 90 g. It contains the optical measuring system and the evaluation electronics. The suction head is centered in the probe shield. The diameter of its circular aperture measures 2 (standard), 4, 6, or 8 mm depending on the skin area and purpose of the examination. The probe is applied perpendicularly to the skin under constant pressure with an elastic spring.

The main unit contains a vacuum pump, a pressure sensor, and an electronic circuit controlling the pump and the analog/digital data conversion. The load of the vacuum can be regulated between 20 and 500 mbar, and the rate of pressure increase or decrease can be selected between 10 and 100 mbar/s. The suction interval (on-time) as well as the relaxation interval (off-time) can be varied between 0.1 and 60 s. Within one measurement cycle as many as 99 measurement repetitions are possible, each with a duration limited to a maximum of 320 s.

The Cutometer is designed for operation with an IBM-compatible personal computer (interface RS-232-C). The delivered standard software allows automatic calculation and storage of measuring results. In addition, data concerning the volunteer or patient, skin area, date and time of measurement, relative humidity, external temperature, and other information can be entered. All data are saved on an ASCII data file for external statistical evaluation. During the measurement process a corresponding curve is displayed showing the depth of the skin penetration into the probe and its retraction as the skin recovers during the relaxation period.[1]

MEASURING MODES

Two measuring techniques are available, a stress–strain mode and a time–strain mode. In the stress–strain mode, the vacuum is increased over a selected period of time, and the deformation (in mm) is displayed as a function of negative pressure (in mbar). In the time–strain mode, which is mostly used to study viscoelastic properties of human skin, a selected vacuum is applied for a chosen time period. The skin deformation is shown as a function of time. The Cutometer offers four different measuring possibilities as the result of the choice of either immediate or slow linear increase and decrease of the applied vacuum. Applying mode 1 (-_), the skin is drawn into the probe with constant negative pressure, which is immediately cut off after a certain time period. In mode 2 (/\), deformation and retraction of the skin are achieved by a linearly increasing and decreasing vacuum. Mode 3 (-\) begins with a constant abrupt strain of the skin followed by a slow decrease of pressure, and mode 4 (/_) combines a linear increase of vacuum and a sudden total release of the skin.[1]

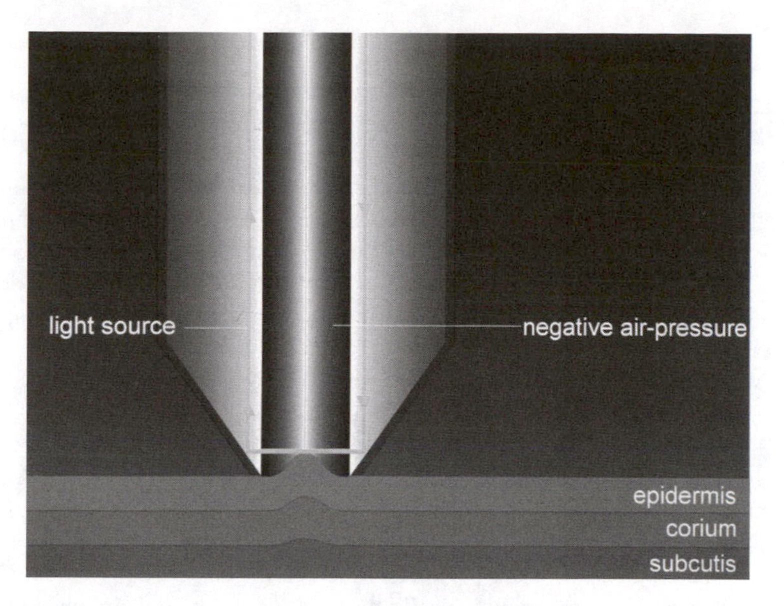

FIGURE 7.1 Schematic view of the optical measuring system within the probe. (From Courage and Khazaka Electronic GmbH, Cologne, Germany. With permission.)

FIGURE 7.2 Cutometer SEM 575. (From Courage and Khazaka Electronic GmbH, Cologne, Germany. With permission.)

MEASURING RESULTS

Figure 7.3 shows a typical curve obtained in the time–strain mode (mode 1). The deformation curve is composed of rapid deformation representing a purely elastic section, followed by a viscoelastic section, and finally by a purely viscous section. The same elements can be differentiated in retraction of the tissue.[2]

Following the nomenclature proposed by Agache et al.,[3] the deformation parameters used in most viscoelastic studies on human skin are U_e, U_v, U_f, U_r, and U_a. U_e is defined as immediate deformation, or skin extensibility; U_v as the delayed distension reflecting the viscoelastic contribution of the skin; U_f as the final skin deformation; and U_r as the immediate and U_a as the final retraction after removal of the vacuum. The Cutometer program offers a cursor mode for the measurement of these parameters directly from the graphical display.

All these parameters are a function of skin thickness. Therefore, it is not meaningful to compare their absolute values between test sites or subjects. Instead, the experimental values should be standardized for skin thickness, which is preferably determined by ultrasound. If data on skin thickness are not available, ratios of the above parameters instead of the parameters themselves can be determined. Certain ratios of these parameters are biologically meaningful, do not depend on skin thickness, and can be compared between sites and subjects. U_v/U_e is the ratio between delayed and immediate deformation and indicates the relative contributions of the viscoelastic plus viscous and the elastic distension to the total deformation. The value of this ratio increases with decreasing elasticity. U_a/U_f is the ratio of total retraction to total deformation, which is called gross elasticity. U_r/U_f, as the ratio of immediate retraction to the total deformation, is called biological elasticity. The ratio U_r/U_e, proposed by Agache et al.,[3] closely resembles U_r/U_f and is also used as a measure of elastic recovery. High values of these ratios — maximum = 1 (100%) — indicate a high level of elasticity.

DERMATOLOGICAL AND COSMETIC APPLICATIONS

Age and anatomical region are important factors influencing the viscoelasticity of human skin and have been studied extensively using the Cutometer. Biological as well as actinic aging is accompanied by a decrease of the biological elasticity of the skin indicated by the decreasing values of the U_r/U_f and U_r/U_e ratios.[4–7] On the other hand, an increase in U_v/U_e, the ratio between viscoelastic and elastic extension, indicates a greater contribution of the viscous part to the total extension with age.[7,8]

Reflecting the differences in structure of the papillary and reticular dermis, significant differences of the viscoelastic properties of skin of different anatomical regions were found in various studies. Comparison of the biomechanical qualities of vulvar and forearm skin, for example, showed that U_v/U_e and U_r/U_f were both significantly lower in vulvar than in forearm skin.[5,6] Differences of the viscoelasticity parameters of three different anatomical skin areas, i.e., eyelid, temple, and forearm, are graphically shown in Figure 7.4. However, significant sex-dependent biomechanical differences were not observed during the studies of physiological influences on skin elasticity.[4,5,7]

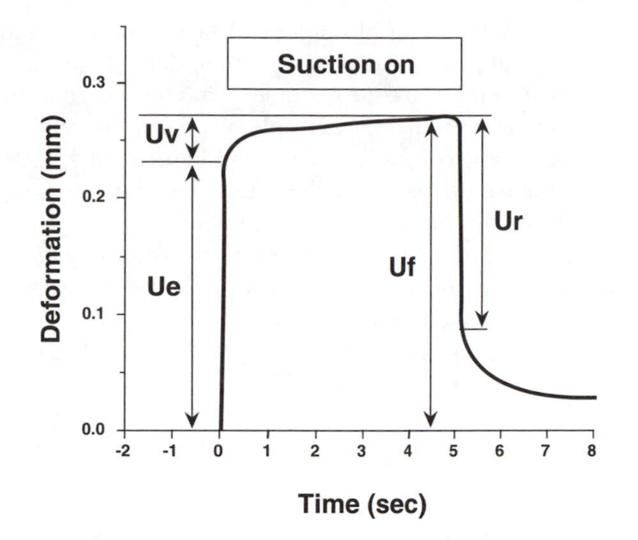

FIGURE 7.3 Typical graphical registration of a strain–time curve obtained on vulvar skin. 2 mm probe, on-time 5 s, off-time 3 s, load 500 mbar. The parameters used to describe the deformation are as follows: U_e = instantaneous deviation length; U_v = viscoelastic and viscous deviation length (creep); U_f = final deviation length; and U_r = instantaneous recovery length.

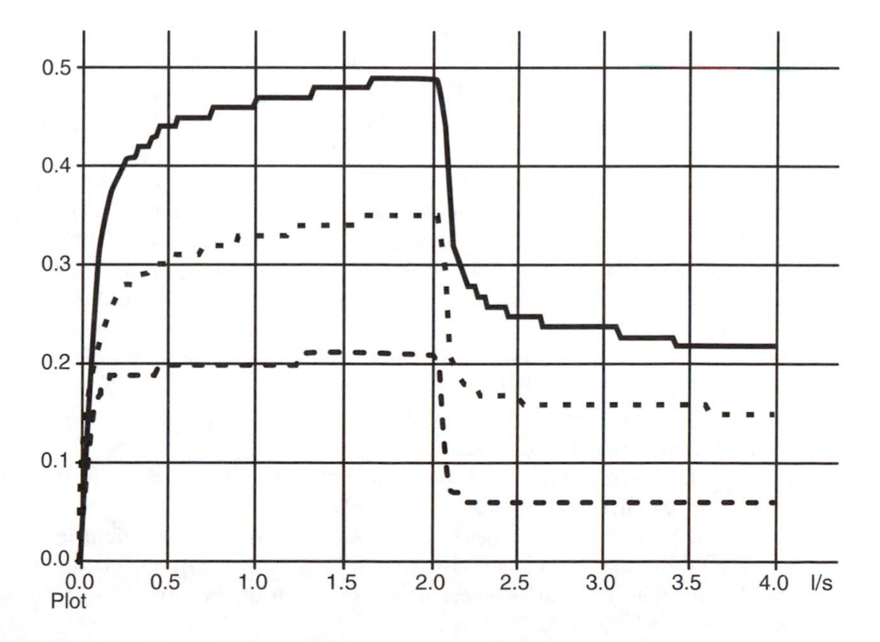

FIGURE 7.4 Three different curves of a 50-year-old person taken under the eye, at the temple, and on the forearm (from top to bottom). Time–strain mode 1 (-_); probe arperture = 2 mm, load = 500 mbar, on-time = 2 s, off-time = 2 s. (From Courage and Khazaka Electronic GmbH, Cologne, Germany. With permission.)

In addition to the influence of physiological factors on the viscoelasticity of human skin, pathological changes underlying various dermatological diseases have been studied in recent years. Thus, the Cutometer has been used to study mechanical properties of skin affected by, for example, psoriasis,[9] systemic sclerosis,[10–12] Ehlers–Danlos syndrome,[13] striae distensae,[14] and hypertroph scars.[15] In addition, objective and quantitative evaluation of the effectiveness of therapeutic strategies has been made utilizing this bioengineering technique.

Furthermore, changes of biomechanical properties following the application of ultraviolet radiation,[16] laser treatment,[17] reconstructive dermal substitutions,[18] and liposuction [19] have been studied. Studies regarding the activity and safety of dermatocosmetics, such as emollients, moisturizers, chemical peelings, and antiaging creams, are another important application.[20,21]

REFERENCES

1. Courage and Khazaka Electronic GmbH, Cutometer SEM 575® Instruction for Use and Customer Information, 1998.
2. Elsner, P., Skin elasticity, in *Bioengineering of the Skin: Methods and Instrumentation,* Berardesca, E., Elsner, P., Wilhelm, K.-P., Maibach, H.I., Eds., CRC Press, Boca Raton, 1995, 53.
3. Agache, P.G. et al., Mechanical properties and Young's modulus of human skin *in vivo*, *Arch. Dermatol. Res.,* 269, 221, 1980.
4. Barel, A.O., Courage, W., and Clarys, P., Suction method for measurement of skin mechanical properties: the Cutometer®, in *Handbook of Non-Invasive Methods and the Skin,* Serup, J. and Jemec, G.B.E., Eds., CRC Press, Boca Raton, 1995, 335.
5. Elsner, P., Wilhelm, D., and Maibach, H.I., Mechanical properties of human forearm and vulvar skin, *Br. J. Dermatol.*, 122, 607, 1990.
6. Wilhelm, K.-P., Cua, A.B., and Maibach, H.I., *In vivo* study of age-related elastic properties of human skin, in *Noninvasive Methods for the Quantification of Skin Function,* Frosch, P.J. and Kligman, A.M., Eds., Springer, Berlin, 1993, 190.
7. Cua, A.B., Wilhelm, K.-P., and Maibach, H.I., Elastic properties of human skin: relation to age, sex, and anatomical region, *Arch. Dermatol. Res.,* 282, 283, 1990.
8. Nishimura, M. and Tsuji, T., Measurements of skin elasticity with a new suction device — relation to age, anatomical region, sun-exposure and comparison with diseased skin, *Jpn. J. Dermatol.*, 1111, 1992.
9. Dobrev, H.P., *In vivo* study of skin mechanical properties in Psoriasis vulgaris, *Acta Derm. Venereol.* (Stockholm), 80, 263, 2000.
10. Dobrev, H.P., *In vivo* study of skin mechanical properties in patients with systemic sclerosis, *J. Am. Acad. Dermatol.,* 40, 436, 1999.
11. Enomoto, D.N. et al., Quantification of cutaneous sclerosis with a skin elasticity meter in patients with generalized scleroderma, *J. Am. Acad. Dermatol.*, 35, 381, 1996.
12. Nikkels-Tassoudji, N. et al., Computerized evaluation of skin stiffening in scleroderma, *Eur. J. Clin. Invest.*, 26, 457, 1996.
13. Henry, F. et al., Mechanical properties of skin in Ehlers-Danlos syndrome, types I, II, and III, *Pediatr. Dermatol.*, 13, 464, 1996.
14. Pierard, G.E. et al., Tensile properties of relaxed excised skin exhibiting striae distensae, *J. Med. Eng. Technol.*, 23, 69, 1999.

15. Fong, S.S., Hung, L.K., and Cheng, J.C., The cutometer and ultrasonography in the assessment of postburn hypertrophic scar — a preliminary study, *Burns,* 23(Suppl. 1), 12, 1997.
16. Habig, J. et al., Einfluß einmaliger UVA- und UVB-Bestrahlung auf Oberflächenbeschaffenheit und viskoelastische Eigenschaften der Haut in vivo, *Hautarzt,* 47, 515, 1996.
17. Koch, R.J. and Cheng, E.T., Quantification of skin elasticity changes associated with pulsed carbon dioxide laser skin resurfacing, *Arch. Facial Plast. Surg*., 1, 272, 1999.
18. van Zuijlen, P.P. et al., Graft survival and effectiveness of dermal substitution in burns and reconstructive surgery in a one-stage grafting model, *Plast. Reconstr. Surg.,* 106, 615, 2000.
19. Henry, F. et al., Mechanical properties of skin and liposuction, *Dermatol. Surg*., 22, 566, 1996.
20. Fischer, T. et al., Instrumentelle Methoden zur Bewertung der Sicherheit und Wirksamkeit von Kosmetika, *Akt. Dermatol.,* 24, 243, 1998.
21. Greif, C. et al., Beurteilung einer Körperlotion für trockene und empfindliche Haut, *Kosmet. Med.,* 19, 24, 1998.

8 Hardware and Measurement Principles: The Gas-Bearing Electrodynamometer and Linear Skin Rheometer

Paul J. Matts

CONTENTS

Measuring Principles 99
Hardware 102
  Instrument Hardware and Control 102
  Calibration of the LSR 104
Software 105
Taking and Interpreting LSR Measurements 105
  Reproducibility and Measurement Site 105
  LSR Measurement Case Study 106
    Study Protocols 106
    Results and Interpretation 107
Discussion 108
Acknowledgments 109
References 109

MEASURING PRINCIPLES

The Gas-Bearing Electrodynamometer (GBE[1–5]) has been used for the last 20 years to obtain sensitive measurements of the stratum corneum. The Linear Skin Rheometer (LSR[6]) is a new instrument for measuring the mechanical properties of the stratum corneum that incorporates all of the measurement principles of the GBE but none of its components. A force-controlled miniature DC servo, gearing, and lead screw replace the magnet/solenoid arrangement of the GBE. Error resulting from conversion of an electrical signal to a mechanical force is automatically compensated for. Consequently,

0-8493-7521-5/02/$0.00+$1.50

this control renders the need for a friction-free bearing redundant. The original Linear Variable Differential Transformer (LVDT) has been replaced with a unit with a sensitivity of 0.01% and force is now measured by a calibrated 100 g load beam. The function generator, signal conditioner, and storage oscilloscope have been replaced by user-friendly software run by a small portable computer. The new design offers greater inherent accuracy than the GBE and requires minimal servicing. The new LSR instrument has been shown to provide sensitive measurements of stratum corneum mechanics. A simple case study is presented in which the LSR was used to measure the mechanical responses of the stratum corneum to two topical moisturizing treatments of differing relative hydration performance (as determined by impedance measurements using the Nova® DPM9003). The relative performance of the two products as measured by the LSR compared favorably with corresponding impedance data, indicating the ability of the LSR to differentiate varying degrees of stratum corneum plasticization in response to hydration.

In measuring subtle changes in the mechanical properties of human stratum corneum, it is essential to use instrumentation that does not induce overly large surface displacement (either parallel or perpendicular to the skin surface). Excessive surface displacement will involve underlying tissue to an unacceptable degree, effectively rendering the measurement void.

With this in mind, one of the instruments to have been most widely used over the last 20 years to obtain sensitive measurements of the stratum corneum, is the Gas Bearing Electrodynamometer (GBE[1–5]). It is able to apply a sinusoidal loading stress of less than 5 g parallel to the skin surface, with a resulting displacement of less than 1 mm in each direction. These force/displacement values are, indeed, sufficiently small to allow investigation of stratum corneum mechanics. This is achieved by suspending an armature in a gas bearing to create near friction-free movement. Changes in the magnetic field generated by a surrounding coil cause the armature to oscillate at a known frequency and amplitude. The coil is activated by a sinusoidal signal from a low-frequency function generator or from a suitable software trigger. The armature of the instrument is typically attached to the skin surface by a stiff wire probe bent to 90° at its free end. A small plastic stub is usually cemented to the free end of the probe and used to attach the probe to the skin surface using a circular piece of double-sided adhesive tape. Displacement of the armature is measured by a sensitive LVDT, mounted coaxially with the coil. Coil and LVDT outputs (force and displacement) are amplified and then supplied for analysis to either a storage oscilloscope or a computer equipped with suitable software. Equipment used in a "classic" GBE workstation is shown in Figure 8.1.

Results of force and displacement measurements of skin are typically displayed as a hysteresis loop (Figure 8.2). Analysis of the gradient of the loop (force/displacement or displacement/force) yields derivatives of the dynamic spring rate (DSR) usually expressed as g/mm (a measure of the force required to stretch or compress the skin per unit extension), mm/N or μm/g (measures of stretching or compression of the skin in response to a given applied force). Such analysis yields derivative information about the elastic properties of the skin. Analysis of the phase lag between force and displacement responses yields derivative information about the viscous properties of the skin.

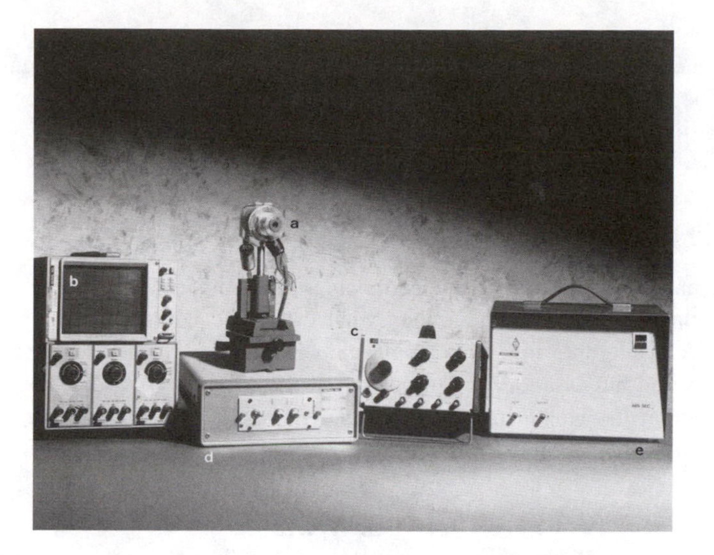

FIGURE 8.1 Equipment used in a GBE workstation: (a) GBE probe, (b) storage oscilloscope, (c) function generator, (d) signal conditioner, (e) air compressor. (From Matts, P.J. and Goodyer, E., *J. Cosmet. Sci.*, 49, 321, 1998. With permission.)

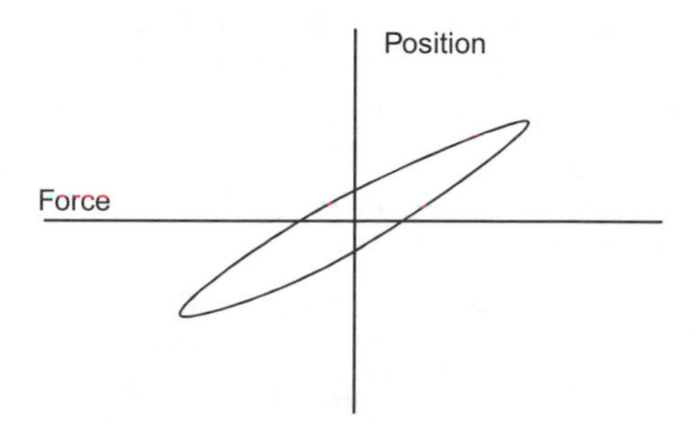

FIGURE 8.2 Typical position vs. force hysteresis loop produced by one GBE measurement cycle. (From Matts, P.J. and Goodyer, E., *J. Cosmet. Sci.*, 49, 321, 1998. With permission.)

More than 20 years of experience with the GBE within the author's laboratories has led to the belief that the GBE measurement principle is still among the best available for measuring changes in the mechanical properties of the human stratum corneum *in vivo*. Subtle but important changes in skin mechanics in response to the application of moisturizing formulas (dubbed "softness" by Maes et al.[3]) have been measured, as have changes in skin "tightness" due to surfactant damage. The author's experience has, however, also highlighted the drawbacks of employing the original Hargens GBE instrument in a modern laboratory. These are as follows:

1. Importantly, the instrument employs an "open-loop" method of control; i.e., during calibration of the instrument, and in subsequent routine operation, one assumes that an applied current equals a given force. As the GBE is calibrated on one point only (3 g), linearity is not guaranteed over the whole measurement range of the GBE. In addition, calibration drift over time is certain.

2. The components needed to run the GBE are bulky and dated (function generator, signal conditioner, storage oscilloscope, compressed gas/air).
3. The probe components are fragile and, in the author's experience, break easily and require excessive servicing when used routinely (for example, the fine copper wires connecting the armature to the body of the probe).
4. The air bearing employed in the probe design is inherently susceptible to misalignment, soiling, and malfunction.

In recent years, the cost of precision has improved greatly. It is now possible to achieve using conventional technology what was achieved previously only through Hargens'[1] considerable ingenuity. A new instrument has been designed and built that retains and builds on all the principles of the original GBE, but contains *none* of the components. This instrument, designated the Linear Skin Rheometer (LSR[6]), is described in the following sections.

HARDWARE

Instrument Hardware and Control

A schematic diagram of the new instrument is shown in Figure 8.3. A force-controlled miniature DC servo [Maxon Type 106328 (0.75 W, 9 V) by Maxon UK], gearing, and lead screw now replace the GBE solenoid arrangement, and drive the LSR probe. The original Schaevitz 050 HR LVDT in the GBE has been replaced with a unit of linearity 0.3% (15 μm) and sensitivity 0.01% (0.5 μm) (Type DFg5 by Solartron Metrology Ltd, U.K.). The force exerted on the probe is now measured directly by a calibrated load beam (Type F301 by Novatech Ltd, U.K., rated 100 g) with an overall repeatability of ±0.03% (±30 mg). The load beam is mounted vertically within the instrument casing and acts horizontally in the direction of motion.

All components fit into one casing measuring 20.0 × 14.8 × 6.9 cm and the whole unit weighs 1.7 kg. The probe housing itself is a lightweight machined perspex chuck mounted on a low-friction swivel assembly allowing 360° movement (analogous to the GBE). This is protected from damage during routine usage by a metal collar. The chuck contains wire grips to allow the wire probes to be inserted or withdrawn by a simple, firm push or pull. A single lead connects the unit to a PC via a 25-pin D-type connector. Power for the LSR unit is taken from the PC via the connector. The unit can be seen in Figure 8.4.

An IBM (or compatible) PC is used to control the movement of the probe and to log force and displacement data. Both force and displacement are monitored continuously at a rate of 1 kHz using a 12-bit ADC plug-in card (National Instruments MIO16). The motor is controlled with an analog output signal also generated by the PC. The desired force/time cycle, which is normally a single sinusoid, is calculated initially and then stored in memory as a table of values. The actual force applied to the probe is compared with the desired value in the table 1000 times a second. A feedback loop is used to control the motor, which moves the load cell in a way that minimizes any discrepancy. The force applied thus follows the desired force/time cycle extremely closely. The control loop uses an algorithm with proportional and integral terms, whose relative weighting can be varied.

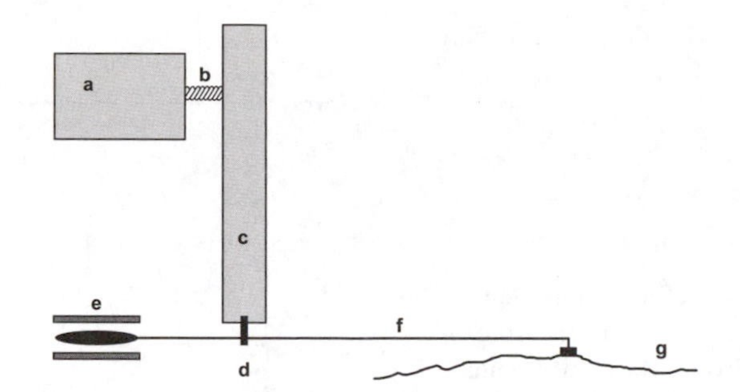

FIGURE 8.3 Schematic diagram of the LSR sensing head: (a) miniature motor, (b) lead screw, (c) load cell, (d) load cell sensing head, (e) LVDT, (f) probe, (g) skin surface. (From Matts, P.J. and Goodyer, E., *J. Cosmet. Sci.*, 49, 321, 1998. With permission.)

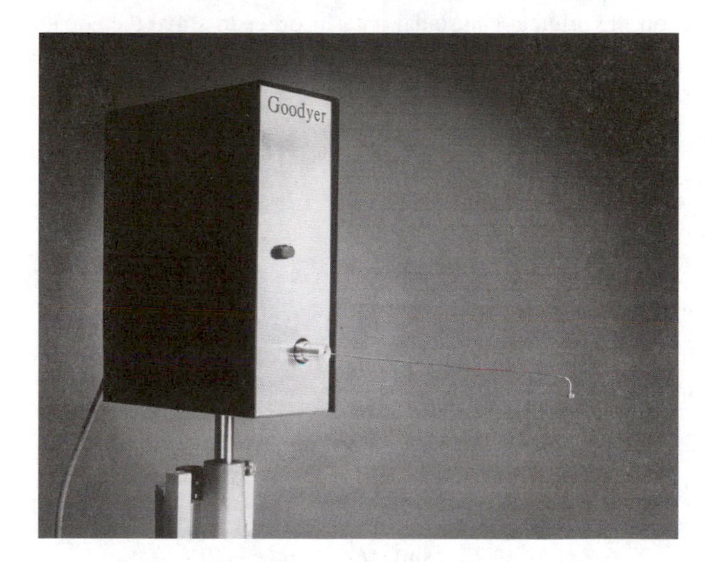

FIGURE 8.4 The Linear Skin Rheometer. (From Matts, P.J. and Goodyer, E., *J. Cosmet. Sci.*, 49, 321, 1998. With permission.)

The PC logs all the force and displacement values over a complete measurement cycle, which is usually set at 0.33 Hz, thus generating 3000 pairs of points over a 3-s cycle. Two waveform plots are then obtained (Figure 8.5). Three parameters may be obtained from these curves:

F_{max} = the peak force that is applied to the skin surface
P_{max} = the peak displacement occurring as a result of that force
T = the phase shift between the two signals

The DSR of the stratum corneum is given simply by the formula F_{max}/P_{max}. Derivatives are calculated and expressed as g/mm, mm/N, and µm/g.

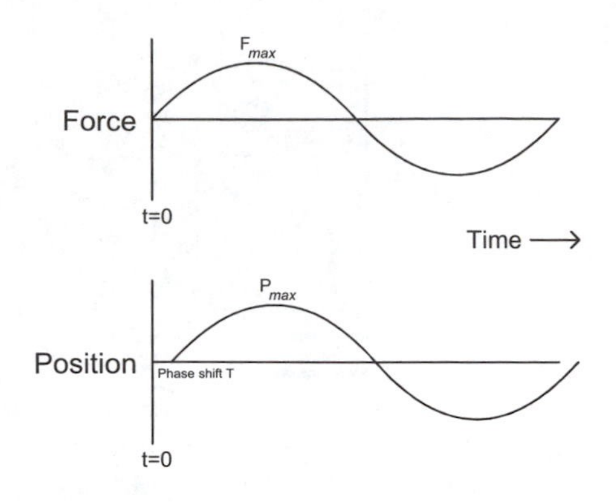

FIGURE 8.5 Waveform plots demonstrating a complete LSR measurement cycle. (From Matts, P.J. and Goodyer, E., *J. Cosmet. Sci.*, 49, 321, 1998. With permission.)

The viscous component of the stratum corneum is often inferred by calculating the area of the ellipse shown in Figure 8.2. A more rigorous approach is to perform a regression on the original sinusoidal data in order to solve the equations:

$$F = F_{\max} \sin(t) \tag{8.1}$$

$$P = P_{\max} \sin(t + T) \tag{8.2}$$

where

F = instantaneous force, $F_{\max}$ = peak force, t = time for one complete cycle in seconds

P = instantaneous displacement, $P_{\max}$ = peak displacement, T = phase shift in radians

Having solved for these equations, it is then a straightforward problem to solve the integral over one cycle that represents the area of the ellipse:

$$\int_0^{2\pi} F_{\max} \sin(t) P_{\max} \cos(t + T) \tag{8.3}$$

The LSR software solves the above equations for both elastic and viscous components of the data. These are subsequently displayed, directly after measurement.

Note: Units of µm/g (1/DSR, a measure of stretching or compression of the stratum corneum in response to a given applied force) will be used as a convenient expression of skin softness in the rest of this chapter.

Calibration of the LSR

A simple calibration jig has been designed and built that allows rapid, absolute calibration of *actual* force and displacement (Figure 8.6). Displacement is measured by a 10 µm resolution Digimatic Indicator [Mitutoyo (UK) Ltd, Warwick, U.K.] traceable to NAMAS calibration standards. Force is measured by a 10 g load beam (Maywood Load Beam type 49034, Maywood Instruments Ltd, Basingstoke, U.K.) with <10 mg accuracy, also traceable to NAMAS calibration standards.

The calibration factors are expressed in terms of ADC input units to achieve 1 g force, and 1 mm displacement. This method means that the calibration factors take account of the signal conditioning and data acquisition electronics as well as the precision transducer characteristics.

FIGURE 8.6 LSR calibration jig. (From Matts, P.J. and Goodyer, E., *J. Cosmet. Sci.*, 49, 321, 1998. With permission.)

SOFTWARE

The LSR software runs on a standard IBM (or compatible) PC of at least 386 33 MHz speed, and is sourced in C programming language. The closed-loop control employed by the LSR is achieved as follows. Because the LSR control loop is a sampled data system, it is essential that a fast real-time clock be generated that triggers the measurement of data samples and updates the control signal output. All IBM PCs have as standard a user interrupt (on interrupt vector 0x1c) called the TIMER TICK, which is available for programmers to use as a regular timing source. This timing signal is generated from the 4.192 MHz system clock via an Intel 8253 programmable interval timer. The BIOS presets this timer to its full scale of 65536 and, therefore, the TIMER TICK interrupt is normally 18.188 Hz. This is far too slow for a sampled data system.

This problem is overcome by reprogramming the 8253 divider to give the desired frequency during the measurement cycle, in this case 1 kHz. In order, however, not to disrupt important internal functions such as monitoring disk drive heads, the timing of interrupt 0x8 (the interrupt number actually triggered by the 8253 output) must be restored. This may be achieved by not using interrupt 0x1c, but replacing the BIOS interrupt function at 0x8 with the control program itself. The original timing is derived within the new interrupt 0x8 by installing a simple counter and calling the BIOS interrupt at the correct interval. In this way, a fast timer is generated that allows data sampling at 1 kHz but which does not harm other internal PC operations.

TAKING AND INTERPRETING LSR MEASUREMENTS

Reproducibility and Measurement Site

For direct comparison with the GBE reproducibility data obtained by Maes et al.,[3] the reproducibility of the LSR was estimated by the same method. In the study,

40 consecutive identical measurements were performed on the back of the hand of a female volunteer. The resulting coefficient of variation for this measurement set was only 2.9% ($n = 40$; mean = 281.8 μm/g; SEM = 1.3 μm/g). This demonstrates the very good reproducibility of the measurement technique and compares very favorably with the value of 3% obtained by Maes et al.[3] for the GBE.

The variation measured is almost certainly due to movement of the subject during the probe cycle. This has always been the main source of error in these types of sensitive measurements, and various means have been employed to minimize subject movement during readings (e.g., use of a pre-cast plaster mold by Maes et al.[3]). However, like Cooper et al.,[4] the author has found that the use of no restraint is preferable and employs a simple sloping table on which subjects rest their hands.

LSR Measurement Case Study

Study Protocols

To determine the ability of the LSR to measure sensitive changes in stratum corneum mechanics in response to simple hydration, the following study was performed. Two moisturizing formulas of differing hydration performance (products A and B; hydration performance was determined by impedance measurements using a Nova Dermal Phase Meter 9003, see below) were applied to the back of the hands of 13 female subjects (aged 18 to 35). The dorsal surface of the hand was chosen for mechanical measurements (1) to conform to previous measurement sites using the GBE[3] and (2) because it is relatively simple to immobilize the hand effectively. The study was performed in a controlled-environment chamber (temperature 20 ± 1°C; relative humidity 45 ± 5%). The plastic stub on the end of the LSR wire probe was attached to skin on the back of the hand via a circular piece of double-sided tape (5 mm diameter). LSR measurements were then performed in triplicate. Baseline measurements were performed before product application. Test products were then applied at a rate of 2 $\mu l/cm^2$ to the entire back of the hand according to a predetermined randomization schedule. LSR measurements were performed at 1, 3, and 6 h after product application. As the whole dorsal surface of each hand was used for product treatment, inclusion of an untreated control was not possible. Results were, therefore, expressed as mean difference from initial pretreatment baseline.

Hydration performance of products A and B was assessed by randomized application at the same rate as above (2 $\mu l/cm^2$) to 5 × 5 cm sites on the volar forearms of 12 female subjects (aged 18 to 35); the volar forearm was chosen as the site for hydration measurements because of its smooth, hairless morphology and its utility as a standard in this type of testing.[7] Each forearm also contained an untreated 5 × 5 cm control site. The study was performed within a controlled-environment chamber (temperature 20 ± 1°C; relative humidity 45 ± 5%). Impedance measurements were performed using a Nova Dermal Phase Meter 9003 with the standard measuring probe DPM 9103 (Nova Instruments) at 1, 2, 4, and 6 h after application, and results expressed as mean difference from untreated control.

Results and Interpretation

The results of the study using the Nova DPM 9003 to measure the hydration efficacy of products A and B can be seen in Figure 8.7. Both products induced significant increases ($p < 0.05$; paired t-test vs. untreated control) in apparent stratum corneum hydration (as measured by impedance changes) up to, and including, 6 h after application. Moreover, product A increased stratum corneum hydration significantly more ($p < 0.05$; paired t-test) than product B at all time points up to and including 6 h after application. Water exerts considerable influence on the mechanical properties of the human stratum corneum because of its complex interactions with keratin.[8,9] This plasticization of the stratum corneum, an essentially viscoelastic material, has been described as skin "softening."[2,3] In the case of topical application of a moisturizing formula, the extent of this softening effect is directly related to the ability of the product to deliver and maintain increased water concentrations within the stratum corneum. This is usually achieved by the delivery of humectant compounds such as glycerol and/or use of occlusive lipidic films. Indeed, in the case of products A and B, product A might be expected to leverage greater increase in stratum corneum hydration as a result of its higher glycerol content (4% w/w glycerol in A, in contrast to 3% w/w in B) and formulation (gel network, in contrast to a simple oil-in-water emulsion in B). For products A and B, therefore, one would expect to be able to measure (1) absolute significant increases in softness for both treatments and (2) differing relative changes in skin softness for both treatments in accordance with their apparent hydration performance.

Results of the study using the LSR to measure the effects of products A and B on stratum corneum mechanics are presented in Figure 8.8. Both products induced significant increases ($p < 0.05$; paired t-test vs. pretreatment baseline) in skin softness at all time points up to and including 6 h after application. Moreover, product A induced greater increases in skin softness than product B throughout the time

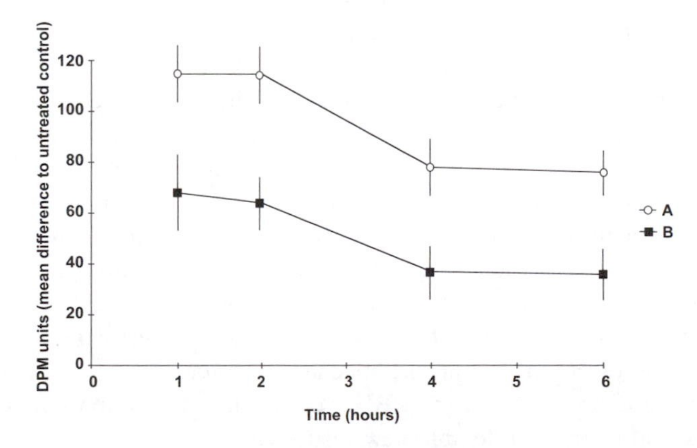

FIGURE 8.7 Relative hydration performance of products A and B (as measured by impedance). (From Matts, P.J. and Goodyer, E., *J. Cosmet. Sci.*, 49, 321, 1998. With permission.)

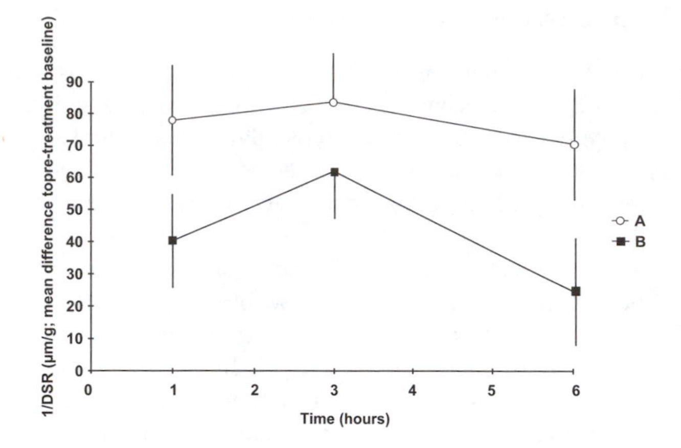

FIGURE 8.8 Relative performance of products A and B as measured by LSR. (From Matts, P.J. and Goodyer, E., *J. Cosmet. Sci.*, 49, 321, 1998. With permission.)

course, significantly so ($p < 0.05$; paired t-test) at 6 h after application. These results compare favorably with the relative hydration profiles of the two products (Figure 8.7). The LSR is, thus, able to measure subtle changes in stratum corneum mechanics in response to hydration and to distinguish between the effect of topical application of moisturizing products with differing relative hydration performance.

DISCUSSION

The control system used for the LSR provides the instrument with an inherently more accurate and reliable measurement capability because it employs closed-loop (feedback) control. The GBE, in contrast, uses an open-loop method of control where a predetermined current is applied to the solenoid and assumed to be transformed into the desired force. Because no determination of the actual force generated is made at the time of measurement, it is difficult to know with certainty the true force applied to the skin. While open-loop techniques can be used successfully in perfectly stable environments, no account can be taken of instantaneous fluctuations in such a system (notably, in this case, subject movement). With the LSR closed-loop system, the true force applied to the skin is measured at a rate of 1 kHz and corrective action taken within 1 ms to restore that measured force to the required value. This system helps ensure that the test sequence is reliable, repeatable, and can dynamically adjust for the inevitable variations that occur during *in vivo* testing. Put another way, because this system allows instantaneous compensation of error resulting from the conversion of an electrical signal to a mechanical force, the need for the friction-free gas-bearing arrangement of the GBE is eliminated. This allows the deployment of a compact, efficient, and flexible new instrument.

ACKNOWLEDGMENTS

The author wishes to thank Mr. Eric Goodyer (E & C Consultancy, Hathern, Leicestershire, U.K.) and Dr. Paul Stevens (Paul Stevens Mechanical Design, Tunbridge Wells, Kent, U.K.) for their significant skill and expertise in the design of the LSR software and hardware. Much of the material included in this chapter is taken from a paper published originally within the *Journal of Cosmetic Science* (Matts, P.J. and Goodyer, E., A new instrument to measure the mechanical properties of human stratum corneum *in vivo*, *J. Cosmet. Sci.*, 49, 321, 1998) to whom the author is grateful for permission to reproduce relevant text and graphical material.

REFERENCES

1. Christensen, M.S., Hargens, C.W. III, Nacht, S., and Gans, E.H., Viscoelastic properties of intact human skin: instrumentation, hydration effects and the contribution of the stratum corneum, *J. Invest. Dermatol.*, 69, 282, 1977.
2. Hargens, C.W. III, The Gas Bearing Electrodynamometer (GBE) applied to measuring mechanical changes in skin and other tissues, in *Bioengineering and the Skin*, Marks, R. and Payne, P.A., Eds., MTP Press, Hingham, MA, 1981, 113.
3. Maes, D., Short, J., Turek, B.A., and Reinstein, J.A., *In vivo* measuring of skin softness using the Gas Bearing Electrodynamometer, *Int. J. Cosmet. Sci.*, 5, 189, 1983.
4. Cooper, E.R., Missel, P.J., Hannon, D.P., and Albright, G.B., Mechanical properties of dry, normal, and glycerol-treated skin as measured by the Gas Bearing Electrodynamometer, *J. Soc. Cosmet. Chem.*, 36, 335, 1985.
5. Hargens, C.W. III, The Gas Bearing Electrodynamometer, in *Handbook of Non-Invasive Methods and the Skin*, Serup, J. and Jemec, G.B.E., Eds., CRC Press, Boca Raton, FL, 1995, 353.
6. Matts, P.J. and Goodyer, E., A new instrument to measure the mechanical properties of human stratum corneum *in vivo*, *J. Cosmet. Sci.*, 49, 321, 1998.
7. Rogiers, V., Derde, M.P., Verleye, V., and Roseeuw, D., Standardised conditions needed for skin surface hydration measurements, *Cosmet. & Toiletries*, 105, 73, 1990.
8. Lèvêque, J.L., Escoubez, M., and Rassneur, L., Water-keratin interaction in human stratum corneum, *Bioeng. Skin*, 3, 227, 1987.
9. Lèvêque, J.L., Water–keratin interactions, in *Bioengineering of the Skin: Water and the Stratum Corneum*, Elsner, P., Berardesca, E., and Maibach, H.I., Eds., CRC Press, Boca Raton, FL, 1994, 13.

9 Hardware and Measuring Principles: The Dermaflex A

Jørgen Serup

CONTENTS

Introduction 111
Measuring Principle and Hardware 111
Preconditioning of Individuals and Measurement 113
The Parameters Expressed by the Device 113
Applications and Interpretation 114
References 114

INTRODUCTION

The Dermaflex A (Figure 9.1) was originally developed as a prototype by the Department of Dermatology, Rigshospital, Copenhagen, Denmark and the Institute of Technical Engineering (ATV), Glostrup, Denmark. Its purpose is to measure objectively the stiffness of the skin in scleroderma (Figure 9.2) in concert with other methods for measurement of skin thickening and change of color in this disease.[1,2] It soon became applied to assessment of psoriasis and histamine wheals exemplifying edematous states (Figure 9.3).[3]

MEASURING PRINCIPLE AND HARDWARE

The Dermaflex is a suction chamber device. The circular aperture of the chamber is 10 mm in diameter, constructed with two sharp, concentric rings to avoid creeping of skin during negative pressure and a surrounding flat bases prepared for the application of a double-sided adhesive ring sealing the probe to the skin. A concave electrode with an outlet tube to the suction system is located in the top of the chamber. This electrode and the surface of the skin exposed in the suction chamber constitute a capacitative system with the distance between skin surface and top electrode the

0-8493-7521-5/02/$0.00+$1.50

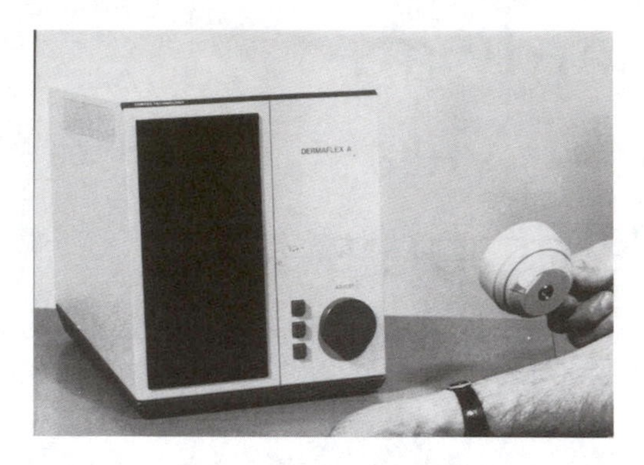

FIGURE 9.1 The Dermaflex A. (Courtesy of Cortex Technology, Hadsund, Denmark.)

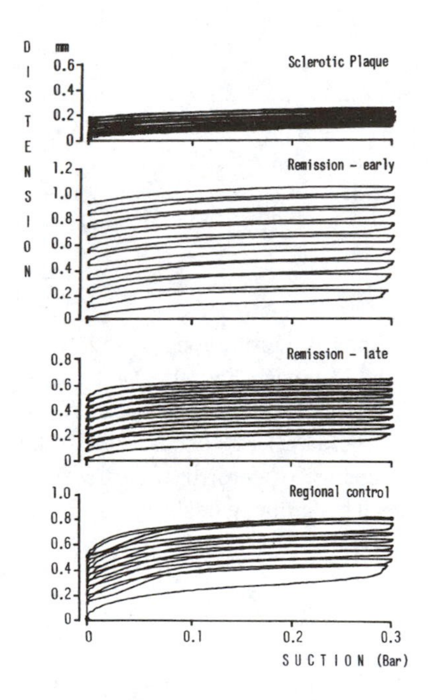

FIGURE 9.2 Stress–strain curves in sclerotic plaque of morphea vs. regional control of clinically uninvolved skin. The distensibility (strain after first suction) is decreased in all stages of the disease. In early remission, the hysteresis or creeping is increased due to some mucinous edema.

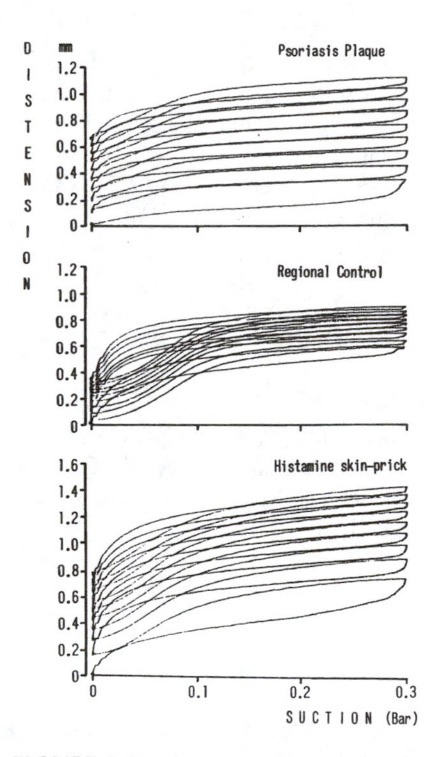

FIGURE 9.3 Stress–strain curves in psoriasis vs. regional control of clinically uninvolved skin. Increased hysteresis of a histamine wheal raised in uninvolved skin illustrates the influence of tissue water on distensibility, particularly reflected in the hysteresis.

variable of the system. This is measured in millimeters with high precision on the order of magnitude of about 0.05 mm. The suction outlet is connected to a negative-pressure reservoir or tank, controlled by a vacuum pump. Because of the reservoir, a preset negative pressure is immediately effected at the initiation of recording in the suction chamber. There is some low resistance in the tube between the reservoir and the probe chamber.

The negative pressure, the duration of suction, and the number of repetitions of suction and release of pressure cycles can be selected. Suction pressure greater than 300 mbar may be unpleasant or painful. The electrical constants of the capacitative measuring system were chosen so that the measured distance and elevation are not dependent on the skin surface hydration state.

The circular opening of the chamber was chosen to standardize the start of recording and to eliminate the influence of Langer line orientation on baseline recording, albeit the elevated skin will show some ellipsoid deformation, which, however, does not disturb the result and reproducibility. The Dermaflex automatically sets the zero elevation immediately at the initiation of negative pressure.

PRECONDITIONING OF INDIVIDUALS AND MEASUREMENT

Individuals should be placed comfortably in a position defined in the study protocol. It is particularly important that the joint positioning be specified, and the measured site located precisely within a few millimeters. Lanugo hairs do not interfere with measurement, but terminal hairs may be a problem. It is important that the probe at the initiation of measurement be positioned neutrally and vertically over the measured site, and that the probe is not supported but remains in position as a result of its own weight. The reason is that the recorded elevation is a delta value, i.e., the difference between the zero surface state and the same surface elevated under the influence of suction.

THE PARAMETERS EXPRESSED BY THE DEVICE

The parameters expressed by the Dermaflex A are as follows:

- Distensibility, i.e., elevation in mm after the first suction. Resilient distensibility (not displayed) is the elevation in mm following the first suction, after the skin has retracted toward its original state.
- Elasticity, i.e., percentage retraction after one suction; 100% is full recovery to zero state.
- Hysteresis or creeping, i.e., the delta distension after repeated suction cycles.

The Dermaflex A appears quite independent of subcutaneous binding. With constant suction, the probe can be elevated millimeters or centimeters without disturbing the distensibility reading. With no suction or a weak suction, even small maneuvers of the probe exert major variation of the displayed strain values, illustrating the important and vulnerable zero state situation, which in the measurement requires much attention and insight to standardize properly.

APPLICATIONS AND INTERPRETATION

The principle is that of a stress–strain exercise. Reduced distensibility is easily understood as skin stiffening exemplified by scleroderma, and increased distensibility as softening exemplified by edematous skin conditions such as inflammation. Poor elasticity percentage is understood as poor turgor, as exemplified by dehydrated dermis, and/or reduced ability of the skin to recover to normal state following a deformation, as exemplified by aged skin.

The reproducibility of the method, body site variations, and findings in scleroderma, Pierini atrophy of morphea, psoriasis, and histamine wheals are described in the original publications.[1–4] Anatomical site differences in 22 sites, comparison of genders, and the influence of gravity and diurnal changes were studied in various studies that provide detailed information.[5–7] Skin elasticity following tissue expansion with expansion of the subcutaneous space was studied in relation to breast cancer surgery.[8] Dr. G.B.E. Jemec and co-workers conducted systematic studies on the Dermaflex A, its precision and variables, the effects of superficial hydration and care products, and finally the use of the Dermaflex A in plastic surgery.[8–15] (Also see Chapter 17 of this book.)

It is noteworthy that there is a high degree of correlation between the different elasticity parameters, and it appears that all essential information is given in the distensibility and the elasticity percentage describing the essential clinical traits of the skin, namely, the ability to stretch and retract to the original state. It should be borne in mind that the first suction comes closest to illustrating the original skin mechanics, and repeated suctions start with a zero drift because of the incomplete recovery from the previous strain, i.e., the creeping dependent on the hydration state of the viscoelastic skin tissue. Gniadecka and Serup[16] have reviewed the Dermaflex A in a recent book chapter.

REFERENCES

1. Serup, J. and Northeved, A., Skin elasticity in localized scleroderma (morphoea). Introduction of a biaxial *in vivo* method, and measurement of tensile distensibility, hysteresis, and resilient distension of normal and diseased skin, *J. Dermatol.* (Tokyo), 12, 52–62, 1985.
2. Serup, J., Localized scleroderma (morphoea). Clinical, physiological, biochemical and ultrastructural studies with particular reference to quantitation of scleroderma (thesis), *Acta Dermatovenereol.* (Stockholm), Suppl. 66, 1–66, 1986.
3. Serup, J. and Northeved, A., Skin elasticity in psoriasis. *In vivo* measurement of tensile distensibility, hysteresis and resilient distension with a new method. Comparison with skin thickness as measured with high-frequency ultrasound, *J. Dermatol.* (Tokyo), 12, 318–324, 1985.

4. Serup, J., Decreased skin thickness of pigmented spots appearing in localized scleroderma (morpheoa). Measurement of skin thickness by 15 MHz pulsed ultrasound, *Arch. Dermatol. Res.*, 276, 135–137, 1984.
5. Malm, M., Samman, M., and Serup, J., *In vivo* skin elasticity of 22 anatomical sites, *Skin Res. Technol.*, 1, 61–67, 1995.
6. Gniadecka, M., Gniadecka, R., Serup, J., and Søndergaard, J., Skin mechanical properties present an adaptation to human upright position, *Acta Dermatovenereol.* (Stockholm), 74, 188–190, 1994.
7. Wickman, M., Olenius, M., Malm, M., Jurell, G., and Serup, J., Alterations in skin properties during rapid and slow tissue expansion for breast reconstruction, *Plast. Reconstr. Surg.*, 90, 945–950, 1992.
8. Jemec, G.B.E. and Serup, J., Epidermal hydration and skin mechanics. The relationship between electrical capacitance and the mechanical properties of human skin *in vivo*, *Acta Dermatovenereol.* (Stockholm), 70, 245–247, 1989.
9. Jemec, G.B.E., Jemec, B., Jemec, B.I.E., and Serup, J., The effect of superficial hydration on the mechanical properties of human skin *in vivo*: implications for plastic surgery, *Plast. Reconstr. Surg.*, 85, 100–103, 1990.
10. Overgaard Olsen, L. and Jemec, G.B.E., The influence of water, glycerin, paraffin oil and ethanol on skin mechanics, *Acta Dermatovenereol.* (Stockholm), 73, 404–446, 1993.
11. Jemec, G.B.E., Gniadecka, M., and Jemec, B., Measurement of skin mechanics. A study of inter- and intra-individual variation using the Dermaflex A, *Skin Res. Technol.*, 2, 164–166, 1996.
12. Petersen, L.K. and Jemec, G.B.E., Plasticizing effect of water and glycerin on human skin *in vivo*, *J. Dermatol. Sci.*, 19, 48–52, 1999.
13. Kynemund, L., Jemec, G.B.E., and Wulf, H.C., Moisturisers for psoriatic skin — do gross morphological differences matter? *Skin Pharmacol. Appl. Phys.*, 14, 20–26, 2001.
14. Jemec, G.B.E. and Wulf, H.C., Correlations between greasiness and plasticising effect of moisturizers, *Acta Dermatovenereol.* (Stockholm), 78, 1–3, 1998.
15. Jemec, G.B.E. and Wulf, H.C., The plasticising effect of moisturizers on human skin *in vivo*: a measure of moisturising potency? *Skin Res. Technol.*, 4, 88–93, 1998.
16. Gniadecka, M. and Serup, J., Suction method for measurement of skin mechanical properties: the Dermaflex, in *Handbook of Noninvasive Methods and the Skin*, Serup, J. and Jemec, G.B.E., Eds., CRC Press, Boca Raton, 1995, 329–334.

10 Hardware and Measuring Principles: The DermaLab

Jørgen Serup

CONTENTS

Introduction..117
Measuring Principle and Equipment...118
Measuring with the DermaLab..120
Results and Their Interpretations 120

INTRODUCTION

The DermaLab (Figure 10.1) is a new suction chamber method for measurement of skin elasticity developed by Cortex Technology, Hadsund, Denmark. The system can be mounted with special probes to measure transepidermal water loss and skin hydration as well. The elastometer device has been developed especially to overcome the problem of baseline setting normally creating startup noise in the recording. The system is designed to simplify and preset factors normally somewhat difficult to measure with high precision and to measure factors that can be sensed technically with high precison and to nominate the latter as measuring end points. Skin elevation or distensibility is a fixed delta value of 1.5 mm, and the applied vacuum is also fixed; however, the vacuum is not applied immediately but, rather, gradually built up to a plateau over some 30 s. The DermaLab operates with controlled pretension load before baseline setting followed by the 1.5 mm stress–strain exposure under the influence of a standard pressure as described above. The device is mounted with an especially lightweight measuring probe. The results are expressed by the manufacturer as an empirical Young's modulus termed *elast* and as the time to achieve the fixed skin elevation of 1.5 mm.

0-8493-7521-5/02/$0.00+$1.50

FIGURE 10.1 The DermaLab skin elasticity measurer. The complete DermaLab system also includes systems and probes for measurement of transepidermal water loss and epidermal hydration. (Courtesy of Manufacturer Cortex Technology, Hadsund, Denmark.)

MEASURING PRINCIPLE AND EQUIPMENT

The probe consists of a polyethylene cylindrical suction chamber with height of 3.8 mm and a circular aperture of diameter 10 mm (Figure 10.2). The probe has a flat 7.5-mm concentric base that is sealed on the skin surface with double-sided adhesive tape. Diodes and a circular lens system for optical monitoring of skin elevation are mounted in the probe cylinder wall at two fixed positions: level 1 at

FIGURE 10.2 The DermaLab suction chamber probe. The probe is especially lightweight, 10 g. The outer dimensions are height 15 mm and diameter 25 mm. The cylindrical suction chamber is 10 mm in diameter and 3.8 mm in height, with optical sensors mounted in the chamber 1.0 and 2.5 mm above the opening toward the skin.

1.0 mm and level 2 at 2.5 mm above the base of the probe. The lens system focus is in the middle of the axis of the cylindrical chamber. A continuous skin surface elevation curve during suction is, thus, not measured. The main body of the equipment includes a base and a display for settings of a number of repeated suctions and relaxations. The suction period is displayed as a result, whereas the relaxation time is a preset variable. Suction time read out as a result may in practice vary between a few seconds in very soft skin and 1 to 5 min in stiff and sclerotic skin.

Results may be printed out or loaded into a computer. The display will show the end vacuum pressure RES1 in pascals and, of course, the main result, the "elast" in megapascals (MPa). In its manual the manufacturer provides the following explanation of the parameters displayed:

- ELAST. The calculated elasticity modulus for the actual measurement sequence based on the first measurement cycle.
- RES1. Indicates the differential vacuum necessary to elevate the skin from detection level 1 to detection level 2 in the first suction cycle. The unit is Pa.
- N (upper in display). Indicates the differential vacuum necessary to elevate the skin from detection level 1 to level 2 in the last actual suction cycle. The unit is Pa.
- TIM1. The elevation time corresponding to the first suction cycle (RES1).
- N (lower in display). The elevation time corresponding to the last actual suction cycle.

The device is specially constructed with resistance in the tube system to apply the vacuum slowly over a period of perhaps 30 to 60 s until a plateau vacuum in the chamber is maintained in equilibrium with the capacity of the pump. The negative-pressure curve in the chamber is not known exactly and is supposed to be nonlinear. The DermaLab has no vacuum tank and no barometric control of the buildup of negative pressure. The pretension control bringing the skin from biological baseline (in principle, there is no touching and no probe held against the surface) to experimental baseline at level 1 at 1.0 mm above the probe base is thought to unfold microanatomical foldings and to harmonize the orientation of the dermal fiber networks at a level of negative pressure believed to be subclinical.

Skin elasticity E displayed as elast in MPa is, according to the manufacturer, a calculated Young's modulus where

$$\Delta X = \Psi \cdot p = \frac{r^4}{E \cdot S^3}$$

ΔX = deviation middle of surface, Ψ = elasticity constant for measured object estimated from civil engineering, p = negative pressure, r = radius of surface measured, s = thickness of object measured.

Constant Ψ is set by manufacturer to 0.5; *r*, defined by the chamber geometry, is 5 mm; *s* is set by manufacturer to standard skin thickness 1.0 mm. This leads to

$$E = 0.3125 \cdot \frac{\Delta p}{\Delta X}$$

The Young's modulus originates from engineering of solid and isotropic materials of some measurable thickness with well-defined mechanical properties, and the modulus is normally not considered applicable to anisotropic materials such as skin. Skin is a complex viscoelastic material with layers of different mechanical resistance. The actual Ψ and the actual thickness of the skin are not known and are not entered as variables. Thus, the modulus given by the DermaLab is an empirical modulus in conflict with prerequisites known in engineering.

MEASURING WITH THE DERMALAB

An advantage of the probe is its light weight. It is important that the probe be fixed gently to the skin surface by double-sided adhesive tape and left undisturbed during recording. As with any other elasticity method, the positioning of the test individual must be carefully standardized with respect to joint position, etc. After a variable suction period needed for the skin to elevate the fixed distance of 1.5 mm, the equipment automatically stops recording and reads out the *E* and the time for elevation to occur. Results may be printed or loaded into a computer. The equipment is easy to operate, and the procedure is described in detail in manufacturer's manual.

Range of operation: There is little experience with applications in cosmetology and dermatology, and no publications have yet appeared. Obviously, very soft skin will swiftly elevate, and stiff and sclerotic skin may never reach the elevation of 1.5 mm unless suction is applied for several minutes, which may lead to prominent edema and even subepidermal blister formation. Thus, the DermaLab appears applicable only to relatively normal skin within a reasonable average range so that measurement can be concluded in about 30 s or less.

Obviously, a lengthy vacuum period results in water accumulation in the measured skin, seen as a wheal-like circular elevation above the surrounding skin, and thus a major measurement artifact may be introduced; see below.

RESULTS AND THEIR INTERPRETATIONS

The manufacturer outlines the following measuring end points:

- *The empirical modulus* E *as end point.* The higher the modulus, the stiffer is the skin; see equations. The limitations of this calculated modulus were discussed above.
- *Suction time as end point.* This may work well if the suction period is relatively short. The longer the suction time, the stiffer is the skin. It shall be borne in mind that the vacuum is built up in the chamber slowly until

a plateau is reached, following an unknown curve; linearity cannot be expected. It should also be considered that water and edema will accumulate in the skin under suction, thicken the skin, and contribute to its elevation especially in stiffer skin requiring a longer time to reach the fixed elevation of 1.5 mm. This is expected to influence mechanical skin properties during suction and recording and to create an artifactual situation where the measured object undergoes a significant change as a result of the recording procedure. This disturbance will influence all end points registered with the DermaLab, as it will with other systems with longer suction periods. To overcome the problems of prolonged suction in sclerotic and stiff skin, the DermaLab may be delivered with a special probe with reduced height between level 1 and level 2.

- *Vacuum needed to elevate (RES1).* This might only be meaningful with very short suction times since after perhaps 30 s the RES1 will be constant and will simply illustrate the suction capacity of the pump. The manometer is mounted close to the pump, and not in the chamber.

Optional end point interpretation and nomenclature (proposed by the author of this chapter):

- *"Stiffness" expressed as* E *and given in arbitrary units.* The value is higher in sclerotic and stiff skin, which reaches the elevation of 1.5 mm only following a higher vacuum and a longer suction period unless the optional probe is used. It should not be called a modulus, and results should be simply given in arbitrary units. The essential limitations are discussed above.
- *"Softness" expressed as reciprocal* E *(1/*E*) and expressed in arbitrary units.* The reciprocal E should be higher in soft skin; see above. This end point might be especially interesting for uses in cosmetology. Units are arbitrary.

Further studies on instrument validation, comparison with other instruments, usefulness, and applications to practical situations should be conducted to finally establish the value of the DermaLab in experimental and clinical research. The DermaLab appears especially suited for the measurement of smoothing and softening effects of cosmetic products on skin.

11 Hardware and Measuring Principles: The Dermagraph in Patients with Systemic Sclerosis and in Healthy Volunteers

HansJörg Häuselmann, Karl Huber, Burkhart Seifert, and Beat Michel

CONTENTS

Summary 124
Background 124
Measuring Mechanical Skin Properties in Health and Disease 124
Systemic Sclerosis 125
Manual Grading of Skin Elasticity in Patients with Systemic Sclerosis 126
Grading of Skin Elasticity Using Different Measuring Devices 126
Methods 126
The Dermagraph 126
Measuring Skin Distensibility and Relaxation with the Dermagraph 126
Determination of Vacuum Probe in Healthy Volunteers and Patients with Systemic Sclerosis 129
*Inter*rater Reliability in Patients with Systemic Sclerosis 129
Reference Values in Healthy Volunteers 129
Results 130
Mean *Intra*rater Reliability with Different Weights of Vacuum Probe 130
*Inter*rater Reliability of Maximal Skin Distensibility in Patients with Systemic Sclerosis 130
Reference Values in Healthy Male and Female Volunteers 131

0-8493-7521-5/02/$0.00+$1.50

Discussion 132
Acknowledgments 135
References 135

SUMMARY

This chapter describes the testing of a novel device, the Dermagraph®, for measuring the distensibility of normal and diseased skin by applying a vacuum onto the skin at 22 defined body areas. *Intra*rater testing resulted in an average weight of the vacuum device of approximately 400 g with a reliability of 0.77 for the distensibility of normal skin. *Inter*rater reliability for the distensibility of normal skin in nearly 100 healthy volunteers at 22 body locations was 0.55 and therefore comparable with the value of the earlier used Sclerimeter® (0.57). The average *intra*rater reliability of the Dermagraph in healthy volunteers was 0.71 and therefore better than with the earlier used prototype Sclerimeter with a mean value of 0.61.[1] There were statistically significant differences between the values of distensibility among both genders, showing higher values at 7 of 22 skin locations in women with the Dermagraph. The *inter*rater reliability in relevant skin locations in ten patients with systemic sclerosis (hand, feet, and face at an early or intermediate stage) showed a mean value of 0.81. The mean *intra*rater reliability of these three relevant skin locations in patients was 0.84, much better than the correspondent value of 0.33 with the Sclerimeter.

The Dermagraph has a standardized and validated weight of the vacuum probe, is handier, and has the ability to produce a printout, comparable to an electrocardiography device. The Dermagraph is a simple and easy-to-use device for measuring parameters of skin elasticity and can therefore be used for longitudinal measurements in patients with systemic sclerosis. With respect to practicability and reliability the Dermagraph can be compared with the Cutometer, except that the Cutometer measures much smaller areas of skin[16] and to the authors' knowledge has never been validated in patients with early and intermediate stages of systemic sclerosis.

BACKGROUND

Measuring Mechanical Skin Properties in Health and Disease

The thickening of skin combined with its decreased distensibility and tethering are hallmarks of systemic sclerosis. Several skin parameters such as area of diseased skin and decreased tethering, seem to correlate well with morbidity (risk of internal organ disease) and rapid deterioration of the disease process and therefore can be used as representative parameters[2–7a] of disease activity and outcome in systemic sclerosis.[7b]

To date, the most often used clinical techniques for measuring skin tethering or distensibility are manually performed skin scores, e.g., modified Rodnan total skin thickness score. In various studies their variability was approximately 20%, even if the examiners were carefully instructed on how to perform the manual skin measurements. As a consequence, easier-to-use manual scores were suggested.[8]

Measuring of skin elasticity goes back to the 1960s, when the first reports on skin measurements by devices were published. But thus far no method or device has been introduced in routine clinical practice for patients with systemic sclerosis, mostly because the techniques were too complicated to use in a day-to-day setting and no comparative trials with the manual scores or validation process of the device itself had been undertaken.[9–29]

Recently, devices that measure skin distensibility by applying a vacuum, comparable to manual skin testing, have been introduced in normal and diseased skin and have also been validated with respect to reliability (*inter*rater reliability; several examiners measure the same patient) and precision (*intra*rater reliability; same examiner, repeated measurements in same patient). Together with the Rüetschi AG, a company manufacturing medical high-precision products, the authors developed a prototype of a device for measuring skin distensibility at the Department of Rheumatology and Physical Medicine at the University Hospital in Zurich and called it the Sclerimeter.[1] Former studies of comparative testing against some published manual skin scores of Harrison,[47] Brennan,[43] Pope,[48] and Silman[8] showed better values for *intra*rater as well as *inter*rater reliability. In comparison with the published devices with the highest reliabilities, such as the SEM 474 Cutometer®,[46] the Sclerimeter showed similar values. These facts stimulated further improvements in the handiness and reliability of the first prototype "Sclerimeter," which led to the "Dermagraph," a device that can easily be used in an inpatient (research) as well as an outpatient setting (Dermatologist and Rheumatologist/Internist) in clinical routine.

Therefore, the objectives for this study were threefold:

1. Exploration of the most appropriate weight of the vacuum device in normal and diseased (systemic sclerosis) skin with respect to measuring precision by determination of the *intra*rater reliability.
2. Measuring the *intra-* and *inter*rater reliability of the measuring process in patients with systemic sclerosis by using the most appropriate weight of the vacuum probe.
3. Finally, acquiring reference values for skin distensibility and relaxation in a population of normal healthy people of both genders and different ages.

Systemic Sclerosis

Systemic sclerosis is a disease that affects the connective tissue of the locomotor system and of internal organs. As a consequence, skin, joints, tendons, and internal organs such as as lungs, intestines (e.g., esophagus), heart, and kidneys can be damaged. Pathogenetically, the disease process leads to a generalized fibrosis and sclerosis of the connective tissue, mediated by an increased synthesis of normal and abnormal components of the connective tissue, mostly of collagen type I and III.

To date, systemic sclerosis has been classified according to manual measurement of the area of diseased skin and other organ involvement (see classification criteria of the American College of Rheumatology, ACR, 1980).[34] Only recently, workers have tried to use other parameters of the disease process to classify systemic sclerosis using other clinical, serologic, and genetic markers.[33] There are published reports suggesting to classify patients with various disease severity and location into subgroups, mainly in the form of limited vs. diffuse disease with significantly different prognoses.[7a,36]

Other research groups differentiate the disease according to the amount of abnormal skin into a limited form (mostly sclerosis of the fingers), an "intermediate" form (skin of proximal extremities is involved), and a "diffuse" form (extension of skin involvement to the torso).[37–40] The classification of Clements et al.[2] with a semiquantitative Skin-Score uses not only the area of affected skin but also its semiquantitative grading of severity. All body areas are treated equally in terms of severity and are summarized into a total skin score with a maximal score of 30.

Manual Grading of Skin Elasticity in Patients with Systemic Sclerosis

Two fingers of the examiners are used to evaluate the degree of tethering of the skin by elevating the skin from its surface. According to the different authors the scale ranges from zero (normal tethering) to three or four (maximally reduced tethering). The examination is performed at defined areas of the skin (e.g., ten body areas used by Clements) by adding the local score number to a total skin score.[2,41–44]

Grading of Skin Elasticity Using Different Measuring Devices

This has been reviewed earlier by the authors' group, and the reader is referred to Reference 1. Basically, there are methods to aspirate the skin from its surface[9–18] to compress the skin toward the underlying resistance,[19,20] or to extend the skin.[21–29]

METHODS

The Dermagraph

The Dermagraph device (Figure 11.1) is a further development of the first prototype, the Sclerimeter.[1] There are no fundamental differences between the two devices; an easier-to-handle vacuum probe (Figure 11.2) was manufactured, and a dot plot printer was installed. The distensibility of the skin is measured by applying a constant vacuum. The amount of aspirated skin is measured in millimeters with a measuring device, which is situated within the aspirating probe. The measuring curve of the distended and relaxed skin over time is printed on a 55-mm-wide paper strip. The initiation of the vacuum is automatically induced as soon as the vacuum probe is in tight contact with the skin, producing a vacuum. There is no need for a pedal as in the Sclerimeter. The diameter of the vacuum probe is 20 mm. Because the precision of the 5-mm vacuum probe for the skin of the fingers was suboptimal, further evaluation of the 5-mm vacuum probe was halted.

Measuring Skin Distensibility and Relaxation with the Dermagraph

The measuring process lasts for 10 seconds (s). During 6 s a constant vacuum, which is shown on the printout, is applied followed by 4 s of skin relaxation (Figure 11.3).

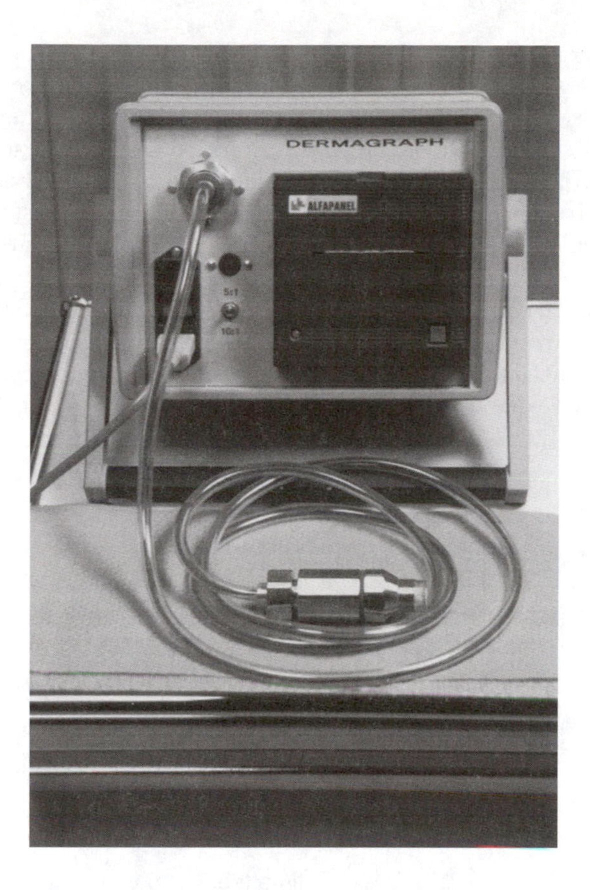

FIGURE 11.1 Dermagraph with vacuum probe in front.

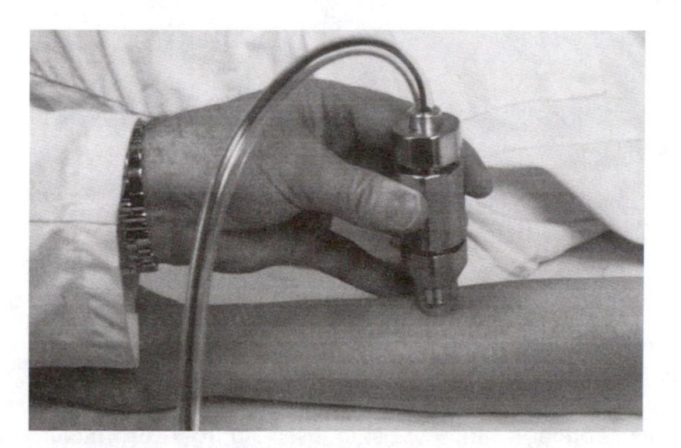

FIGURE 11.2 Vacuum probe applied perpendicularly onto the ventral lower arm of a healthy volunteer.

The aspiration or tethering of the skin over time is not constant but, because of the plasticity of the skin, it follows the mathematical equation of a hyperbole. After 6 s, the vacuum is stopped, resulting in an initial phase of fast relaxation followed by a slow phase (elastic retraction). The relaxation period has been set to 4 s.

The whole process (distention and relaxation process as well as vacuum over time) is printed onto a paper strip (Figure 11.4). The following measuring parameters have been taken from an earlier publication of Agache et al.[45] and are shown in units of millimeters with a precision of a tenth of a millimeter (see Figure 11.3). The skin distensibility after maximal vacuum (U_e), the distention at a maximal vacuum for 6 s (U_f), and the residual distention, 4 s after stopping the vacuum (U_x). The three measuring parameters are printed immediately after the curve. Several body areas can be measured without any delay, as the parameters are stored. As soon as the

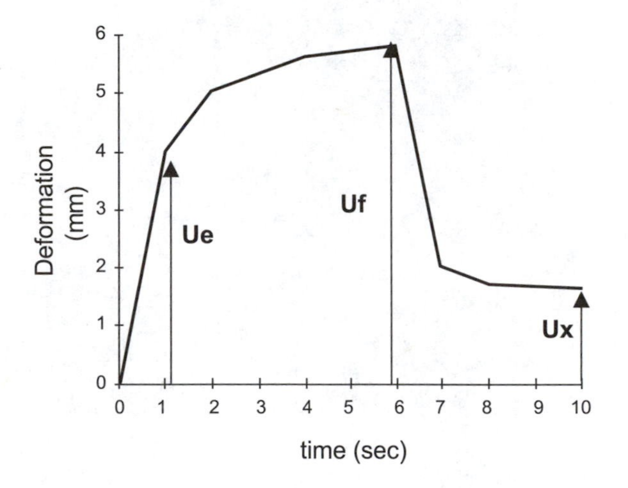

FIGURE 11.3 Schematic representation of measuring process with the Dermagraph. The parameters initial skin distensibility U_e, maximal distensibility U_f, and residual distensibility U_x are defined and marked by arrows.

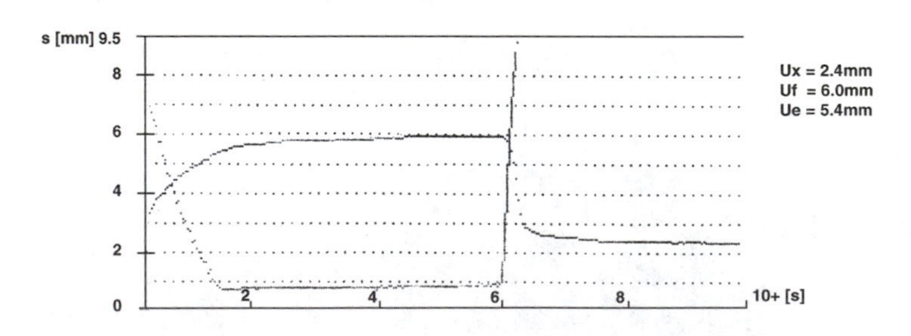

FIGURE 11.4 The applied vacuum and skin distensibility over time are printed on a paper strip. The absolute values of initial skin distensibility U_e, maximal distensibility U_f, and residual distensibility U_x are printed onto the paper strip in millimeters with the precision of a tenth of a millimeter.

machine sends two beeps, the measuring process must be terminated and the next skin area can be measured without delay.

According to the earlier publication[1] the precise localizations of the 20 measured body areas are in concordance with the manually measured skin tethering published by Clements et al.[2] They represent face, torso, abdomen, upper back, upper and lower arms, hands, thighs, calves, and feet. The exact definition of the localizations and the positioning of the patient have been published elsewhere.[1]

Determination of Vacuum Probe in Healthy Volunteers and Patients with Systemic Sclerosis

To evaluate the most appropriate weight of the device in terms of optimal precision of the measuring process, the *intra*rater reliability of this device was tested in ten patients with systemic sclerosis at the Department of Rheumatology and Physical Medicine in Zurich (eight women and two men, 37 to 74 years old), as well as in ten healthy volunteers (six women and four men, 16 to 81 years old).

Five different weights of the vacuum probe were chosen to measure each of the 22 defined skin locations (see below): 88, 176, 225, 442, and 619 grams (g). With the exception of the temperature, no other parameter was controlled for. All measurements were done at an ambient temperature between 19 and 23°C. The usual parameters at all 22 body areas as mentioned above were measured with respect to *intra*rater reliability of the five different weights. In addition, all parameters were calculated for the three most relevant measured body areas in patients with systemic sclerosis (hand, foot, and face). After these tests were performed and the best, e.g., most reliable, weight of the vacuum probe was defined, all following examinations were performed with this single best vacuum probe.

*Inter*rater Reliability in Patients with Systemic Sclerosis

Three raters measured ten patients in groups of three at all 22 body areas. The selection of the raters with respect to their order of measuring the patients was randomized. Two raters were experienced with the handling of the Dermagraph; for the third rater the device was new except for a 15-minute training program before the test. The vacuum probe used has been chosen according to the best results of *intra*rater variability testing (442 g).

Reference Values in Healthy Volunteers

In 96 healthy volunteers (51 men, 45 women, 16 to 81 years of age), the same skin parameters of distensibility, relaxation, and elasticity as mentioned above were measured at all 22 body areas with the definitive vacuum probe of 442 g. Mean values and standard deviation were calculated. *P* values were calculated for significant differences between volunteers and patients with systemic sclerosis as well as for differences between the genders of the healthy volunteers.

RESULTS

Mean *Intra*rater Reliability with Different Weights of Vacuum Probe

With respect to different weights of the vacuum probe of the Dermagraph, examinations of *intra*rater reliability showed very good values for the maximal (U_f) and initial (U_e) distensibility irrespective of the weight. The mean value of the *intra*rater reliability for U_f of all measured body areas in ten patients with systemic sclerosis was between 0.847 (with a vacuum probe weighing 225 g) and 0.745 (with 619 g). The mean value for the three most relevant body areas in this disease, hand, feet, and face, was between 0.872 (with 225 g) and 0.807 (with 619 g). With regard to the ten healthy volunteers, the reliability was lower. The mean value of all 22 measured body areas was between 0.729 (with 619 g) and 0.609 (with 176 g); the mean value of *intra*rater reliability of the three body areas (hand, feet, and face) was between 0.654 (with 619 g) and 0.510 (with 176 g).

Values of *intra*rater reliability of U_x (residual distention 4 s after stopping the vacuum) showed a much higher dependency on the weight of the vacuum probe, because reliability clearly dropped when using a probe weighing only 88 g. In patients with systemic sclerosis, the mean value for *intra*rater reliability of all 22 body areas was between 0.837 (using 442 g) and 0.638 (with 88 g). The corresponding values for the three most relevant body areas were between 0.720 (with 442 g) and 0.312 (with 88 g). Even lower mean values could be shown in the ten healthy volunteers: between 0.696 (with 225 g) and 0.369 (with 88 g) for all 22 body areas and between 0.479 (with 619 g) and 0.239 (with 88 g) if only the three most relevant areas were measured.

The corresponding mean reliability values for the different weights with respect to the skin parameters "relaxation" ($U_f - U_x$) and "elasticity" according to the definition of Agache ($U_f/U_f - U_x$) were between the excellent values for maximal skin distension (U_f) and the lowest values for U_x (data not shown). To eventually define the most appropriate, e.g., most reliable, weight of the vacuum probe, all values were subject to a ranking order considering all mean reliability values of U_f and U_x of all 22 body areas and the most relevant three body areas, hand, feet, and face. This resulted in four different ranking orders of the five weights. Summarizing the ranking orders, the weight of 442 g was clearly positioned best followed by 225 g of the vacuum probe. Three out of four times the weight of 442 g ranked first.

*Inter*rater Reliability of Maximal Skin Distensibility in Patients with Systemic Sclerosis

*Inter*rater reliability for maximal skin distension U_f in patients with systemic sclerosis showed better results if considering only the two experienced raters. Values from all three raters were between 0.882 (right hand) and 0.033 (back left side) (Figure 11.5). Of the 22 reliability values, 11 were over 0.5, 9 of them over 0.6, and the rest lower than 0.5. Considering only the two raters familiar with the Dermagraph,

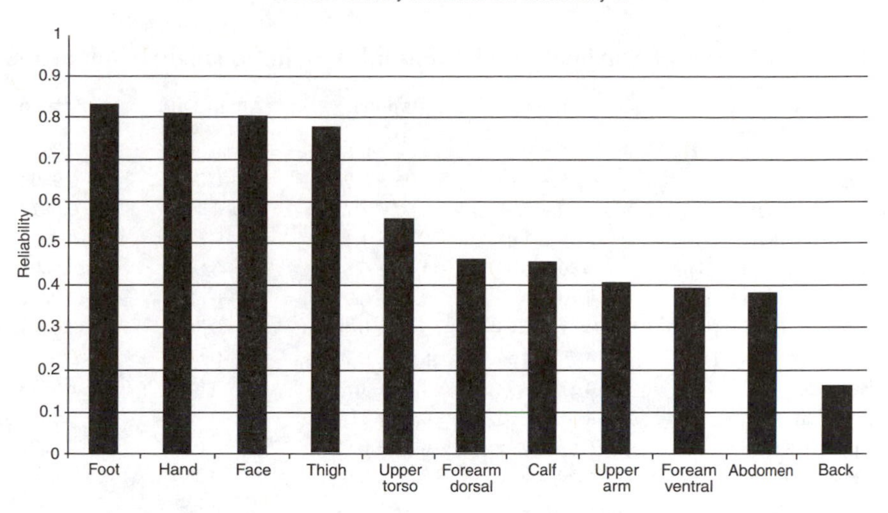

FIGURE 11.5 *Inter*rater reliabilities of maximal skin distensibility U_f at 11 body areas in patients with systemic sclerosis (three raters) are shown in rank order.

*inter*rater reliability was between 0.830 (right foot) and 0.170 (back, right side). Of the 22 values, 19 were over 0.5, 14 over 0.6, the rest below 0.5.

The best *inter*rater reliability values in patients with systemic sclerosis were calculated from measurements at the three most relevant body areas, hands, feet, and face. The corresponding excellent values were (all three raters) between 0.882 (right hand) and 0.741 (left hand). Four of the six values were over 0.8. Interestingly, the familiarity of the rater with the Dermagraph was less important at these three relevant body areas, because the variability values of the two experienced raters were very similar to all three raters (between 0.888 and 0.690), and four out of the six values were above 0.5.

Reference Values in Healthy Male and Female Volunteers

The highest mean reference values for maximal skin distensibility U_f (between right and left side of the body) in healthy volunteers of both genders were between 6.06 mm (abdomen in men) and 6.38 mm (upper arms in women); values are shown in Table 11.1. Mean standard deviation (between right and left side of the body) was between 0.38 mm (calves in men) and 1.34 mm (hands in women) (Table 11.1). There were significant differences (P value < 0.05) for maximal skin distensibility in 12 of 22 body areas between the two genders (Table 11.1).

Mean reference values from healthy volunteers of both genders in terms of the parameter relaxation ($U_f - U_x$) were between 2.19 mm (feet in men) and 4.64 mm (abdomen in women); values are shown in Table 11.2. It was only possible to show significant differences between the two genders at four skin locations, always showing higher values in women than in men.

TABLE 11.1
Reference Values of Maximal Skin Distensibility U_f in 96 Healthy Volunteers

Body Areas	Men	Women	Mean Diff.	*P* Value
Face right	5.24 ± 0.89	5.04 ± 0.80	+0.20	0.2629
Face left	5.05 ± 0.94	5.06 ± 0.96	–0.01	0.9802
Upper arm right	5.79 ± 0.96	6.50 ± 1.24	–0.71	**0.0023**
Upper arm left	5.58 ± 1.06	6.25 ± 1.11	–0.67	**0.0033**
Lower arm ventral right	4.59 ± 0.69	5.45 ± 0.88	–0.86	**<0.0001**
Lower arm ventral left	4.50 ± 0.49	5.18 ± 0.77	–0.68	**<0.0001**
Lower arm dorsal right	4.77 ± 0.43	5.13 ± 0.58	–0.36	**0.0008**
Lower arm dorsal left	4.79 ± 0.54	5.37 ± 0.75	–0.59	**<0.0001**
Hand right	4.22 ± 0.98	4.84 ± 1.32	–0.62	**0.0095**
Hand left	4.51 ± 1.03	4.79 ± 1.37	–0.28	0.2502
Upper torso right	5.66 ± 0.61	6.26 ± 0.93	–0.60	**0.0003**
Upper torso left	5.50 ± 0.56	5.85 ± 0.64	–0.35	**0.0059**
Abdomen right	6.12 ± 0.76	6.27 ± 0.83	–0.15	0.3610
Abdomen left	6.01 ± 0.60	6.12 ± 0.81	–0.11	0.4481
Thigh right	4.63 ± 0.58	4.64 ± 0.43	–0.01	0.9197
Thigh left	4.61 ± 0.58	4.63 ± 0.44	–0.02	0.8889
Foot right	4.51 ± 0.70	4.63 ± 1.06	–0.12	0.5072
Foot left	4.26 ± 0.60	4.64 ± 0.99	–0.38	**0.0233**
Calf right	4.61 ± 0.39	4.69 ± 0.43	–0.08	0.3581
Calf left	4.48 ± 0.37	4.54 ± 0.33	–0.07	0.3601
Back right	4.83 ± 0.54	5.12 ± 0.50	–0.29	**0.0079**
Back left	4.84 ± 0.55	5.10 ± 0.52	–0.26	**0.0211**

Mean references values for the parameter "elasticity" according to the definition of Agache ($U_f/U_f - U_x$) in healthy volunteers were between 1.37 (abdomen in women) and 2.14 (foot in women). Mean low values, e.g., high elasticity (lower than 1.6) were measured at the abdomen in women (1.44), at the back in men (1.50), and at the back in women (1.52). High values, e.g., low elasticity, were measured at the extremities (Table 11.3). Significant differences between the two genders were seen at eight body areas. Importantly, all measurements with regard to maximal skin distensibility, relaxation, and elasticity did not show any significant age dependency.

DISCUSSION

The Dermagraph seems to be very reliable for measuring skin distensibility, relaxation, and elasticity at body areas that are relevant for patients with systemic sclerosis, namely, the face, hands, and feet, irrespective of being familiar with or especially trained for this device, a fact that underlines the easy handling of the Dermagraph. The reliability tests demonstrate relevant differences between the reliability values of maximal skin distension between patients and healthy volunteers. One possible explanation for this could be that the underlying area is neither homogeneous because of extensor tendons nor exactly horizontal to allow easy positioning of the apparatus

TABLE 11.2
Reference Values for Skin Relaxation $U_f - U_x$ in 96 Healthy Volunteers

Body Areas	Men	Women	Mean Diff.	*P* Value
Face right	3.02 ± 0.75	2.93 ± 0.95	+0.08	0.6275
Face left	3.01 ± 0.86	3.02 ± 0.97	–0.01	0.9385
Upper arm right	3.20 ± 0.98	3.63 ± 1.44	–0.43	0.0889
Upper arm left	3.09 ± 1.10	3.35 ± 1.10	–0.26	0.2561
Lower arm ventral right	2.53 ± 0.52	2.88 ± 0.81	–0.35	**0.0136**
Lower arm ventral left	2.38 ± 0.41	2.53 ± 0.61	–0.15	0.1663
Lower arm dorsal right	2.48 ± 0.44	2.65 ± 0.61	–0.17	0.1261
Lower arm dorsal left	2.54 ± 0.61	2.91 ± 0.89	–0.36	**0.0206**
Hand right	2.15 ± 0.83	2.98 ± 1.28	–0.83	**0.0002**
Hand left	2.51 ± 0.95	2.94 ± 1.30	–0.43	0.0639
Upper torso right	3.47 ± 0.85	4.03 ± 1.07	–0.56	**0.0052**
Upper torso left	3.20 ± 0.84	3.65 ± 0.90	–0.44	**0.0148**
Abdomen right	4.34 ± 0.96	4.72 ± 1.16	–0.38	0.0831
Abdomen left	4.28 ± 0.76	4.56 ± 1.07	–0.29	0.1328
Thigh right	2.90 ± 0.59	2.68 ± 0.57	+0.21	0.0778
Thigh left	2.86 ± 0.60	2.66 ± 0.58	+0.20	0.1010
Foot right	2.23 ± 0.58	2.33 ± 0.97	–0.11	0.5146
Foot left	2.17 ± 0.71	2.35 ± 0.84	–0.19	0.2417
Calf right	2.86 ± 0.36	2.97 ± 0.61	–0.11	0.2672
Calf left	2.73 ± 0.37	2.80 ± 0.43	–0.07	0.3910
Back right	3.31 ± 0.60	3.45 ± 0.62	–0.14	0.2697
Back left	3.27 ± 0.60	3.45 ± 1.90	–0.18	0.1581

in a perpendicular direction. In addition, patients sometimes have difficulties relaxing their periorbital muscles when the Dermagraph is measuring skin distensibility at the face. Because of the sclerosis of the skin, some of these difficulties are less pronounced in patients with systemic sclerosis, which affects mostly both hands, the face, and the feet. According to the reliability results it seems to be important to provide precise instructions to examiners who are measuring normal or little-affected skin.

With respect to gender differences, the values of maximal distensibility U_f were again higher in women with normal skin, as already shown in the earlier study testing the first prototype Sclerimeter.[1] The same tendency could be shown for the parameter relaxation $U_f - U_x$. In contrast, the parameter elasticity $U_f/U_f - U_x$ did not show a gender specific difference but a significant difference with respect to skin localization in both genders, e.g., high elasticity at the torso and low elasticity at the extremities.

Surprisingly, a significant age dependency of measured skin parameters could not be demonstrated with the Dermagraph in almost 100 healthy volunteers 16 to 81 years of age.

Comparing the actual data with the earlier published values of the Sclerimeter, a clearly better average (of all 22 body areas) *intra*rater reliability of the Dermagraph (0.71 vs. 0.61) is demonstrated. The mean *inter*rater reliability was about the same

TABLE 11.3
Reference Values for Elasticity According to the Definition of Agache $U_f/U_f - U_x$ in 96 Healthy Volunteers

Body Areas	Men	Women	Mean Diff.	*P* Value
Face right	1.79 ± 0.32	1.83 ± 0.44	–0.04	0.6176
Face left	1.76 ± 0.33	1.80 ± 0.50	–0.05	0.5787
Upper arm right	1.91 ± 0.39	1.95 ± 0.46	–0.04	0.6538
Upper arm left	1.90 ± 0.37	1.97 ± 0.39	–0.07	0.3635
Lower arm ventral right	1.85 ± 0.24	1.96 ± 0.29	–0.11	**0.0410**
Lower arm ventral left	1.93 ± 0.27	2.11 ± 0.32	–0.18	**0.0043**
Lower arm dorsal right	1.96 ± 0.30	1.99 ± 0.29	–0.03	0.6376
Lower arm dorsal left	1.94 ± 0.31	1.94 ± 0.35	+0.01	0.9336
Hand right	2.10 ± 0.48	1.76 ± 0.42	+0.34	**0.0004**
Hand left	1.94 ± 0.45	1.74 ± 0.37	+0.19	**0.0239**
Upper torso right	1.69 ± 0.30	1.61 ± 0.24	+0.08	0.1544
Upper torso left	1.79 ± 0.32	1.66 ± 0.28	+0.13	**0.0440**
Abdomen right	1.45 ± 0.18	1.36 ± 0.22	+0.09	**0.0391**
Abdomen left	1.43 ± 0.19	1.38 ± 0.19	+0.05	0.2145
Thigh right	1.62 ± 0.19	1.78 ± 0.25	–0.17	**0.0004**
Thigh left	1.65 ± 0.24	1.78 ± 0.26	–0.14	**0.0091**
Foot right	2.11 ± 0.43	2.17 ± 0.54	–0.06	0.5392
Foot left	2.11 ± 0.54	2.10 ± 0.48	+0.004	0.9722
Calf right	1.63 ± 0.19	1.61 ± 0.19	+0.02	0.5609
Calf left	1.67 ± 0.20	1.66 ± 0.23	+0.01	0.8037
Back right	1.49 ± 0.21	1.52 ± 0.22	–0.03	0.5264
Back left	1.50 ± 0.18	1.51 ± 0.22	–0.01	0.8606

Note: The lowest possible value of 1 denotes highest elasticity.

(0.55 vs. 0.57), but the average value of 0.81 in the three most relevant body areas in patients with systemic sclerosis, hand, foot, and face, was much higher than with the earlier Sclerimeter (0.47). In summary, the Dermagraph shows a more reliable measuring process of skin distensibility in the most relevant body areas, hand, foot, and face, for patients with systemic sclerosis.

To the authors' knowledge the only comparable device to measure skin distensibility with a vacuum is the SEM 474 Cutometer, which measures immediate and late distensibility as well as immediate and definite retraction of the skin.[46] In 19 patients with a late form of systemic sclerosis Enomoto et al.[46] measured an excellent *intra*rater reliability of 0.94 and an *inter*rater reliability of 0.94 for maximal skin distensibility at 74 body areas. The study of Enomoto is not directly comparable to this one because of the different patient population (only consideration of patients with a late stage of systemic sclerosis). The Dermagraph has been developed to examine the skin of patients with systemic sclerosis early on and prospectively over time with the option to correlate early skin disease with disease outcome (potential complications of internal organs). Therefore, of most interest were patients with an early or intermediate stage of systemic sclerosis (examination on *inter*rater reliability) or healthy volunteers (half of people in examinations of *intra*rater reliability). To compare the two devices directly, an examination with the Cutometer must be undertaken in patients with early and intermediate stages of systemic sclerosis, as well.

The Dermagraph can be used in longitudinal studies in patients with systemic sclerosis with or without testing new therapeutic agents by in-house rheumatologists, dermatologists, and internists (transplantation units) as well as for physicians seeing outpatients with systemic sclerosis and similar skin diseases. Whether the Dermagraph can be helpful in other diseases, such as reflex algodystrophy, still has to be shown. The most relevant and important task ahead is to show whether the Dermagraph and other devices offer an advantage in predicting disease outcome in patients with systemic sclerosis. This must be done in prospective studies comparable to the low vs. high penicillamine trial.[7b]

ACKNOWLEDGMENTS

For acquiring normal reference values, the authors appreciate the kind help of the staff of the Institute for Physical Therapy and Rehabilitation GmbH, Jannhöhe 3, 23701 Eutin, Germany, as well as the staff of the manufacturing company of the Dermagraph, Rüetschi AG in Yverdon and Murten, Switzerland.

REFERENCES

1. Häuselmann HJ, Renfer K, Seifert B, Brühlmann P, Seitz J, Ruetschi Ch, and Michel BA, Skin bioengineering methods in the monitoring of scleroderma, *Curr. Prob. Dermatol.*, 1998, 26, 134–144.
2. Clements PJ, Lachenbruch PA, Ng SC, Simmons M, Sterz M, and Furst DE, Skin score: a semiquantitative measure of cutaneous involvement that improves prediction of prognosis in systemic sclerosis, *Arthritis Rheum.*, 1990, 33, 1256–1263.
3. Farmer RG, Gifford RW Jr, and Hines EA Jr, Prognostic significance of Raynaud's phenomenon and other clinical characteristics of systemic scleroderma: a study of 271 cases, *Circulation*, 1960, 21, 1088–1095.
4. Barnett AJ and Coventry DA, Scleroderma: 1. Clinical features, course of illness and response to treatment in 61 cases, *Med. J. Aust.*, 1969, 1, 992–1001.
5. Steen VD and Medsger TA, Jr, Epidemiology and natural history of systemic sclerosis. *Rheum. Dis. Clin. North Am.*, 1990, 16, 1–10.
6. Steen VD and Medsger TA Jr, Osial TA Jr, Ziegler GL, Shapiro AP, and Rodnan GP, Factors predicting development of renal involvement in progressive systemic sclerosis, *Am. J. Med.*, 1984, 76, 779–786.

7a. Medsger TA Jr, Steen VD, Ziegler G, and Rodnan GP, The natural history of skin involvement in progressive systemic sclerosis (abstr.), *Arthritis Rheum.*, 1980, 23, 720–721.

7b. Clements PJ, Hurwitz E, Wong WK, Seibold JR, Mayes M, White B, Wigley F, Weisman M, Barr W, Moreland L, Medsger TA Jr, Steen VD, Martin RW, Collier D, Weinstein A, Lally E, Vaya J, Weiner SR, Andrews B, Abeles M, and Furst DE, Skin thickness score as a predictor and correlate of outcome in systemic sclerosis: high dose vs. low dose penicillamine trial, *Arthritis Rheum.*, 2000, 43, 2445–2454.

8. Silman A, Harrison M, Brennan P, and the ad hoc International Group on the Assessment of Disease Outcome in Scleroderma, Is it possible to reduce observer variability in skin score assessment of scleroderma? *J. Rheumatol.*, 1995, 22, 1277–1280.
9. Grahame R and Holt PJL, The influence of ageing on the *in vivo* elasticity of human skin, *Gerontologia* (Basel), 1969, 15, 121–139.
10. Bluestone R, Grahame R, Holloway V, and Holt PJL, Treatment of systemic sclerosis with D-penicillamine, *Ann. Rheum. Dis.*, 1970, 29, 153–158.
11. Piérard GE and Lapière CM: Physiopathological variations in the mechanical properties of skin, *Arch. Dermatol. Res.*, 1977, 260(3), 231–239.
12. Piérard GE, Piérard-Franchimont C, and Lapière CM, Connective tissue compartments in scleroderma. Study of the structure and the biomechanical properties, *Dermatologica*, 1985, 170(3), 105–113.
13. Piérard GE, Histological and rheological grading of cutaneous sclerosis in scleroderma, *Dermatologica*, 1989, 179(1), 18–20.
14. Serup J and Northeved A, Skin elasticity in localized scleroderma (morphaea). Introduction of a biaxial *in vivo* method for measurement of tensile distensibility, hysteresis and resilient distension of diseased and normal skin, *J. Dermatol.*, 1985, 12(1), 52–62.
15. Bjerring P, Skin elasticity measured by dynamic admittance: a new technique for mechanical measurements in patients with scleroderma, *Acta Derm-Venereol. Suppl.* (Stockholm), 1985, 120, 83–87.
16. Cua AB, Wilhelm KP, and Maibach HI, Elastic properties of human skin: relation to age, sex and anatomical region, *Arch. Dermatol. Res.*, 1990, 282, 283–288.
17. Cozzani E, Gnone M, Gaddia G, and Cipriani C, A Non-invasive method for measuring the visco-elastic properties of the skin in healthy subjects and in subjects with scleroderma, *G. Ital. Dermatol. Venereol.*, 1989, 124(9), 403–406.
18. Piérard GE, Letawe C, Dowlati A, and Piérard-Franchimont C, Effect of hormone replacement therapy for menopause on the mechanical properties of skin, *JAGS*, 1995, 43, 662–665.
19. Parot S and Bourlière F, A new technique for the measurement of the compressibility of the skin and subcutaneous tissue. Role of sex, age and site of measurement, *Gerontologia*, 1967, 13(2), 95–110.
20. Falanga V and Bucalo B, Use of a Durometer to assess skin hardness, *J. Am. Acad. Dermatol.*, 1993, 29, 47–51.
21. Christensen MS, Hargens CW III, Nacht S, and Gans EH, Viscoelastic properties of intact human skin: instrumentation, hydration effects, and the contribution of the stratum corneum, *J. Invest. Dermatol.*, 1977, 69(3), 282–286.
22. Thacker JF, Iachetta FA, and Allaire PE, *In vivo* extensometer for measurement of the biomechanical properties of human skin, *Rev. Sci. Instrum.*, 1977, 48(2), 181–185.
23. Gunner CW, Hutton WC, and Burlin TE, The mechanical properties of skin *in vivo* — a portable hand-held extensometer, *Br. J. Dermatol.*, 1979, 100(2), 161–163.
24. Gunner CW, Hutton WC, and Burlin TE, An apparatus for measuring the recoil characteristics of human skin *in vivo*, *Med. Biol. Eng. Comput.*, 1979, 17(1), 142–144.
25. Manschot JF and Brakkee AJ, The measurement and modelling of the mechanical properties of human skin *in vivo*, I. The measurement, *J. Biomech.*, 1986, 19(7), 511–515.
26. Manschot JF and Brakkee AJ, The measurement and modelling of the mechanical properties of human skin *in vivo*, II. The model, *J. Biomech.*, 1986, 19(7), 517–521.
27. Kalis B, De Rigal J, Léonard F, Lévêque JL, Riche O, Le Corre Y, and De Lacharriere O, *In vivo* study of scleroderma by non-invasive techniques, *Br. J. Dermatol.*, 1990, 122, 785–791.

28. Ballou SP, Mackiewicz A, Lysikiewicz A, and Neuman MP, Direct quantitation of skin elasticity in systemic sclerosis, *J. Rheumatol.*, 1990, 17, 790–794.
29. Venturini R, Biocca M, Rispoli E, and De Pascalis V, Methods of the mechanical properties of the skin, *Boll. Soc. Ital. Biol. Sper.*, 1974, 50(3), 125–130.
30. Moll I, Unsere dynamische Haut, in *Dermatologie*, Jung EG, Ed., Hippokrates Verlag, Stuttgart, 1989, 17–29.
31. Leonhardt H, *Histologie, Zytologie und Mikroanatomie des Menschen. Taschenlehrbuch der Gesamten Anatomie*, Vol. 3, Georg Thieme Verlag, Stuttgart, 1981, 301–317.
32. Elsner P, Skin elasticity, in *Bioengineeering of the Skin: Methods and Instrumentation*, Berardesca E, Elsner P, Wilhelm KP, and Maibach HI, Eds., CRC Press, Boca Raton, FL, 1995, 53–64.
33. Holzmann H, Schlieter A, and Ramirez-Bosca A, Proposal of a revised classification of systemic sclerosis, *Eur. J. Dermatol.*, 1993, 3, 262–265.
34. Subcommittee for Scleroderma Criteria of the American Rheumatism Association Diagnostic and Therapeutic Criteria Committee: Preliminary Criteria for the Classification of Systemic Sclerosis (Scleroderma), *Arthritis Rheum.*, 1980, 23, 581–590.
35. Kelley WN, Harris ED, Ruddy S, and Sledge CB, Eds., *Textbook of Rheumatology*, 4th ed., WB Saunders, Philadelphia, 1993, 1113–1143.
36. LeRoy EC, Black C, Fleischmajer R, Jablonska S, Krieg T, Medsger TA Jr, Rowell N, and Wolheim F, Scleroderma (systemic sclerosis): classification, subsets and pathogenesis, *J. Rheumatol.*, 1988, 15, 202–205.
37. Giordano M, Valentini G, Migliaresi S, Picillo U, and Vatti M, Different antibody patterns and different prognosis in patients with scleroderma with various extent of skin sclerosis, *J. Rheumatol.*, 1986, 13, 911–920.
38. Barnett AJ, Miller MH, and Littlejohn GO, A survival study of patients with scleroderma diagnosed of 30 years (1953–83): the value of a simple cutaneous classification in the early stages of the disease, *J. Rheumatol.*, 1988, 15, 276–283.
39. Masi AT, Classification of systemic sclerosis (scleroderma): relationship of cutaneous subgroups in early disease to outcome and serological reactivity, *J. Rheumatol.*, 1988, 15, 894–898.
40. Arbeitsgruppe der Arbeitsgemeinschaft Dermatologischer Forschung (ADF): Klinik der progressiven systemischen Sklerodermie (PSS). Multizentrische Untersuchung an 194 Patienten, *Hautarzt*, 1986, 37, 320–324.
41. Rodnan GP, Lipinsky E, and Lucksick J, Skin thickness and collagen content in progressive systemic sclerosis and localized scleroderma, *Arthritis Rheum.*, 1979, 22, 130–140.
42. Kahaleh MB, Suttany GL, Smith EA, Huffstutter JE, Loadholt CB, and LeRoy EC, A modified scleroderma skin scoring method, *Clin. Exp. Rheumatol.*, 1986, 4, 367–369.
43. Brennan P, Silman A, Black C, Bernstein R, Coppock J, Maddison P, Sheeran T, Stevens C, and Wolheim F, Reliability of skin involvement measures in scleroderma, *Br. J. Rheumatol.*, 1992, 31, 457–460.
44. Zachariae H, Bjerring P, Halkier-Sorensen L, Heickendorff L, and Sondergaard K, Skin scoring in systemic sclerosis: a modification — relations to subtypes and the aminoterminal propeptide of type III procollagen (PIIINP), *Acta Derm. Venereol.* (Stockholm), 1994, 74, 444–446.
45. Agache PG, Monneur C, Lèvêque JL, and De Rigal J, Mechanical properties and Young's modulus of human skin *in vivo*, *Arch. Dermatol. Res.*, 1980, 269, 221–232.

46. Enomoto D, Kekkes J, Bossuyt P, Hoekzema R, and Bos J, Quantification of cutaneous sclerosis with a skin elasticity meter in patients with generalized scleroderma, *J. Am. Acad. Dermatol.*, 1996, 35(3), 381–387.
47. Harrison A, Lusk J, and Corkill M, Reliability of skin score in scleroderma, *Br. J. Rheumatol.*, 1993, 32, 170.
48. Pope JE, Baron M, Bellamy N, Campbell J, Carette S, Chalmers I, Dales P, Hanly J, Kaminska EA, Lee P, Sibley J, and Stevens A, Variability of skin scores and clinical measurements in scleroderma, *J. Rheumatol.*, 1995, 22, 1271–1276.

12 Hardware and Measuring Principles: The Durometer

Marco Romanelli and Vincent Falanga

CONTENTS

Introduction 139
The Measuring System 139
Accuracy and Reproducibility of the Measurements 141
Medical Applications 141
  Scleroderma 141
  Lipodermatosclerosis 142
  Neuropathic Foot 143
Conclusions 144
References 145

INTRODUCTION

The mechanical properties of the human skin are complex; although the skin is readily accesible, it is always a difficult task to identify simple and noninvasive measurement methods suitable for a specific skin parameter. This chapter describes a novel and simple instrument called a durometer, which is now used to assess and quantify the degree of skin hardness. The durometer is an engineering instrument widely employed in industry to measure the hardness of metals, plastic, rubber, and other nonmetallic materials. The instrument has been applied in the last few years in several medical conditions where skin hardness is a clinical feature of the disease and where there is a need for quantification of hardness for prognostic and therapeutic reasons.

THE MEASURING SYSTEM

The first instrument used to assess skin hardness was a Rex durometer (model 1700, Rex Gauge Company, Inc., Glenview, IL). This instrument is the international standard for measuring the hardness of rubber, plastic, nonmetallic materials, and soft

0-8493-7521-5/02/$0.00+$1.50

tissue such as skin. The durometer is a portable hand-held device, which is provided with a calibrated gauge that registers linearly the relative degree of hardness on a scale of units divided from 0 to 100 (Figure 12.1). This feature is the result of a spring-loaded interior that senses hardness by applying an indentation load on the specimen. The bottom of the durometer has a small, dull indentor which is retractable and is responsible for the measurements registered on the gauge (Figure 12.2). The instrument is rested by gravity against the skin and must be held in a vertical position (Figure 12.3). Initial hardness is defined as that recorded within 1 s of firm contact of the durometer with the skin. For many materials, this initial reading remains unaltered after the first second of measurements (near perfect elasticity). Imperfect elasticity is referred to as plasticity and is observed when continued contact of the durometer with the surface leads to a higher reading. The degree of this plasticity is referred to as creep, and in animal tissue the amount of creep is minimal. The type of durometer used in dermatology to assess skin hardness is Model 0 which is provided with an 822 g springload. To eliminate operator error, the company has improved the first model with a specified amount of weight (400 g), located on top of the instrument, which is used to depress the indentor without any additional pressure. This constant load principle provides a more consistent reading to the operator every time the gauge is applied to the skin. For measurements on the skin, the durometer is used at 25°C and four consecutive readings are usually taken at the same site. Between readings the durometer is reset to 0 and measurements are made with the patient supine, thus avoiding muscle tension or contraction.

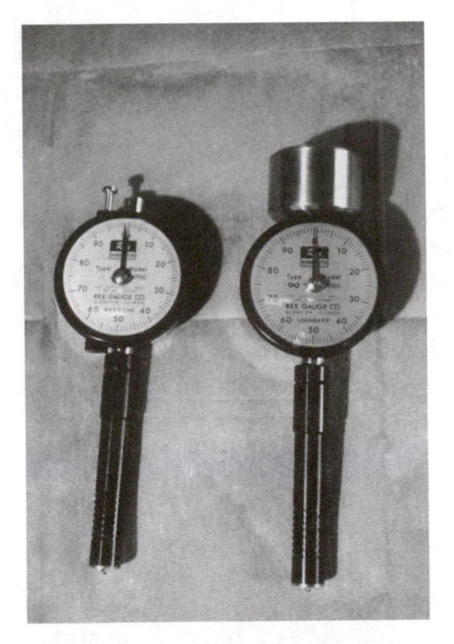

FIGURE 12.1 Durometer model 1700 (Rex Gauge), type 00, with (right) and without (left) constant load weight.

FIGURE 12.2 Bottom of the durometer with the retractable indentor responsible for detecting the degree of hardness.

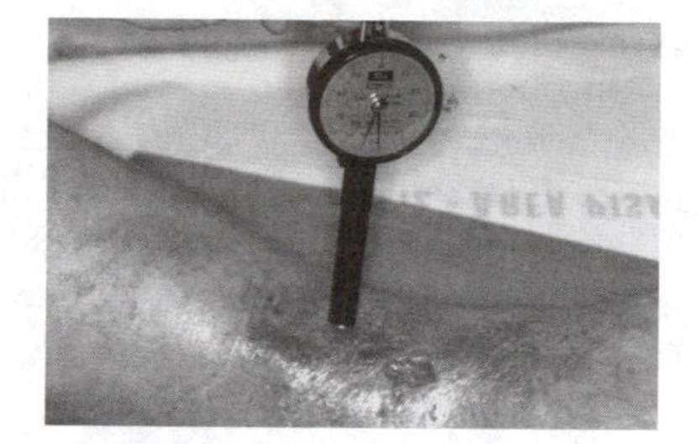

FIGURE 12.3 Durometer rested by gravity over an area of lipodermatosclerosis in a patient with venous insufficiency and a chronic wound.

ACCURACY AND REPRODUCIBILITY OF THE MEASUREMENTS

Durometer readings obtained with four consecutive determinations at the same site in each subject and in different clinical conditions differed constantly by less than 5%.[1,2] The durometer was tested for interobserver and intraobserver variability in a study[3] where patients with morphea, a localized form of scleroderma, were investigated for skin induration. The durometer was used in combination with a skin scoring system called MSS (Modified Skin Score).[4] The interobserver variability in the durometer score was only 0.5% and was even lower than the interobserver variability of the MSS (0.5 vs. 2.2%). The intraobserver variability, which is the variation in readings obtained by the same observer, was estimated as 0. The correlation between the total MSS and the durometer score was approximately 0.5. Insensitive durometer readings were found in different areas of the skin, such as the forehead and dorsal digit, where subcutaneous tissue is less represented. The hypothesis of underlying bone or tendon structure giving higher readings at these sites was raised as a possible explanation of this failure to discriminate between normal and indurated skin.[1]

MEDICAL APPLICATIONS

SCLERODERMA

Sclerodermatous skin is characterized by induration of the lower dermis and directly underlying tissue, which could be affected diffusely as in systemic sclerosis or in a more localized involvement as in morphea. The degree and extent of skin hardness in scleroderma are used as a prognostic value for the disease, and are generally assessed subjectively using a clinical skin scoring system. Objective measurements of skin involvement in scleroderma have included the use of ultrasonography, skin elastometers, transcutaneous oxygen tension measurements, and also such more-sophisticated techniques as magnetic resonance imaging. All these methods were able to provide useful information about the extent of skin involvement but are generally considered not practical for sequential measurements in clinical practice. Falanga and Bucalo[1] were the first to describe the use of a durometer in patients with systemic sclerosis. They investigated six anatomic sites in 12 patients and in 12 control individuals using the durometer in combination with a skin severity score. A direct relationship between the skin severity score and durometer readings was present for all anatomic sites except the forehead: the higher the skin score, the greater the amount of deflection measured with the durometer (p = 0.0032, p = 0.0002, and p = 0.0005 for skin scores of 1, 2, and 3, respectively). Moreover, the durometer was able to differentiate between increased levels of skin hardness (Table 12.1). In this study no statistically significant relationships were found among body weight, skin score, diameter of anatomic sites measured, and durometer readings. The use of the durometer was not associated with any pain or discomfort. Same results of increased skin hardness in morphea were obtained by Aghassi et al.[2] in a study where the durometer was also used in combination with a laser Doppler perfusion imager. In that study the authors found a relationship between age and durometer readings over the ventral aspect of the forearm in normal controls. Skin hardness rises from birth, reaches a constant level by puberty, and then remains steady until age 65 years, when the skin begins to lose its firm quality progressively. Women had lower durometer readings compared with men, and the instrument was able to define contiguous areas of different degrees of sclerosis in morphea lesions.

The efficacy of treatment has been investigated in patients with localized scleroderma by means of durometer readings in two different studies. Seyger et al.[5] evaluated the effect of low-dose methotrexate (15 mg/week) in nine patients with widespread morphea in a 24-week trial using the durometer and other clinical assessments to monitor the therapeutic regimen. A total of seven patients of nine showed a decrease of durometer readings at the end of the trial, with mean values changing from 101 ± 13.9 before treatment, to 89.7 ± 10.3 after treatment ($p = 0.07$). The greatest improvement in durometer score among the sites examined was in the arm region ($p = 0.03$). In another study, Karrer et al.[6] evaluated the efficacy of topical photodynamic therapy by means of durometer readings and clinical score in five patients with progressive localized scleroderma who had failed to respond to other therapies. In all patients the therapy was highly effective as measured by durometer score, which showed a mean reduction of 20% after a treatment period of 6 months.

Lipodermatosclerosis

Chronic lipodermatosclerosis is characterized by hyperpigmented indurated skin on the medial aspect of the leg and is common in patients with venous insufficiency. The severity of induration of lipodermatosclerosis has been associated with poor ulcer healing. Nemeth et al.[7] reported a correlation between the degree of skin induration in lipodermatosclerosis and ulcer healing. They used a clinical score to assess skin induration and found that the lipodermatosclerosis was less severe in patients who had ulcers that healed than in those who did not. Durometer testing of skin hardness was performed in a group of patients with lipodermatosclerosis with and without venous ulcers.[8] A clinical scoring for hardness by a blinded observer was used in combination with durometer measurements at a skin site midway between the upper and lower margin of lipodermatosclerosis. Measurements of transcutaneous oxygen pressure ($TcpO_2$) at the same site were also performed. Compared with normal skin in control subjects, increased skin severity scores were associated with higher durometer readings ($p < 0.01$). This direct correlation between clinical scores and durometer readings was linear ($r = 0.9621$). The durometer was

TABLE 12.1
Durometer Readings in Patients with Lipodermatosclerosis and Venous Ulcers Who Had Different Skin Scores

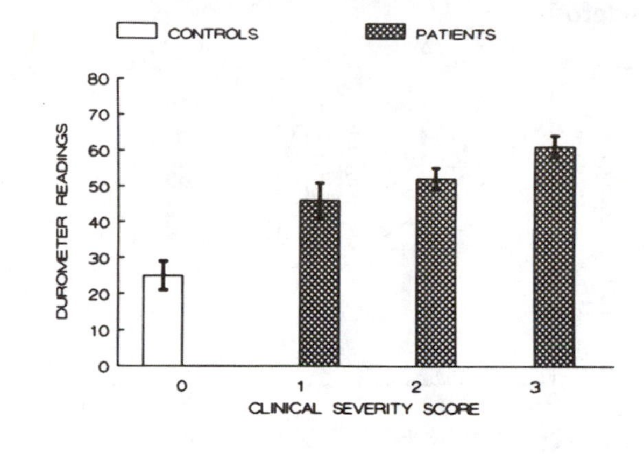

Source: Romanelli, M. and Falanga, V., *J. Am. Acad. Dermatol.*, 32, 188, 1995. With permission.

able to differentiate between increased levels of skin hardness. Thus, a skin severity score of 2 reflected a higher durometer reading than that of a skin score of 1 ($p = 0.0016$). An inverse and statistically significant relationship ($r = 0.431$; $p = 0.0079$) between $TcpO_2$ measurements and durometer readings was found in both patients and control subjects (Table 12.2). The correlation between $TcpO_2$ measurements and durometer readings found in that study confirms the hypothesis of poor oxygen diffusion through dense and thickened skin as occurs in lipodermatosclerosis. LeBlanc et al.[9] used the durometer to assess the area of lipodermatosclerosis around venous ulcers and to obtain a map of skin induration. Their hypothesis was that maximum skin induration correlated with the presence of ulcers. The authors found that venous ulcers are located in skin that is most affected by lipodermatosclerosis. Durometer measurements showed a progressive and linear decrease from the top of the ulcer's edge to the knee ($r = 0.925$). In the same study, the authors tested 14 patients with pitting edema and 8 controls without edema. There were no significant differences in durometer readings in patients with minimal or more severe edema ($p > 0.05$) regardless of anatomic location.

Neuropathic Foot

Patients with diabetes with peripheral polyneuropathy are at increased risk for foot ulceration. One of the key events in the pathogenetic pathway to neuropathic ulceration is hyperkeratosis, which develops in areas of increased pressure. Nonenzymatic glycosilation and autonomic dyshidrosis have been implicated in the pathogenesis of skin hardening. Piaggesi et al.[10] measured skin hardness at three different sites on the foot of a selected group of patients with diabetes, avoiding areas with bony prominences. Skin hardness was more pronounced in neuropathic patients than in non-neuropathic patients and controls ($p < 0.01$). Among neuropathic patients, those

TABLE 12.2
Relation between $TcpO_2$ Measurements and Durometer Readings (scattergram) or Clinical Severity Score (inset) in Patients with Lipodermatosclerosis

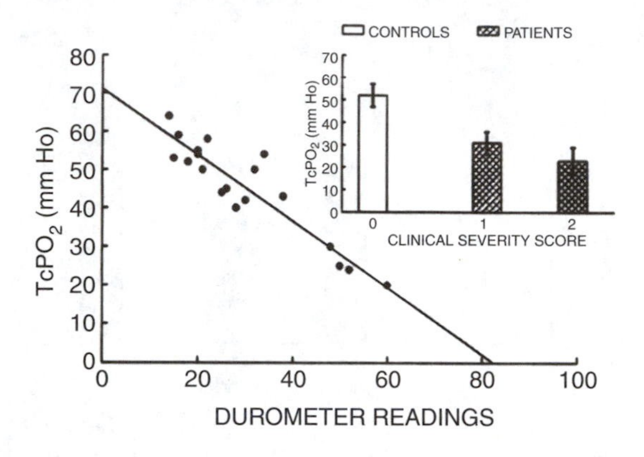

Source: Romanelli, M. and Falanga, V., *J. Am. Acad. Dermatol.*, 32, 188, 1995. With permission.

with metatarsal hyperkeratosis had significantly higher durometer readings compared with those without hyperkeratosis ($p < 0.001$). Subjects were also evaluated for neuropathy by determining the vibration perception threshold (VPT) with a biothesiometer. This instrument is an electrical device provided with a vibrating probe, which is applied to the test site and to which an increasing vibrating intensity is applied. The lowest of three intensities of vibration defines the VPT. In that study, a significant positive correlation was found between durometer readings and VPT at the malleolus ($r = 0.516$) and at the allux ($r = 0.624$) in neuropathic patients (Table 12.3).

CONCLUSIONS

Durometry is a technique for measuring skin hardness. The method takes advantage of the ability of the durometer to sense and record hardness on a linear scale, and it is unique in its simplicity and ease of use. Measurements with the durometer are highly reproducible in the same subject and in persons with the same degree of skin induration. Data obtained indicate that the durometer may be insensitive when used in areas of skin where the subcutaneous tissue is not well represented. At this stage the instrument is used as a standard noninvasive tool to measure skin hardness in systemic sclerosis. The authors believe that durometry could be applied in other dermatologic diseases to assess skin involvement and to monitor the efficacy of treatment.

TABLE 12.3
Relation between Vibration Perception Threshold at Malleolus (VPT-M) and Skin Hardness Measurement at Heel (DMT-H) in Patients with Neuropathic Diabetes

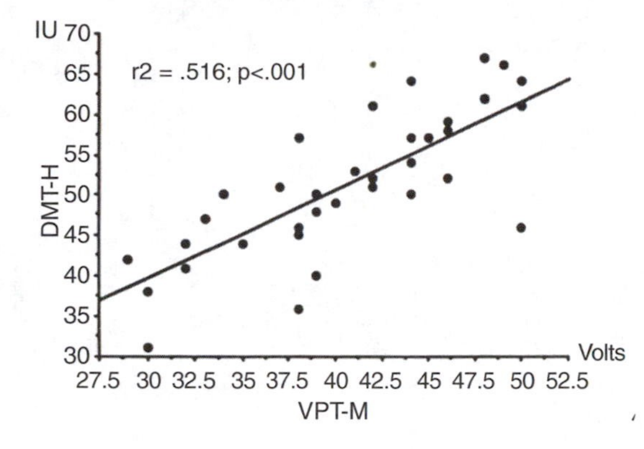

Source: Piaggesi, A. et al., *J. Diabet. Complications*, 13, 129, 1999. With permisison.

REFERENCES

1. Falanga, V. and Bucalo, B., Use of a durometer to assess skin hardness, *J. Am. Acad. Dermatol.*, 29, 47, 1993.
2. Aghassi, D., Monoson, T., and Braverman, I., Reproducible measurements to quantify cutaneous involvement in scleroderma, *Arch. Dermatol.*, 131, 1160, 1995.
3. Seyger, M.M.B. et al., Reliability of two methods to assess morphea: skin scoring and the use of a durometer, *J. Am. Acad. Dermatol.*, 37, 793, 1997.
4. Zachariae, H. et al., Skin scoring in systemic sclerosis: a modification — relations to subtypes and the aminoterminal propeptide of type III procollagen (PIIINP), *Acta Derm-Venereol.* (Stockh.), 74, 444, 1994.
5. Seyger, M.M.B. et al., Low-dose methotrexate in the treatment of widespread morphea, *J. Am. Acad. Dermatol.*, 39, 220, 1998.
6. Karrer, S. et al., Topical photodynamic therapy for localized scleroderma, *Acta Derm-Venereol.* (Stockh.), 80, 26, 2000.
7. Nemeth, A.J., Eaglstein, W.H., and Falanga, V., Clinical parameters and transcutaneous oxygen measurements for the prognosis of venous ulcers, *J. Am. Acad. Dermatol.*, 20, 186, 1989.
8. Romanelli, M. and Falanga, V., Use of a durometer to measure the degree of skin induration in lipodermatosclerosis, *J. Am. Acad. Dermatol.*, 32, 188, 1995.
9. LeBlanc, N. et al., Durometer measurements of skin induration in venous disease, *Dermatol. Surg.*, 23, 285, 1997.
10. Piaggesi, A. et al., Hardness of plantar skin in diabetic neuropathic feet, *J. Diabet. Complications*, 13, 129, 1999.

13 Hardware and Measuring Principles: The Ballistometer

Peter T. Pugliese and John R. Potts

CONTENTS

Description of the Instrument 147
Principles of Operation 148
Basic Concept of the Measurement 149
Basic Physics of Ballistometry 149
Intrinsic Parameters for Characterizing Materials 151
Determination of Coefficient of Restitution 152
Determination of Cutaneous Absorption Coefficient 152
Determination of Stiffness 153
Measuring with the Ballistometer 154
Methods and Materials 154
Procedure 154
Results 155
Amplitude 155
Stiffness 155
Cutaneous Absorption Coefficient 157
Coefficient of Restitution 157
Discussion 157
Conclusions 158
References 159

DESCRIPTION OF THE INSTRUMENT

The IDRA® (Integrated Dynamic Rebound Analyzer) ballistometer consists of ten main parts: an external interface unit, a 9-v DC power supply, an internal PC interface card, an interconnecting cable, a holding magnet, a rotary transducer, a ballistic hammer, a multipositional mounting plate, a fixed stand, and IDRA control and data processing software (Figure 13.1).

0-8493-7521-5/02/$0.00+$1.50

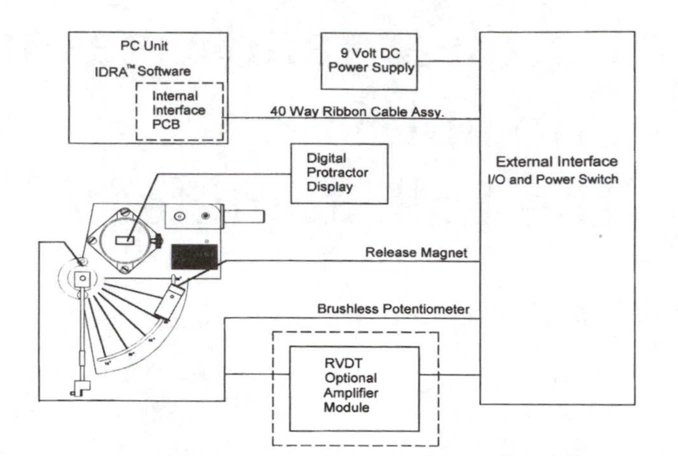

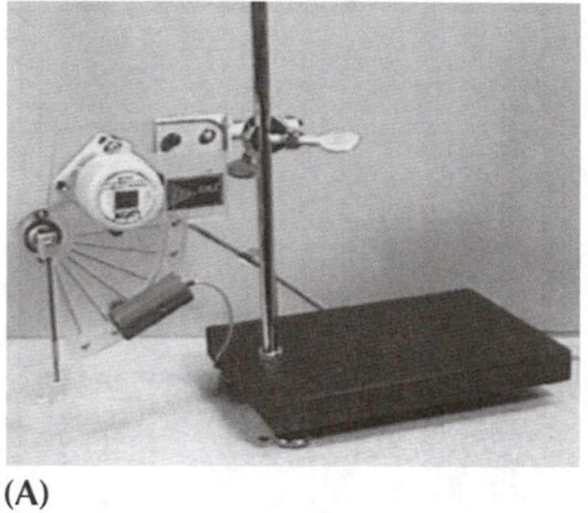

(A)

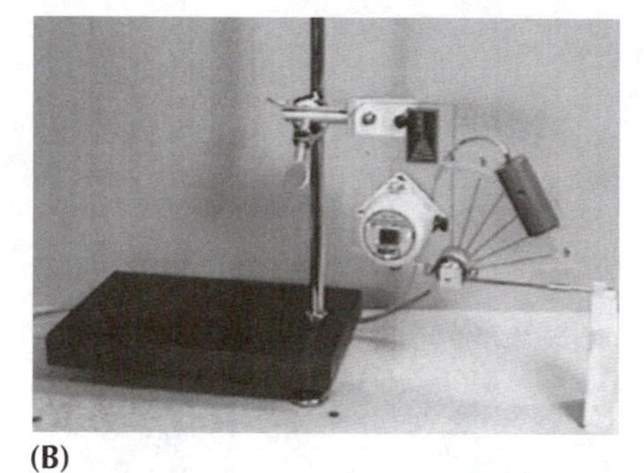

(B)

FIGURE 13.1 Transducer mounting plate; (A) horizontal; (B) vertical. Illustration of the relationship of the parts on the IDRA mounting plate assembly. Shown is a lightweight ballistic hammer, rotary transducer, multipositional mounting plate, magnet for holding and releasing the ballistic hammer at a fixed angle, and a rigid support stand. Also shown is an optional attachment and a digital protractor system, which is used for accurately and conveniently calibrating the angular displacement of the hammer to within 0.1°. Alternatively, calibration of the angular displacement may be done by using the numbered angle markings on the mounting plate, which are spaced at 15° intervals over a quadrant.

PRINCIPLES OF OPERATION

IDRA is a PC-based ballistometer and employs a dynamic technique for assessing intrinsic viscoelastic properties of skin. That is, it is an instrument for determining physical properties that do not depend on the method of measurement. For relatively soft materials, such as skin, the measurement technique involves a lightweight hammer, anchored at one end, that free-falls onto the test surface under gravitational force. The resulting hammer oscillatory displacement–time data are recorded and analyzed, and characteristic physical parameters are determined. Ballistometry is a method of choice for characterizing the viscoelasticity of skin, since it is nondestructive, noninvasive, fast, and easy to use.

Basic Concept of the Measurement

The IDRA ballistometer system accepts two types of suitable rotary transducers: RVDTs (Rotary Variable Differential Transformers) and Brushless Potentiometers (Hallpot® transducers). Both RVDT and Hallpot transducers give highly reproducible DC analogs, with excellent linearity, over a wide range of angular displacement. The Hallpot used in an IDRA system is custom-made for third-party research and development by Elweco, Inc., and incorporates several superior, critical design features such as miniature, ultralow-friction bearings with nonviscous lubricant, very wide sensor bandwidth, and excellent reproducibility and linearity for up to 60° angular displacement.

An RVDT model R30A transducer, manufactured by Lucas, was employed for the studies discussed in this chapter. This transducer has miniature, ultralow-friction bearings and has sufficient bandwidth to permit accurate recording of the dynamic angular displacement–time response of the ballistic hammer, as it contacts and rebounds from the surface of the skin. RVDTs are essentially bipolar devices with four linear regions (in which the output voltage is proportional to the angular displacement) and two null points. A properly calibrated RVDT, in combination with its AC–DC amplifier, produces a DC signal that is directly proportional to the angular displacement of the transducer shaft and attached ballistic hammer. The IDRA system design permits facile alignment of the RVDT optimal null point, the long axis of the hammer, and the angular markings on the mounting plate. The highest precision is achieved by nulling the RVDT at the midpoint of the full angular displacement range.

As shown in Figure 13.1A, the mounting plate is positioned for horizontal impact with a test surface. Vertical impact with a test surface is achieved by simply rotating the mounting plate 90° counterclockwise, as shown in Figure 13.1B. The accuracy, reproducibility, and relevancy of characteristic parameters, determined by ballistometry, depend markedly on the skin elasticity, design of the ballistic hammer, testing and data processing techniques, dynamic measurement range, and the test quadrant selected. Significant surface movement during a measurement cannot be tolerated. A 95% measurement success rate is easily achieved with proper measurement technique. Unacceptable surface movement is automatically detected and flagged in IDRA software by comparing the baseline readings just before the hammer is released to those at steady state.

Basic Physics of Ballistometry

The hammer is a rigid body that rotates about a noncentroidal axis and the basic scalar equation needed to represent the motion is

$$\sum M_o = I_o \cdot \alpha \tag{13.1}$$

where $\sum M_o$ is the algebraic sum of the moments of the external forces about the axis of rotation (o), I_o is the moment of inertia of the hammer about the axis of rotation, and α is the angular acceleration of the hammer. This is a second-order nonlinear differential equation and has no closed-form solution.

For either vertical or horizontal contact with the skin, the potential energy of the hammer can be calculated from the mass of the hammer, the distance from the center of rotation to the hammer's center of gravity, and the angular displacement of the hammer, with respect to the test surface. The horizontal or vertical position of the arm is assumed to be 0°, for reference.

Potential Energy of the Hammer for Horizontal Impact

$$\text{P.E.} = \text{mass} \times \text{distance} \times (1 - \cos(\text{angular displacement})) \tag{13.2}$$

Potential Energy of the Hammer for Vertical Impact

$$\text{P.E.} = \text{mass} \times \text{distance} \times \sin\,(\text{angular displacement}) \tag{13.3}$$

These formulae represent a simplified instrument model. They assume that the relative potential energy of the hammer is zero at the horizontal or vertical position, and that the angle of the hammer in contact with the surface at steady state, i.e., the contact angle, is 0°. In practice, it is more convenient to allow for some measurement-to-measurement variation in the actual contact angle, which is recorded and included in accurate estimates of the characteristic parameters.

The angular displacement of the hammer with time is accurately recorded during a measurement, which permits calculation of the angular velocity, and angular acceleration of the hammer, at various times. The angular velocity of the hammer, at the instant of impact with the test surface, can also be calculated from the formula for the kinetic energy of hammer rotation, with the value of kinetic energy set equal to the potential energy of the hammer at a previous peak displacement.

Kinetic Energy of the Rotating Hammer

$$\text{K.E.} = {}^1/_2 \times \text{moment of inertia} \times (\text{angular velocity})^2 \tag{13.4}$$

The moment of inertia of the hammer may be accurately determined from its oscillatory period during a free swing experiment, by treating the hammer as a compound pendulum. The value of the moment of inertia determined from a free-swing experiment agrees closely with that calculated for the hammer using formulas for the moments of inertia of symmetrical bodies.

The product of its moment of inertia and angular velocity is defined as the angular momentum of the hammer. The average force of the hammer impact on a test surface can be estimated from the formula for angular impulse, which is the change in angular momentum.

Force of Impact of Hammer with Test Surface

$$\text{Force} = \text{moment of inertia} \times (\text{change in angular velocity}) / \\ ((\text{distance from hammer tip to center of rotation}) \times (\text{time for arm to stop})) \tag{13.5}$$

For hammers of sufficient length, to a first approximation, the uniaxial deformation of the test surface can be estimated by multiplying the distance from the hammer tip to the center of rotation and the sine of the maximum angle of deformation.

$$\text{Deformation} = (\text{distance from hammer tip to center of rotation}) \times \sin(\text{maximum angle of deformation}) \quad (13.6)$$

Typical parameter values used for computer calculation of force and deformation, in the IDRA system, include hammer free-swing period = 0.66 s, hammer center of gravity distance = 5.62 cm, hammer mass = 8.75 g, and hammer tip distance = 11.8 cm. All other parameters were determined from program analysis of transducer signals collected with 12 bit A/D conversions under computer interrupt control.

Intrinsic Parameters for Characterizing Materials

Figure 13.2 shows the basic response curve for the IDRA ballistometer system. As shown in Figure 13.2, the oscillatory angular displacement of the hammer is recorded with time. The baseline, shown as a line above and parallel to the time axis, is the position of the hammer when it is resting on the test surface. Four major parameters are defined as measures of viscoelasticity (AMP, COR, CAC, and stiffness) and will be discussed in detail below. Angular displacement–time data above the baseline are used to determine AMP, COR, and CAC, whereas that below the baseline is used to determine stiffness.

The AMP is defined as the amplitude or angular displacement measured with respect to the baseline, and is a measure of elasticity. Fthenakis et al.[1] reported the

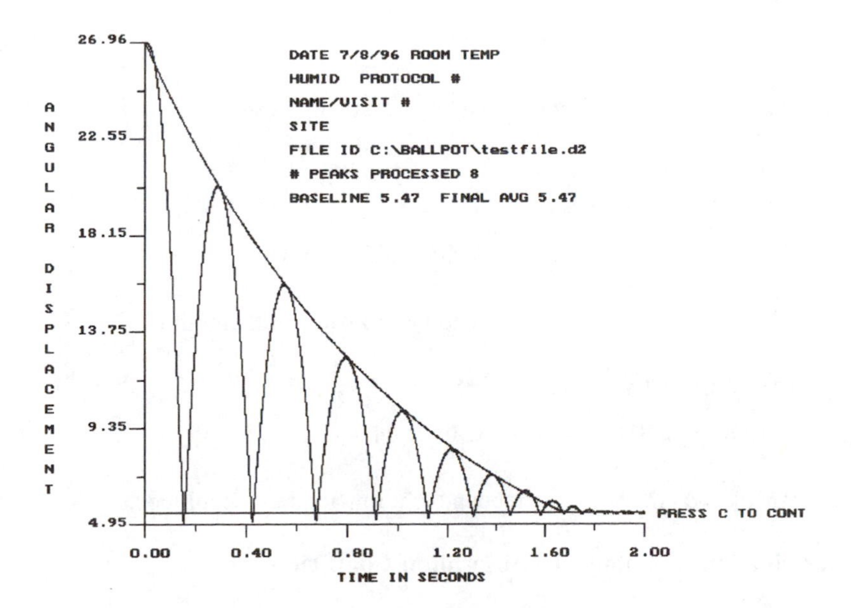

FIGURE 13.2 IDRA basic response curve (test material).

use of AMP to measure changes in elasticity in cosmetic claims studies. The ballistometer employed in these studies was based on principles discussed by Hargens.[2]

The COR, or coefficient of restitution, first applied by Tosti et al.,[3] is a measure of elasticity defined as the ratio of the hammer rebound speed to its speed just before impact. The COR is calculated two ways: (1) as the ratio of rebound and initial hammer speed, and (2) as the square root of the P.E. ratio of two adjacent peaks.

The CAC, or cutaneous absorption coefficient, first applied by Adhoute et. al.,[4] is also a measure of elasticity defined as a dynamic time constant; it presumes that hammer impact energy is lost exponentially with time, and is calculated from peak heights.

Stiffness, first reported by Potts and Pugliese,[5] provides stress–strain information and is defined as the ratio of the hammer impact force to the skin deformation. Stiffness is defined as the force of impact in newtons divided by the deformation in centimeters. Stiffness provides information on the stress–strain behavior of a test material.

For accurate simultaneous determination of COR, CAC, stiffness, and other parameters, an IDRA system employs interrupt-driven, 12 bit A/D data collection, at rates of up to 8000 data points/s, coupled with optimal parameter calculation techniques. The instrument bandwidth is adequate to permit accurate free-fall, contact, and rebound angular displacement measurements to ±0.05° with nonlinearity of 0.5° over a range of 60°.

Determination of Coefficient of Restitution

From Hammer Speed

$$\mathrm{COR}_W = W_R/W_I \tag{13.7}$$

where W_R and W_I are the rebound and impact angular speeds.

From Peak Amplitudes

Vertical impact with skin

$$\mathrm{COR} = \mathrm{SQRT}\ [\sin(\mathrm{AMP}_R)\ /\sin\ (\mathrm{AMP}_I)] \tag{13.8}$$

where AMP_R and AMP_I are the rebound and impact peak amplitude.

Horizontal impact with skin

$$\mathrm{COR} = \mathrm{SQRT}\ [(1 - \cos\ (\mathrm{AMP_R}))\ /(1 - \cos\ (\mathrm{AMP_I}))] \tag{13.9}$$

where AMP_R and AMP_I are the rebound and impact peak amplitude.

Determination of Cutaneous Absorption Coefficient

$$(\mathrm{AMP}_n + \mathrm{BL}) = (\mathrm{AMP}_0 + \mathrm{BL}) * \exp(-Kt_n) \tag{13.10}$$

where K is the cutaneous absorption coefficient, t_n is the time at the nth peak, AMP refers to the angular displacement vs. the baseline at the peaks, and BL the baseline displacement from the horizontal or vertical. K is determined as a one-parameter fit of the exponential equation, above. K is also calculated for peak amplitudes, assuming exponential behavior, using the following formula:

$$K = -(1/t_n) * \log((\mathrm{AMP}_n + \mathrm{BL})/(\mathrm{AMP}_0 + \mathrm{BL})) \tag{13.11}$$

where t_n is the time at the nth peak.

Determination of Stiffness

As already stated, stiffness is defined as the force of impact in newtons divided by the deformation in centimeters. It is the slope of the hammer impact force vs. surface deformation curve, which is shown in Figure 13.3.

The amplitude, impact force, stiffness, CAC, surface deformation, COR calculated from peak amplitudes, and COR calculated from angular speeds are tabulated for sequential rebounds in IDRA as shown in Figure 13.4. These processed data are automatically stored on disk in an IDRA system as an ASCII file, which may be easily imported into commercial spreadsheets for further data processing and study. Raw data for each study are also stored on disk in ASCII format.

The COR and CAC place less demand on instrumentation, since they are generally determined from peak values of angular displacement. CORs can be determined with a precision of 1% and CACs to within 1 to 2%. In contrast, stiffness, which is defined as the force of impact of the ballistic hammer divided by the

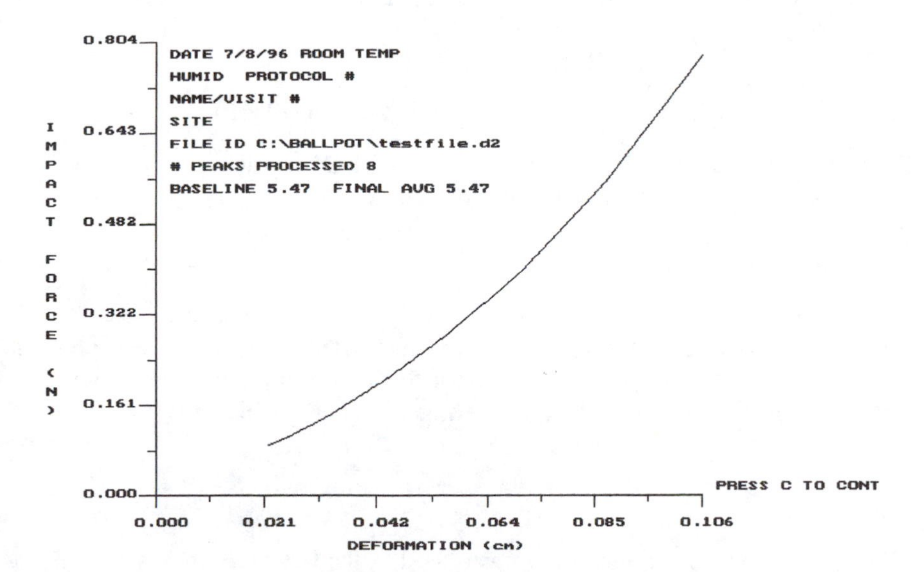

FIGURE 13.3 Impact force vs. surface deformation.

```
DATE 7/8/96 ROOM TEMP
HUMID  PROTOCOL #
NAME/VISIT #
SITE
FILE ID C:\BALLPOT\testfile.d2
# PEAKS PROCESSED = 8
BASELINE 5.47 FINAL AVG 5.47
AVG FIT M = -0.95+/-0.02 B =27.0
AVERAGE CAC = 0.95 STD DEV = 0.02

  AMP  F(N) STIFF CAC  DEFORM COR CORW
14.95 0.804 7.59 0.97 0.106 0.75 0.72
10.47 0.565 6.51 0.96 0.087 0.76 0.77
 7.15 0.397 5.67 0.96 0.070 0.76 0.77
 4.71 0.276 4.96 0.96 0.056 0.76 0.72
 2.96 0.189 4.33 0.96 0.044 0.75 0.75
 1.78 0.128 3.75 0.95 0.034 0.74 0.74
 1.05 0.088 3.29 0.93 0.027 0.75 0.68
 0.56 0.070 3.18 0.92 0.022 0.71 0.61

                                        PRESS C TO CONT
```

FIGURE 13.4 Tabulated data for sequential rebounds.

associated uniaxial surface deformation, requires instrumentation with wide bandwidth and high angular resolution. The data used to determine stiffness is below the baseline with displacements ranging only a few degrees over a time period of milliseconds. Stiffness can be determined with a precision of 5 to 6% for skin, given an instrument calibration technique that is optimal for the simultaneous measurement of all three parameters. With more optimal calibration conditions set for the determination of stiffness, measurement precision can be significantly improved for this parameter.

MEASURING WITH THE BALLISTOMETER

The instrument described above was used in a clinical evaluation of viscoelastic properties of the forearm in 70 female subjects.

METHODS AND MATERIALS

An IDRA ballistometer was used in this study. The study group consisted of 70 female subjects, in good health, and they were instructed not to use skin care products or soap on their forearms for 10 days prior to the test day. All subjects signed an informed consent document before entering the study. Table 13.1 shows the age distribution of the test subjects.

PROCEDURE

The volar surface of the right forearm was used as the test site. Subjects were seated in a comfortable though rigid chair and their right forearm was extended onto a table. Placement of the arm in relation to the ballistometer is critical to obtaining reproducible results. Relaxation of the subject is essential to an accurate measurement. Prior

TABLE 13.1
Age Distribution of Subjects (years)

35–39	40–44	45–49	50–54	55–65
CB	LF	MT	DK	JK
DH	CA	DP	SE	SS
DC	DW	CZ	GS	PP
ED	AF	BS	LP	DV
RF	PB	KE	BM	PL
VC	MM	PK	LC	BW
CB	LW	PJP	RE	PH
CA	VC	LR	MC	MD
MY	SW	SE	RT	PW
DW	LS	BM		YG
VZ	AN	GS		JC
	PB	LB		RR
	PR	DF		GW
	CG	LF		PP
	CR	SS		SN
	KR	JR		
	PB			
	KH			
	KO			

to making the measurement, the subject is instructed in all aspects of the procedure to help achieve relaxation. The zero position of the ballistometer arm is calibrated by placing the hammer against the skin, and the instrument is adjusted to zero following the instructions supplied by the manufacturer. This zero point is taken as the baseline of the graph that will be generated in the computer. The measurement is made by releasing the hammer from the magnetic holder electronically.

RESULTS

The following parameters were generated by the computer program from the input data for each subject: amplitude, stiffness, cutaneous absorption coefficient, and coefficient of restitution.

AMPLITUDE

The amplitude is a measure of elasticity as it relates to the rebound energy of the skin. The amplitude decreases as the subject's age increases as shown in Figure 13.5.

STIFFNESS

Skin stiffness relates to "skin hardness" and is a function of deformation of the skin on impact. It is expressed as a ratio of the impact force to the skin deformation and thus represents a true stress–strain parameter. This is a new measurement value obtained from the ballistometer, first reported by Potts and Pugliese.[5] The stiffness value has an inverse relationship to age until the fourth to fifth decade, after which it increases sharply with age in females. This relationship is shown in Figure 13.6.

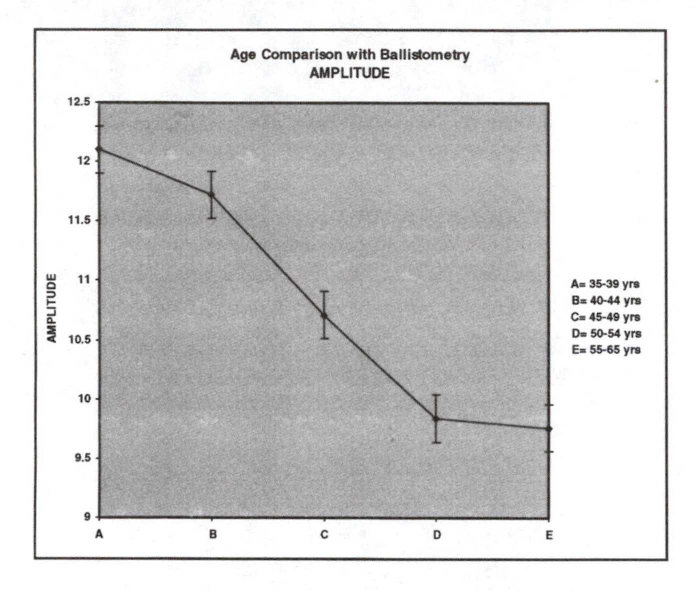

FIGURE 13.5 Amplitude.

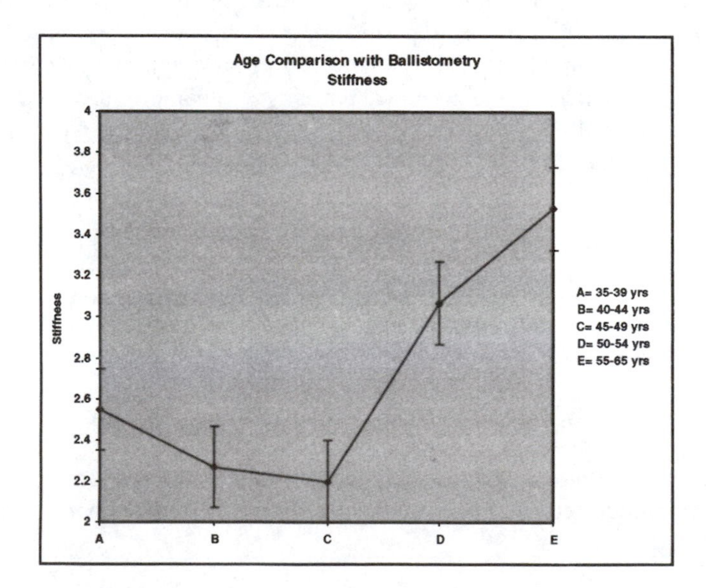

FIGURE 13.6 Stiffness.

Cutaneous Absorption Coefficient

The CAC is a measure of the elasticity defined as a dynamic time constant and presumes that the hammer impact energy is lost exponentially with time. In this study the CAC increases with age, as shown in Figure 13.7.

Coefficient of Restitution

The COR represents the ratio between the hammer rebound speed to the hammer speed just before the impact on the skin. The COR increases directly with age as is shown in Figure 13.8.

DISCUSSION

The three basic parameters reported by previous investigators, amplitude, CAC, and COR, are confirmed in this study. Amplitude and COR decrease with age, while COR increases with age. These three parameters move in the direction one would expect as the subject's age increases. The stiffness parameter, however, shows a biphasic curve when plotted as the subject's age increases.

This phenomenon might be explained if one considers first that all the subjects were female, and second that the ages of the subjects ranged from 35 to 49 years during the period that the stiffness decreased; a period of life for females representing the maximum effect of estrogen on connective tissue. It is well known that estrogen stimulates collagenase production by fibroblasts, a mechanism responsible for endometrial sloughing in the menstrual cycle; however, this action of collagenase

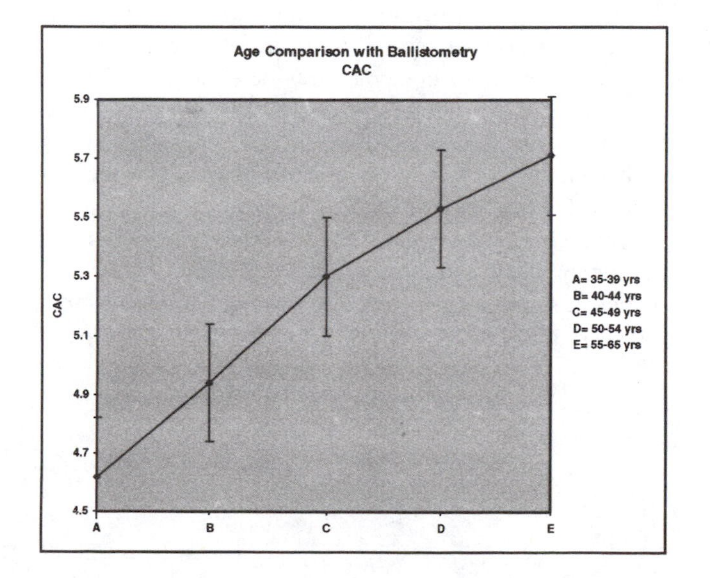

FIGURE 13.7 CAC.

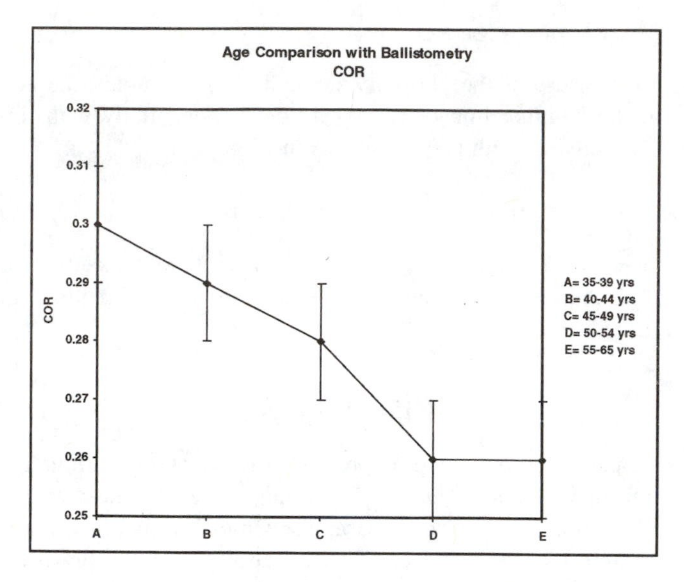

FIGURE 13.8 COR.

is decreased, but does not diminish entirely.[6] Cross-linking of collagen and lack of production of new collagen would contribute to overall stiffness.

CONCLUSIONS

The IDRA ballistometer is described in detail and clinical data presented using this instrument to evaluate viscoelastic properties of the skin of 70 female subjects. The ballistometer is a sensitive and accurate means of determining viscoelastic properties of skin. It is noninvasive, offers a quick and easy evaluation, and has a wide dynamic measurement range. It measures four parameters: amplitude, stiffness, cutaneous absorption coefficient, and coefficient of restitution, three of which have been previously reported. Stiffness is reported in this chapter as an additional parameter. The data obtained in the clinical study on 70 female subjects show a decrease in the values with age for amplitude and the coefficient of restitution, while the cutaneous absorption coefficient shows a direct increase with age (Figure 13.9). Stiffness follows a biphasic curve decreasing to age 50 and increasing sharply after age 50.

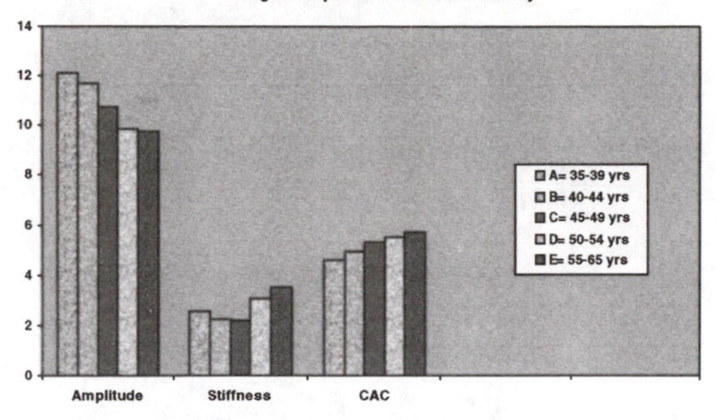

FIGURE 13.9 Age comparison with ballistometry.

REFERENCES

1. Fthenakis, C.G., Maes, D.H., and Smith, W.P., *In vivo* assessments of skin elasticity using ballistometry, *J. Soc. Cosmetic Chem.*, 42, 211, 1991.
2. Hargens, C.W., *Ballistometry*, CRC Press, Boca Raton, FL, 1995.
3. Tosti, A., Compagno, G., Fazzini, M., and Villardita, S., A ballistometer for the study of the plastoelastic properties of skin, *J. Invest. Dermatol.*, 69, 221, 1991.
4. Adhoute, H., Berbis, P., and Privat, Y., Ballistometric properties of aged skin, in *Aging and the Skin*, Balin, A.K. and Kligman, A.M., Eds., Raven Press, New York, 1989.
5. Potts, J.R. and Pugliese, P.T., Ballistometry techniques and instrumentation for the physical characterization of skin *in vivo*, paper presented at 20th Anniv. Symp. Int. Soc. Bioeng. Skin, Miami, FL, February 15, 1996.
6. Fischer, G., Comparison of collagen dynamics in different tissues under the influence of estradiol, *Endocrinology*, 93, 1216, 1973.

14 Hardware and Measuring Principles: The Microindentometer

Christopher J. Graves and C. Edwards

CONTENTS

Introduction 161
Stratum Corneum Mechanical Competence 162
Introduction 162
Review of Indentation Research 163
Measurement Principles 164
Introduction 164
Hardware 165
Indentometer Actuator and Needle 165
Positioning System 166
Mounting Components 166
Force Measurement 168
Computer Control 168
Taking and Interpreting Measurements 169
Position Testing of Needle Relative to Surface 169
Indentation of Human Stratum Corneum 169
Discussion and Pitfalls 172
Further Development 174
Discussion and Conclusion 175
References 176

INTRODUCTION

The stratum corneum provides an important physical barrier that protects the body from desiccation, infection, and minor traumata. The "competence" of the stratum corneum in providing physical protection against injury depends on its biomechanical properties. This protective property is often thought of as "hardness" but is better described as increased stiffness, tensile strength, and mass.

0-8493-7521-5/02/$0.00+$1.50

While impairment of this barrier often leads to dermatitis, many endogenous skin disorders actually reduce the competence of the skin in providing this barrier. The aim of this instrumental technique is to measure the mechanical properties of the stratum corneum in isolation from its underlying structures. This will be of major benefit in studying the causes and treatment of many skin disorders and dermatoses such as atopic dermatitis or ichthyosis where abnormal keratinization is a feature of the disorder.

This chapter describes just one instrumental approach adopted to obtain mechanical information from this layer using dynamic microindentation. A prototype computer-controlled indentometer has been developed that uses a 1-mm-diameter flat-tipped needle to indent the stratum corneum by a preset distance of 20 μm. A piezoactuator is used with a response time of 1 ms. Dynamic indentation serves to isolate the stratum corneum surface response from that of the slower underlying tissues, but produces considerable inertial artifacts. To measure the true mechanical resistance of the skin surface, matched force transducers and then an impedance head were used as a means of canceling out these artifacts. The precise positioning of the indentometer tip to the skin surface and its subsequent indentation is automated by the computer program ensuring both reproducibility and ease of use.

Measurement of changes in the reaction force to indentation will enable characterization of the mechanical properties of the stratum corneum. Thus, changes due to disease or its treatment, and to application of topical agents such as emollients, can be quantified.

STRATUM CORNEUM MECHANICAL COMPETENCE

INTRODUCTION

One of the primary functions of the stratum corneum is to provide a physical barrier protecting against mechanical insult and trauma.[1] Apart from an exceptional ability of the skin to repair itself, the *competence* of the stratum corneum at providing protection against mechanical injury is related to its strength, flexibility, stiffness, and hardness. The *hardness* of a material is defined as resistance to wear, indentation, and scratching,[2] although what is perceived as increase in hardness of the stratum corneum may in fact be a combination of increased stiffness, strength, and mass, as well as true hardness. However, tensile strength, stiffness, and hardness are interrelated properties.[3]

Thus, measurement of stratum corneum hardness is one way that the mechanical competence of stratum corneum can be assessed. In classical engineering, the measurement of hardness is usually carried out by indenting the surface of the material being tested, and this practice has been used for the testing of skin. Schade[4] was the first to use the technique on skin and since then it has regularly been advanced as a technique by other researchers. By measuring the forces involved in the point indentation of stratum corneum, it is believed that one or more parameters relating to strength, stiffness, and hardness can be obtained. This would then provide a measure of stratum corneum mechanical competence.

Biomechanical characterization of the stratum corneum could do much to improve clinical decision making and therapy. Impairment of this physical barrier invariably results in some form of dermatitis.[5,6] Conversely, many endogenous skin disorders reduce the mechanical competence of the skin in providing this barrier.[7] In many disorders, abnormal keratinization or desquamation results in the stratum corneum becoming thickened and stiff, often leading to brittleness and fissuring.[8] In some disorders, severe mechanical fragility can result.[9] It is also likely that it is altered after prolonged occlusion of the skin leading to an increased susceptibility to mechanical (and possibly chemical) trauma. Indeed, changes found in stratum corneum compliance after glove patch occlusion[10] have raised the question whether this is indicative of a less mechanically protective stratum corneum, as perhaps would be expected. To address these problems, a microindentometer device was constructed, based on the principles of an earlier instrument.[11]

This chapter describes the development of a research instrument to characterize the mechanical response of the stratum corneum *in vivo* to microindentations to assess stratum corneum mechanical competence.

REVIEW OF INDENTATION RESEARCH

Many researchers have used indentation as a way to study the mechanical properties of skin. However, very few have used the technique to investigate stratum corneum properties. Rather, the dermis, whole skin, and even muscle have been the focus of attention. Schade[4] used a spherical ball on the end of a pivoted arm. The force for indentation was provided by a 50-g mass. The depth of indentation produced was of the order of millimeters. Schade's method and design has been used by a number of investigators.[12–14] Most later workers have used smaller indentors. A flat disk was first used by Vlasblom[15] and then by Dikstein and Hartzshtark,[16] who continued to use weights to provide the indentation force. A different approach was adopted by Barbenel et al.,[17] who compressed a fold of skin with a predetermined constant deformation and measured the force that was required to do this. Some authors have employed commercially available hardness testers used in the rubber industry. Peck and Glick[18] used one such device called a durometer to carry out *in vitro* measurements on keratin specimens. Falanga and Bucalo[19] and Romanelli and Falanga[20] used a similar durometer device to assess skin hardness in patients with scleroderma. Measurements were taken by resting the device under gravity against the skin of finger pads, forearms, and thighs, and taking four consecutive readings as the skin exhibits creep. Good correlation was found between durometer readings and clinical severity score. A different technique was described by Prall,[21] based on the lowest load able to cause a stylus to visibly scratch the skin. A scratch was distinguished by its ability to scatter light.

A number of authors have published work on measuring the hardness of muscle tissue. Honda[22] used a spherical indentor of 5 mm radius and compared measured values with a theoretical spring and dashpot model. Horikawa et al.[23] devised a hand-held instrument for evaluating muscle hardness consisting of a 1-cm^2 pressure disk attached to a load cell. A separate optical displacement transducer was fixed to the side of the instrument. In operation, pressure is applied manually. The indentation produced and the force required to produce it is continually recorded. Other authors have developed large-scale *in vivo* indentation systems for the investigation of pressure sores,[24] and for the biomechanical characterization of limb tissues in rehabilitation engineering.[25–27]

On a scale that could prove useful for future *in vitro* studies of stratum corneum, a cell-indentation apparatus (cytoindenter) has recently been developed to obtain intrinsic material properties of individual cells attached to a rigid substrate.[28] Meanwhile, researchers investigating the dermal anchoring of skin equivalents have used a spherical indenter in a quasi-static test of tensile properties.[29]

Only one indentation device has been constructed specifically to investigate the pliability and hardness of stratum corneum.[30–32] The device consists of an interchangeable needle mounted on a moving coil vibrator, which was electronically controlled,[33] to give a single indentation movement of variable travel to a maximum of 100 μm, lasting 4 ms. The needle was manually positioned above the skin using a modified microscope translation stage. Needles used had tip diameters of 1, 10, and 20 μm. The original version used a linear variable displacement transducer; however, this was removed at a later stage as it was found to be a source of artifact.[11] *In vitro* samples of thigh skin that had been differentially hydrated were indented. A decreasing resistance to indentation was found with increasing levels of hydration.[32] Successive removal of the stratum corneum using skin surface biopsies[34] produced a reduction in the indentation force registered.[31] The effects of delipidization and hydration on the forearm skin indentation were measured.[11] Delipidization with diethyl ether was found to cause an increase in resistance to indentation; however, the effect had almost disappeared 20 min later. Hydration of forearm skin was found to reduce resistance to indentation.

MEASUREMENT PRINCIPLES

INTRODUCTION

The microindentometer is based on the principle that indentation of the stratum corneum by a needle is opposed by the extensible horny membrane. Hendley et al.[11] stated:

> As the skin is a viscoelastic structure the faster the [indenting] needle moves the smaller is the time available for skin surface elongation to relieve the indenting force and the greater the viscous drag. Therefore, the force registered on initial impact will depend mainly on the properties of the horny layer.

Although the theory is still poorly developed, it is known that viscous effects are proportional to velocity, so if the stratum corneum is viscoelastic, then the force registered by a rapidly moving needle will depend on stratum corneum viscosity as well as stiffness. If the underlying epidermis deforms only a little due to stratum corneum viscosity, then one can consider that the reaction force depends mainly on stratum corneum properties. However, it is possible that a stiffer stratum corneum would cause depression of the epidermis under an area of stratum corneum.

The success of the device depends on its ability to rapidly indent the stratum corneum by an amount that is of an equivalent order of magnitude to its thickness (10 to 15 μm). Impedance measurements of human finger pulp[35] based on the mechanical model attributed to Voigt and Kelvin give a minimum mechanical impedance of around 250 Hz with characteristic stiffening at higher frequencies. This indicates that the needle must be able to move at a velocity greater than 5×10^3 ms^{-1} (10 μm in 2 ms) in order for the skin to present itself as a stiff system.

The authors have chosen to implement an impact response approach to the extraction of stratum corneum properties. This maximizes the possibility of separating the mechanical responses of underlying tissues in the time domain, since it would be expected that some time delay in the response of tissues would be proportional to the distance from the indenting tip. Furthermore, input from a step function by definition imparts a range of frequencies into the skin and enables analysis of the response in the frequency domain. Methods of extracting useful frequency responses from step or point function excitation are well developed.

HARDWARE

A microindentometer device was designed and constructed. In part it was based on the original indentometer device of Guibarra, Nicholls, Hendley, and co-workers, reviewed above. Their device used a moving coil vibrator, incorporating a linear variable displacement transducer (LVDT). It was clear that a far more reliable instrument could be produced by utilizing some of the recent advances made in the field of laser and optics technology. The component parts and operation of the new indentometer shown in Figure 14.1 are described below.

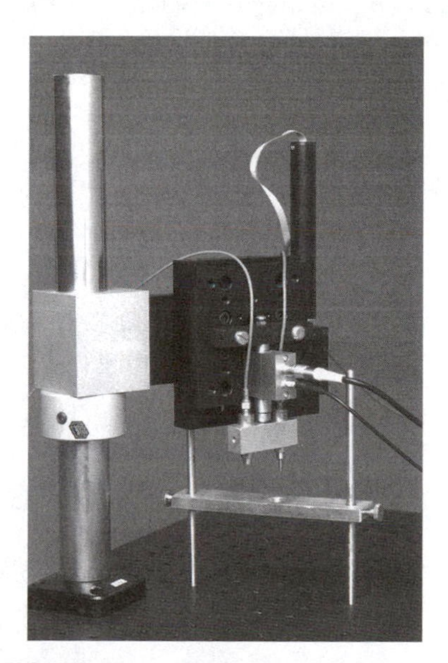

FIGURE 14.1 The indentometer device without arm rest or pressure cuff.

Indentometer Actuator and Needle

The instrument works by rapidly indenting the stratum corneum to a maximum depth of 20 μm and monitoring the force of reaction. A piezoelectric actuator and controller (Photon Control/Melles Griot NGS20ZC and NDA501-75Fc) were used. The actuator can move a component, such as a lens, mirror, laser, or in this case a force transducer and "needle," by distances of up to 20 μm at a speed of 10 μm in 1 ms. This type of actuator will perform the operation repeatedly and without error. Needles of 1 mm diameter and smaller were used (Figure 14.2).

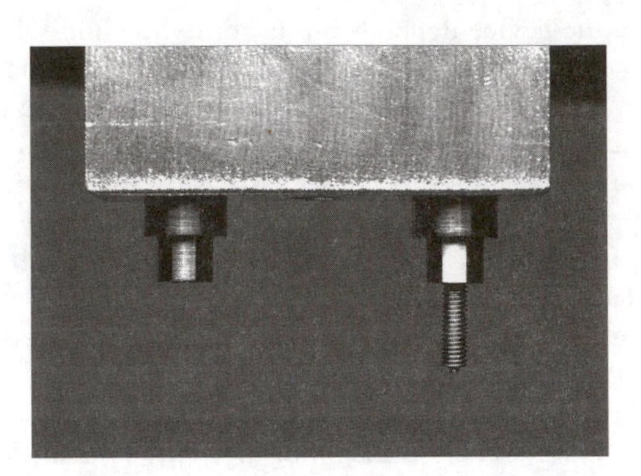

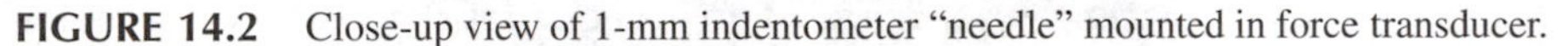

FIGURE 14.2 Close-up view of 1-mm indentometer "needle" mounted in force transducer.

Positioning System

One of the major technical problems with stratum corneum indentometry is accurately positioning the indentometer needle above the stratum corneum before commencement of indentation. This is because it is difficult to measure precisely the location of the part of the skin surface to be indented, in relation to the position of the needle. The indentometer resolved this problem by using an automated probing and indentation action as shown in Figure 14.3. The automated procedure that was used is described as follows:

1. The indentometer moved toward the skin in 10 μm steps performing a 20-μm indentation "jab" after each maneuver.
2. When a predetermined, sufficiently large reaction force was registered,* the needle had reached a position of maximum indentation and the stepping forward ceased.
3. The indentometer moved back 50 μm and made a reference measurement, indenting "free air."
4. Then the indentometer returned to the position of maximum indentation and repeated the previous "jab."
5. Finally, the indentometer stepped back 2 μm, made a 20 μm indentation, and repeated this maneuver 20 times.

All vertical movement was computer controlled and provided by an encoder-driven motorized positioner (37-1104 Ealing Electro-Optics plc).

Mounting Components

The whole device was attached to a damped column (Ealing Optics) shown in Figure 14.1, which was in turn bolted to an aluminum "bread board" (Newport).

* This is calculated from previous measurements to ensure that in the final part of the procedure (step 5), some of the indentations commence with the needle in contact with the skin and some commence clear of it.

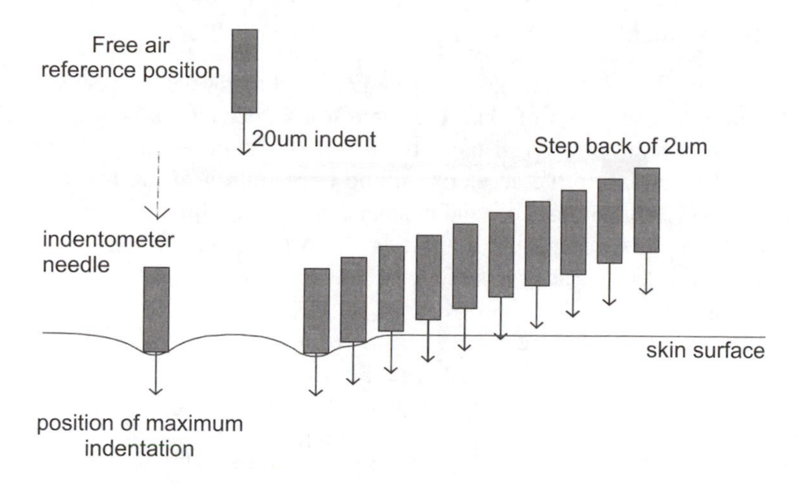

FIGURE 14.3 Schematic diagram of indentometer movement control.

This was mounted on a vibration-isolated concrete slab to minimize the effects of extraneous vibrations on measurements. An arm rest was constructed for immobilizing the forearm in position. A metal stop bar with a 12-mm-diameter hole in it was clamped down over the skin to prevent any movement. To immobilize hands for measurement, a blood pressure cuff was used in place of the arm rest. By inflating the cuff under the hand, the area of skin to be measured was pressed against the hole in the metal bar (Figure 14.4). Sufficient pressure was used to keep the hands still; however, measurements had to be completed within a maximum of 5 min because of the discomfort from blanching.

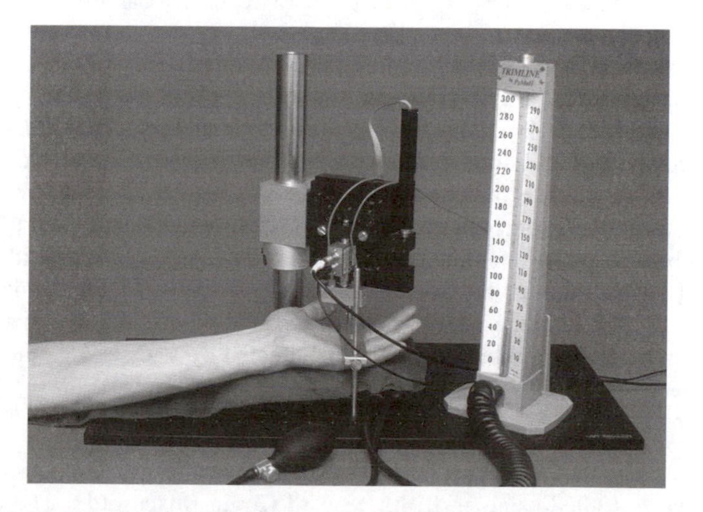

FIGURE 14.4 Indentometer device in use with hand immobilization provided by an inflated blood pressure cuff.

Force Measurement

The reaction force to rapid indentation was measured using two piezoelectric force transducers. On indentation of skin, the force transducer into which the needle is mounted registers not only a resistance force from the skin but also a large inertial force contribution due to the acceleration and deceleration of the transducer. The signal from the second transducer and balanced charge amplifier was used to cancel out this inertial force component by putting the two signals through a differential amplifier (Figure 14.5). The resultant signal was low-pass-filtered.

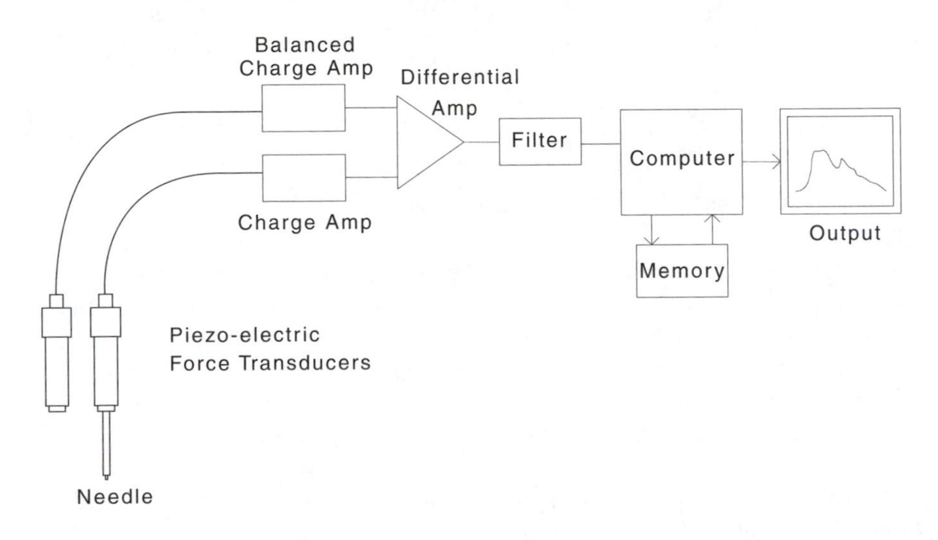

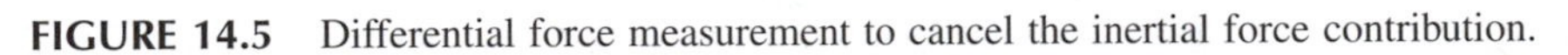
FIGURE 14.5 Differential force measurement to cancel the inertial force contribution.

As it was not possible to balance the two force transducers completely, a further cancellation of residual artifacts was achieved by using the recorded response from indenting free air and subtracting this from each indentation signal measured. The resultant signal was taken as the reaction force measurement. Figure 14.6 shows this further cancellation. The bottom trace is the filtered signal placed in computer memory after indentation of free air and includes residual inertial artifacts. The second-to-bottom trace is the filtered signal from indentation of skin with the same artifacts. The top trace is the indentation signal with the artifacts canceled. The second-to-top trace is the resultant indentation signal (with artifacts canceled) from a previous measurement when the skin was indented by a lesser amount.

Computer Control

An IBM-compatible PC computer was used. The piezoelectric actuator/controller system was controlled using an IEEE PC card (Brain Boxes Ltd.). The motorized positioner was controlled using a DCX controller card and DCX-MC110 DC servo module (Precision Micro Control Corp.). The PC interface and the DCX-PC100 motion controller drivers were used for software control. Data acquisition was

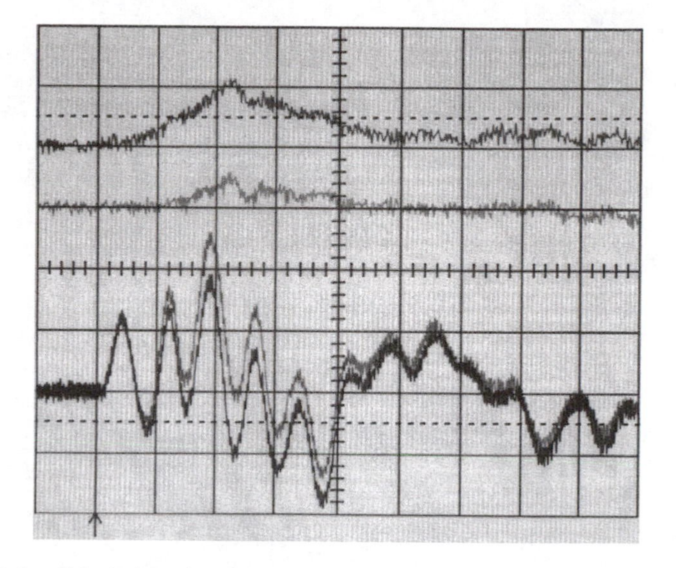

FIGURE 14.6 Skin indentation force signals. Bottom trace: free-air reference signal in memory; second to the bottom trace: signal from indentation of skin; top trace: resultant force signal after subtraction of artifacts; second to the top trace: resultant force signal from a previous measurement.

achieved using a PCI20091W 89 kHz analog input board (Burr Brown), in DMA mode. Integrated software was written in Microsoft QuickBasic v4.5.

TAKING AND INTERPRETING MEASUREMENTS

Two experiments were carried out using the device.

POSITION TESTING OF NEEDLE RELATIVE TO SURFACE

Measurements were made of an *in vitro* test surface (insulating tape), in order to test the automated probing and indentation action. Three successive runs were made using the same surface, but moving the indentor laterally by a millimeter between runs, so as not to indent any spot more than once.

Figure 14.7a shows the results of the first three series of indentations. The position of maximum indentation is marked as 0 µm. As the indentometer stepped away from the material, it eventually cleared the surface and no indentation signal was detected. From this, the position of the surface was deduced and the data replotted relative to the surface. The result of transforming the data in this way is shown in Figure 14.7b. The force reading representing a full 20 µm indent from a starting position just touching the surface is shown at –20 µm on the graph.

INDENTATION OF HUMAN STRATUM CORNEUM

Measurements were carried out on the palms of two volunteers. Sites at the base of the fingers where calluses had formed and sites on the thenar eminence were measured.

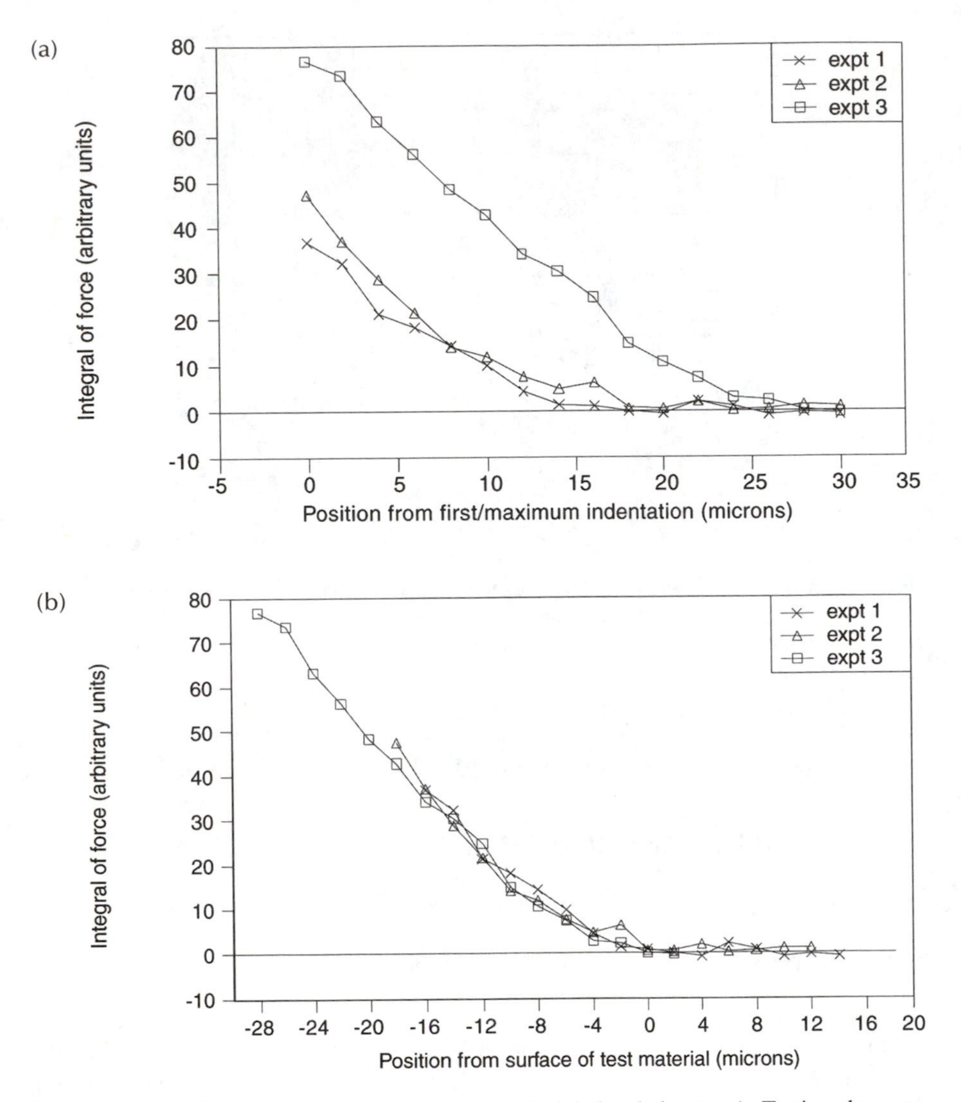

FIGURE 14.7 Indentation of an *in vitro* test material (insulating tape). Testing the automated probing and indentation action for determining position of the needle relative to a surface. (a) Position relative to first (maximum) indentation; (b) position relative to surface of test material.

In both experiments the integral of the force signal was used to represent the resistance to indentation, rather than the peak force. Computer calculation of the peak indentation force is complicated by the fact that the signals still contain some noise ripple. Averaging the signal is problematic: the time overheads are large for a reliable and accurate determination of peak force. Calculating the area under the force curve is a much quicker and simpler way of assessing the magnitude of the resistance to indentation. Analysis of some data from this experiment was carried out using both the force integral and a manual analysis of peak force.

A sample force trace from a 20 μm indentation of the thenar eminence is shown in Figure 14.8. Peak indentation was achieved in about 1 ms. Figure 14.9 provides a comparison of the peak and integral force methods of representing the resistance to indentation. It can be seen that the two traces virtually superimpose.

Data from the indentation of callus and thenar eminence skin are shown in Figures 14.10 and 14.11. Each set of plotted data was the average of two measurement cycles. The callus measured from volunteer "CE" gave a greater resistance force to 20 μm indentation than the callus from volunteer "RH." The response of the thenar stratum corneum of both volunteers was similar, the resistance force to 20 μm indentation being almost identical.

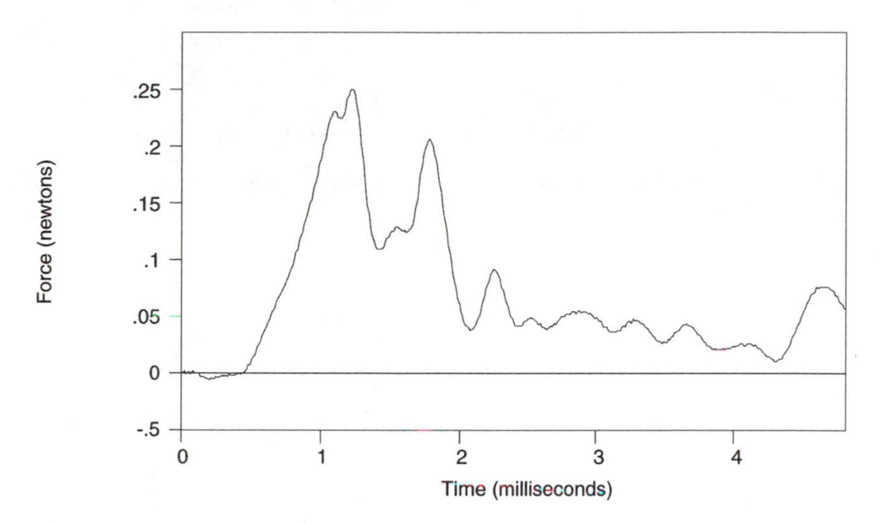

FIGURE 14.8 A sample indentometer force trace from a 20 μm indentation of stratum corneum from the thenar eminence *in vivo*.

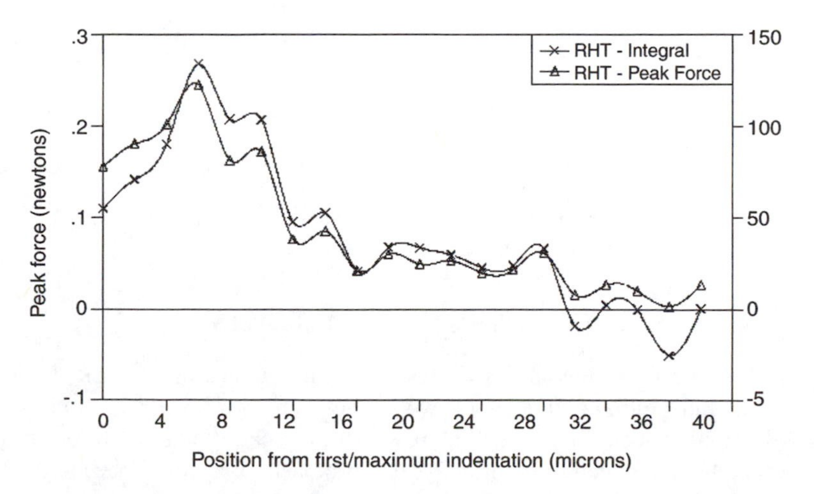

FIGURE 14.9 A 20 μm indentation of human stratum corneum *in vivo*. Comparison of "peak force" and "integral of force."

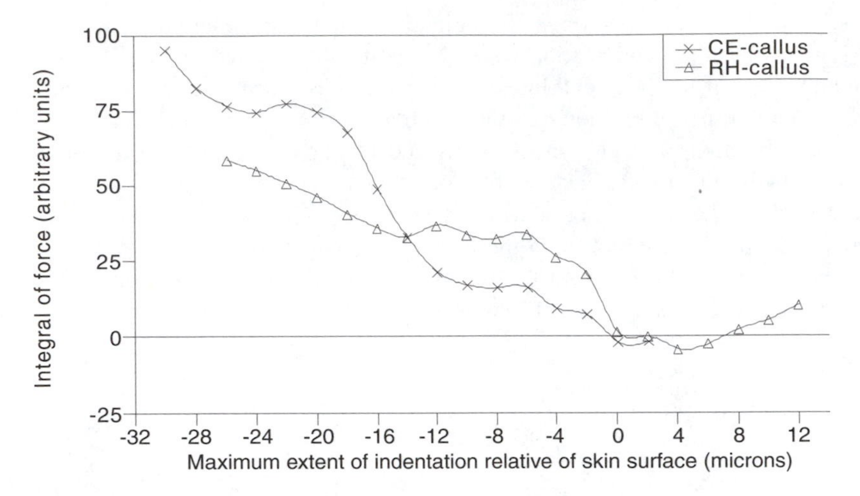

FIGURE 14.10 A 20 μm indentation of human callus stratum corneum *in vivo*. Moving average data from two indentation cycles.

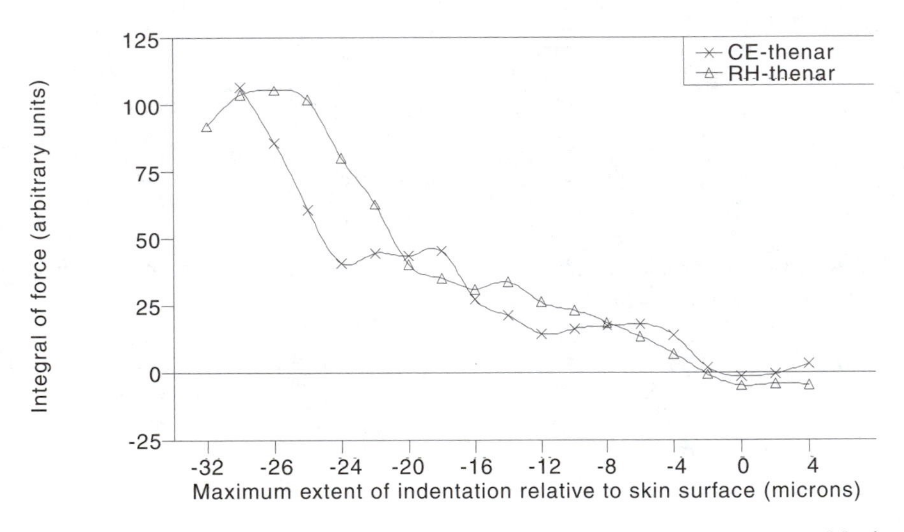

FIGURE 14.11 A 20 μm indentation of human thenar stratum corneum *in vivo*. Moving average data from two indentation cycles.

DISCUSSION AND PITFALLS

The results from indenting an *in vitro* test material demonstrate that the automated probing and indentation routine works well. After transformation, a clear correlation was observed between each trace (Figure 14.7b), indicating that the position of the test surface had been deduced correctly. It can be seen that for experiment 1 (-x-), the predetermined reaction force for maximum indentation (20 μm) was insufficient and was only just sufficient for experiment 2. It was modified for experiment 3.

The superposition of the traces in Figure 14.9 demonstrates the validity of using the area under the force curve (integral) in place of the peak force, to represent the magnitude of the resistance to indentation. A peak force of 0.2 N was approximately equivalent to 100 arbitrary units (force × time).

In both Figures 14.10 and 14.11, at the point when the indentometer no longer indents the skin there is a clear drop in the force readings. The response from the callus of volunteer "CE" appears to be stiffer than that from volunteer "RH." Clearly, the resistance to 20-μm indentation is greater. However, the same was not observed for measurements taken from the thenar eminence. The similar response of the thenar stratum corneum is as expected as both volunteers were male, in their 30s, and in the same type of occupation.

From Figure 14.11 it would appear that once greater than full indentation occurs, there is a rapid increase in the reaction force. However, more data should be collected to gain a greater understanding of how the skin surface responds to the indentation cycle. This would also enable optimization of the stepping and indentation cycle, improving its efficiency in obtaining a prompt reading. The 4 μm delay in the rise of trace from volunteer "CE" (-x-) may be because noise or artifacts resulted in an incorrect deduction of the skin surface position. However, it is also possible that the measurements were affected by a trough in the skin surface. The surface of the stratum corneum consists of many relatively flat plateaus separated by a network of troughs that crisscross the skin surface. If the indenting needle straddles a trough of similar magnitude to its diameter, then the resistance to indentation is likely to be reduced in some way. Consideration was given as to the size of these plateaus. A series of pictures of increasing magnification were taken of positive skin replicas using an electron microscope. From these, it was possible to relate both the size of the plateaus and individual corneocytes to the topography of the skin.

Immobilizing the hand reliably and allowing for no movement or vibration of the instrument is vital for successful measurements. Figure 14.12 illustrates this

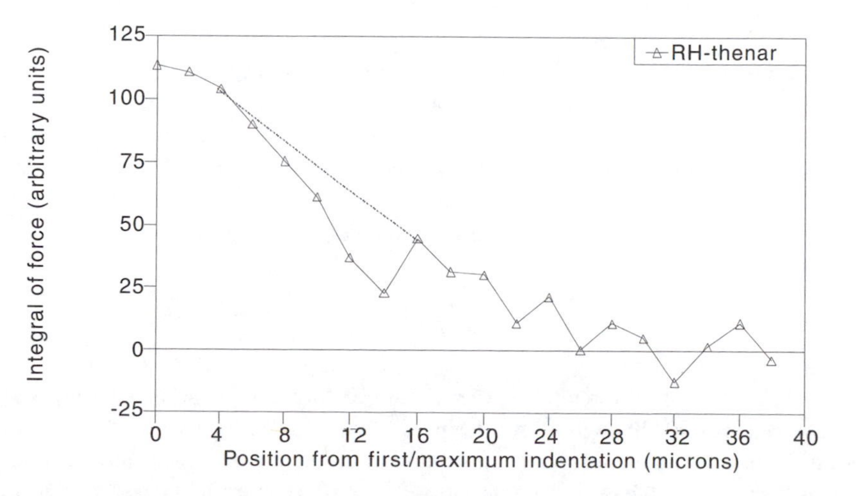

FIGURE 14.12 Repeated 20 μm indentation of human stratum corneum *in vivo*, demonstrating the effect of a temporary drop in supporting cuff pressure.

point well. It shows data from the repeated indentation of the stratum corneum of one volunteer. In this case, the pressure in the supporting cuff was inadvertently allowed to drop, so that the skin surface slowly moved away from its original position. This was realized halfway through the procedure, so when the pressure was returned to its original level, the skin surface moved back to its original position, producing a slight increase in the force measured. The dashed line indicates the trace that would have been obtained had the pressure drop not occurred.

The question arises whether the indentation cycle involving multiple indentations affects the measurement by initially damaging the skin. While it remains a possibility, it is not likely because the speed of indentation (~0.01 ms^{-1}), although fast, probably does not pierce the stratum corneum. However, this needs to be tested in future experiments. Nicholls[32] used a stereomicroscope to examine skin that he had indented by 90 μm at a speed of approx 4.5×10^{-2} ms^{-1}. However, no signs of surface damage were found.

Mention should be made of the elimination of inertial forces by differential measurement. No mention is made in any of the published work on the original indentometer (reviewed above) of accounting for any inertial force contribution. There is no doubt that inertial forces would have affected the measurements made. With this in mind, the validity of the data is in doubt. Some of the trends reported are convincing and it is possible that a harder, stiffer stratum corneum would produce a significantly greater deceleration of the indenting needle and thus a greater inertial contribution. If this is the case, then the original device measured hardness in terms of increased deceleration of the indenting needle, rather than increased resistance to its penetration.

FURTHER DEVELOPMENT

Dynamic indentation, although helping to isolate the stratum corneum surface response from that of softer underlying tissues, produces considerable inertial artifacts, which the above design has difficulty in eliminating.

An impedance head (PCB Piezotronics), which incorporates a load cell and an accelerometer in a single compact package, has been used to replace the matched force transducer arrangement described above. It measures the force of resistance offered by the stratum corneum to indentation and enables cancellation of the inertial forces generated by the acceleration/deceleration of the mass of the indentation head and needle assembly. Initially, inertial mass cancellation was carried out in the time domain using

$$F_s = F_m - m_i a$$

where F_s is the sensed force, F_m is the measured force, m_i is the inertial mass, and a the acceleration. Although with filtering and calibration for free-air "null" indentation representative data have been obtained, it cannot be consistently captured. This is because the dynamic characteristics of the transducers and the overall system are frequency dependent. What is required is a frequency-dependent transfer function to describe the dynamic response of the system. The transfer function for the indentometer impedance head without skin contact is

$$H_I = A/F_I = 1/M_M$$

where A is the acceleration, F_I the indicated input force, and M_M the inertial mass. Figure 14.13 shows the frequency response function (transfer function) for the impedance head obtained using a two-channel digital signal analyzer (Diagnostic Instrumentation D2200). The measured transfer function with the indentometer in contact with the skin is

$$H_M = A/F_M$$

where F_M is the measured force. Figure 14.14 is the frequency response function obtained from 10 μm indentation of palmar skin. The force applied to the skin is $F_S = (F_M - F_I)$. It is now possible to substitute the two frequency-dependent transfer functions using the equations above, to obtain the corrected indentation force. This can be done using fast Fourier transforms and their inverse.

DISCUSSION AND CONCLUSION

Although reasonable inertial compensation was achieved in the arrangement of matched force transducers described above, the matching was not exact and a repeated differential with a free-air indentation was necessary to recover a signal from skin. Furthermore, the arrangement was a little bulky and placed restrictions on which skin site could be indented. The use of an impedance head is more compact

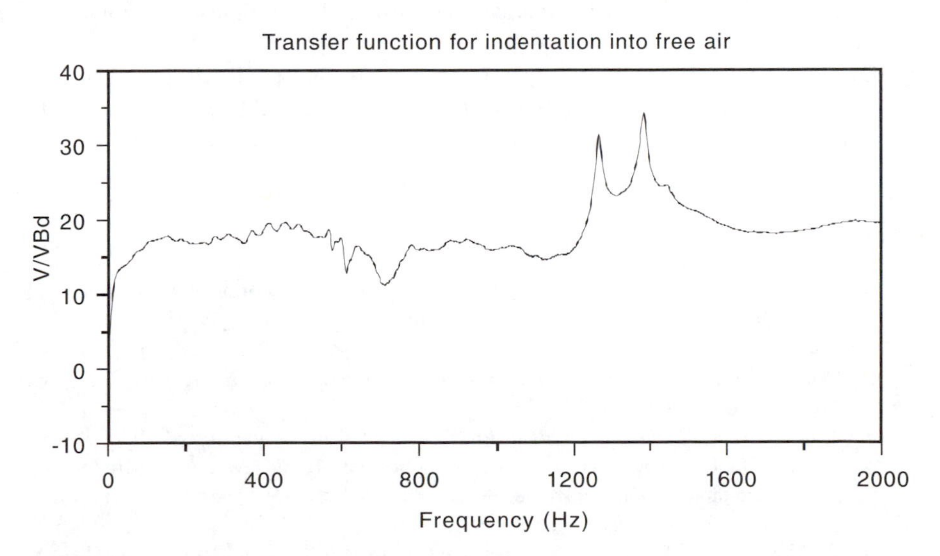

FIGURE 14.13 Transfer function free air.

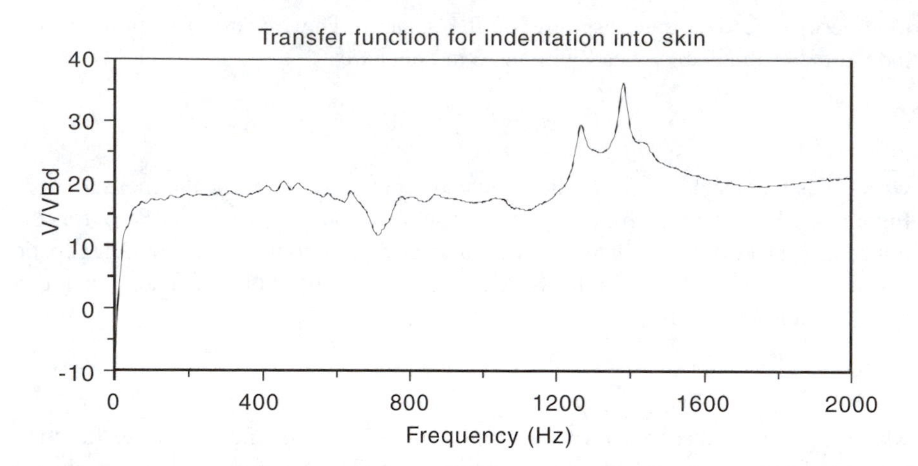

FIGURE 14.14 Transfer function 10 μm indentation of palmar skin.

and should eventually allow dispensing with the free-air measurement immediately prior to skin indentation.

Dynamic point indentation currently rests on the theory expounded by Hendley et al.[11] The current assumption is that a velocity of 0.01 ms^{-1} is sufficient for the viscous properties of stratum corneum to be significant. If it is found that faster indentation is required, then the inertial compensation technique adopted will be even more pivotal to the success of the device.

In conclusion, the ability to measure the reaction force from the precise indentation of the stratum corneum has been demonstrated. Inertial forces have successfully been eliminated using differential measurement. The point of contact with the skin surface can be found with relative ease using the automated stepping and indentation technique.

By measuring changes in the reaction force to indentation, it will be possible to characterize the mechanical properties of stratum corneum and to assess its mechanical competence in disease and treatment.

REFERENCES

1. Jackson, S.M. and Elias, P.M., Skin as an organ of protection, in *Dermatology in General Medicine*, 4th ed., Vol. 1, Fitzpatrick, T.B. et al., Eds., McGraw-Hill, New York, 1993, 241.
2. Houpt, G.L., *Science for Mechanical Engineering Technicians. A Part 2 Course*, McGraw-Hill, London, 1973, 323.
3. Titcomb, G.R.A., *Fundamentals of Engineering Science*, Hutchinson, 1970, 721.
4. Schade, H., Untersuchungen zur Organfunction des Bindegewebes. I. Mittheilung: Die Elasticitätsfunction des Bindegewebes und die intravitale Messung ihrer Störungen, *Z. Exp. Pathol. Ther.*, 11, 369, 1912.
5. Wilkinson, D.S., Dermatitis from repeated trauma to the skin, *Am. J. Ind. Med.*, 8, 307, 1985.

6. Kanerva, L., Physical causes of occupational skin disease, in *Occupational Skin Disease*, 2nd ed., Adams, R.M., Ed., W.B. Saunders, Philadelphia, 1990, 41.
7. Shmunes, E., Predisposing factors in occupational skin diseases, *Dermatol. Clin.*, 6, 7, 1988.
8. Nicholls, S., King, C.S., and Marks, R., Morphological and quantitative assessment of physical changes in the horny layer in ichthyosis, in *The Ichthyoses*, Marks, R. and Dykes, P.J., Eds., MTP Press, Lancaster, 1978, 95.
9. Williams, M.L. and Elias, P.M., From basket weave to barrier: unifying concepts for the pathogenesis of the disorders of cornification, *Arch. Dermatol.*, 129, 626, 1993.
10. Graves, C.J., Edwards, C., and Marks, R., The occlusive effects of protective gloves on the barrier properties of the stratum corneum, *Curr. Prob. Dermatol.*, 23, 87, 1995.
11. Hendley, A., Marks, R., and Payne, P.A., Measurement of forces for point indentation of the stratum corneum *in vivo*: the influences of age, sex, site delipidisation and hydration, *Bioeng. Skin Newsl.*, 3, 234, 1982.
12. Wiltrup, G., *Undersogelser over Elasticitetsforholdene i Subcutis*, A. Busck, Copenhagen, 1924.
13. Kirk, E. and Kvorning, S.A., Quantitative measurements of the elastic properties of the skin and subcutaneous tissue in young and old individuals, *J. Gerontol.*, 4, 273, 1949.
14. Kirk, J.E. and Chieffi, M., Variation with age in elasticity of skin and subcutaneous tissue in human individuals, *J. Gerontol.*, 17, 373, 1962.
15. Vlasblom, D.C., Skin Elasticity, thesis, University of Utrecht, the Netherlands, 1967.
16. Dikstein, S. and Hartzshtark, A., *In vivo* measurement of some elastic properties of human skin, in *Bioengineering and the Skin*, Marks, R. and Payne, P.A., Eds., MTP Press, Lancaster, 1981, 45.
17. Barbenel, J.C., Gibson, F., and Turnbull, F., Optical assessment of skin blood content and oxygenation, in *Bed Sore Biomechanics*, Kenedi, R.M. and Cowden, J.M., Eds., Macmillan, London, 1976, 83.
18. Peck, S.M. and Glick, A.W., A new method for measuring the hardness of keratin, *J. Soc. Cosmet. Chem.*, 7, 530, 1956.
19. Falanga, V. and Bucalo, B., Use of a durometer to assess skin hardness, *J. Am. Acad. Dermatol.*, 29, 47, 1993.
20. Romanelli, M. and Falanga, V., Use of a durometer to measure the degree of skin induration in lipodermatosclerosis, *J. Am. Acad. Dermatol.*, 32, 188, 1995.
21. Prall, J.K., Instrumental evaluation of the effects of cosmetic products on skin surfaces with particular reference to smoothness, *J. Soc. Cosmet. Chem.*, 24, 693, 1973.
22. Honda, T., Development of a system for static measurement of skin-muscle hardness and a fundamental study on its applications [in Japanese], *Nippon Eiseigaku Zasshi* [*Jpn. J. Hyg.*], 45, 860, 1990.
23. Horikawa, M. et al., Non-invasive measurement method for hardness in muscular tissues, *Med. Biol. Eng. Comput.*, 31, 623, 1993.
24. Ferguson-Pell, M.W., Hagisawa, S., and Masiello, R.D., A skin indentation system using a pneumatic bellows, *J. Rehabil. Res. Dev.*, 31, 15, 1994.
25. Zheng, Y.P. and Mak, A.F.T., An ultrasound indentation system for biomechanical properties assessment of soft tissues *in vivo*, *IEEE Trans. Biomed. Eng.*, 43, 912, 1996.
26. Pathak, A.P. et al., A rate-controlled indentor for *in vivo* analysis of residual limb tissues, *IEEE Trans. Rehabil. Eng.*, 6, 12, 1998.
27. Zheng, Y.P. et al., Biomechanical assessment of plantar foot tissue in diabetic patients using an ultrasound indentation system, *Ultrasound Med. Biol.*, 26, 451, 2000.

28. Tabata, N., Tagami, H., and Kligman, A.M., A twenty-four-hour occlusive exposure to 1% sodium lauryl sulfate induces a unique histopathologic inflammatory response in the xerotic skin of atopic dermatitis patients, *Acta Derm. Venereol.* (Stockholm), 78, 244, 1998.
29. Lafrance, H. et al., Study of the tensile properties of living skin equivalents, *Biomed. Mater. Eng.*, 5, 195, 1995.
30. Nicholls, S. et al., Measurement of point deformation (P.D.) of human skin *in vivo* — contribution of the stratum corneum, *J. Invest. Dermatol.*, 70, 227, 1978.
31. Guibarra, E., Nicholls, S., and Marks, R., Measurement of force required for point indentation of stratum corneum. The effects of hydration, *Bioeng. Skin Newsl.*, 2, 29, 1979.
32. Nicholls, S., Development of Physical Methods for Investigation of Stratum Corneum Structure and Function, thesis, University of Wales, Cardiff, U.K., 1981, 129.
33. Byrne, J., A feedback controlled stimulator that delivers controlled displacements or forces to cutaneous mechanoreceptors, *IEEE Trans. Biomed. Eng.*, 66, 1975.
34. Marks, R. and Dawber, R.P.R., Skin surface biopsy: an improved technique for the examination of the horny layer, *Br. J. Dermatol.*, 84, 117, 1971.
35. Thompson, D.E., Hussein, H.M., and Perritt, R.Q., Point impedance characterization of soft tissues, in *Bioengineering and the Skin*, Marks, R. and Payne, P.A., Eds., MTP Press, Lancaster, 1981, 103.

15 Standardization of Skin Biomechanical Measurements

R. Randall Wickett

CONTENTS

Introduction 179
Form of the Data 180
Instrumental Factors 180
Contact Pressure of the Instrument on Skin 181
Pretension 182
Environmental Factors 182
Subject Factors 183
  Influence of Gender 183
  Race 183
  Body Site 183
  Age 183
  Subject Factors and Control Sites 183
Conclusions 184
References 184

INTRODUCTION

Standardization of measurement methods has been a goal of many researchers working on noninvasive measurement of skin function. For example, Pierard[1] stated, "optimization of noninvasive biophysical measurements should benefit from strict standardization of measurements and frequent calibration of devices." While no absolute standards for skin measurements have been published, helpful guidelines have been published for measurement of transepidermal water loss (TEWL)[2] and the electrical properties of skin for assessment of skin hydration. Cook[4] called for standardization of mechanical measurements 20 years ago, and recently the EEMCO group has begun a series of papers discussing guidelines for skin mechanical measurements.[5,6] Rodrigues[6] illustrates the difficulty in standardizing biomechanical

0-8493-7521-5/02/$0.00+$1.50

measurements of the skin by pointing out that that estimates of Young's modulus of elasticity for skin *in vivo* vary by four orders of magnitude.

There are at least two factors that make standardization of mechanical measurements difficult. As anyone reading this book will realize, there are several commonly used devices for determining the mechanical properties of skin that work by quite different principles. For example, the dermal torque meter will obviously require different standards than the ballistometer. The other problem is that different workers are trying to obtain fundamentally different classes of information from the measurements. Experiments designed to test the efficacy of a moisturizer are likely to require significant differences in procedure from those designed to determine the effects of wound treatments on the mechanical properties of scars or the effects of medical interventions on diseases of the skin. For these reasons, this chapter will not attempt to promulgate standards for each type of measurement under each condition. Instead, the chapter discusses the factors that the author considers important to control as carefully as possible and some principles of standardization that may apply to the general problem of measuring the mechanical properties of skin.

FORM OF THE DATA

The many different instruments provide data in various formats. The two commonly used, commercially available instruments, with which the author is most familiar, the Cutometer® and the Dermal Torque Meter® (DTM), can both provide an output of the general form shown in Figure 15.1. This output results when the deforming force (vacuum or torque) is applied for a period of time and then released.

U_e = elastic deformation of the skin due to the application of stress (vacuum or torque) by the instrument
U_v = viscoelastic creep occurring after the elastic deformation
U_f = total extensibility of the skin
U_r = elastic deformation recovery due to stress removal
U_a = total deformation recovery at the end of the stress-off period
R = amount of deformation not recovered by the end of the stress-off period
U_a/U_f = overall elasticity of the skin including creep and creep recovery
U_r/U_e = pure elasticity ignoring viscoelastic creep
U_v/U_e = ratio of viscoelastic to elastic extension called the viscoelastic ratio
U_r/U_f = ratio of elastic recovery to total deformation

INSTRUMENTAL FACTORS

The majority of instruments for measuring the properties of skin *in vivo* supply a specified deforming force to the skin, and the extent of the resulting deformation is measured. It is not really possible to separate contributions from the various skin layers, but the general principle is to minimize the extent of the deformation when attempting to study effects on the stratum corneum and to use larger deformations

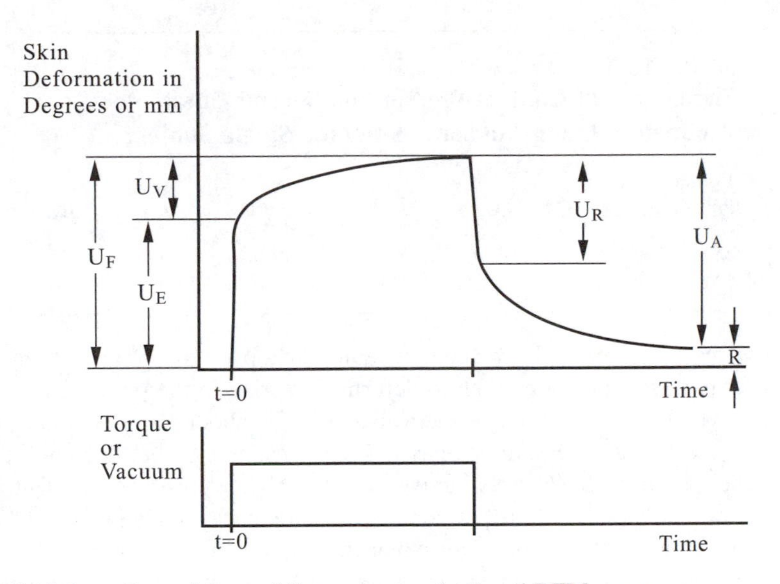

FIGURE 15.1 General form of Cutometer (mode 1) and DTM curves.

when attempting to study the mechanical properties of the dermis. Most studies of the effect of water or moisturizer on skin elasticity with the Cutometer have used the 2-mm-diameter probe.[7–10] The author has found that using a pressure of 200 mbar of negative pressure leads to more sensitivity to moisturizing effects compared with 500 mbar of negative pressure.[11] Pierard et al.[12] performed a systematic study comparing the 2- and 8-mm-diameter Cutometer probes and concluded that the claim that the 2-mm probe is better suited for the study of moisturizers is only justified for pressures of 200 mbar or less. With the DTM the author used the 1-mm ring gap when studying moisturizers[9] or short-term application of water[13] and the 5-mm ring gap when investigating the mechanical properties of scars resulting from treatment of burns with cultured skin substitutes.[14] Although use of lower deformations does lead to more apparent sensitivity of the measurement to effects that should be occurring in the stratum corneum, Diridollou and co-workers[15,16] used a combination of ultrasound and suction measurements to show that both the dermis and even the subcutaneous fat are deformed under 50 mbar of negative pressure with a 3-mm-diameter opening. Thus, it is necessary to keep in mind that at least some contribution to the measurement is coming from lower layers of the skin even when relatively small deformations are used.

CONTACT PRESSURE OF THE INSTRUMENT ON SKIN

Contact pressure is an important variable with virtually all biophysical measurements that are made with skin contact. The effect is easily verified with the Cutometer by making measurements at different contact pressures on the same site. Table 15.1 shows the effects of contact pressure on some of the Cutometer parameters obtained at the same site on a single subject.

TABLE 15.1
The Effects of Contact Pressure on Several Cutometer Parameters Using the Same Site on a Single Subject

Contact Pressure, g	U_e	U_v	U_f	U_a	U_r	R	U_v/U_e
250	0.42	0.27	0.69	0.58	0.46	0.11	0.675
500	0.29	0.22	0.51	0.41	0.30	0.10	0.750

Pressures were varied by placing weights on top of the probe. Pushing the Cutometer probe down just enough to deform the spring to the point that the outer and inner cylinders of the probe are even also results in about 250 g of contact force. Increasing the contact pressure decreased U_e, U_v, U_f, and U_a but did not effect R. The viscoelastic ratio U_v/U_e was increased at the higher pressure. The Cutometer spring allows fairly good control of contact pressure, but practice is required and it is very important to make this point when training operators. It is also important that the person performing the measurements be blind to the treatment so that he or she does not unconsciously influence the measurement through contact pressure. Whenever possible, the same operator should perform all measurements in any given study. With the DTM, the weight of the instrument is usually used to control contact pressure. With other instruments the contact pressure should obviously be controlled as carefully as possible.

PRETENSION

The Cutometer gives the option of applying a brief pretension to the skin. Barel et al.[17] report that using the precondition or pretension mode of the Cutometer can have a significant effect on some of the elastic parameters measured. U_r/U_e and U_r/U_f were found to be higher after pretension was applied indicating that preconditioned skin recovers more of its elastic deformation. If preconditioning is used, it should be reported.

ENVIRONMENTAL FACTORS

The first consideration in performing *in vivo* trials with human subjects is to ensure that the subjects are comfortable and equilibrated to the room conditions. The author's group usually controls temperature at 20 to 22°C and relative humidity at 35 to 45% and allows at least 20 min equilibration time with subjects sitting comfortably in a reclining chair. While mechanical measurements are not as sensitive to skin surface hydration as electrical measurements,[8] they have been shown to have some sensitivity to hydration[7,8,18] so control of relative humidity and equilibration to avoid sweating by the subject are important. If other biophysical measurements are also being made, as they nearly always are, these measurements are usually taken before mechanical assessments. This is especially important with the DTM because the DTM probe is affixed to the skin with double-sided tape.

SUBJECT FACTORS

INFLUENCE OF GENDER

For the most part, studies of the mechanical properties of skin have not found an influence of gender on skin elasticity parameters.[17,19] Pierard et al.[12] investigated the effect of gender on the Cutometer in a study with 100 male and 100 female subjects on forearm skin using both the 2- and 8-mm-diameter probes. The only statistically significant effect of gender was on U_f with the 2-mm-diameter probe. Auriol et al.[7] saw no difference in baseline forearm data but did see a larger effect of short-term hydration in women when using the Cutometer.

RACE

There have been relatively few studies of the effect of race on the mechanical properties of skin. Berardesca et al.[20] reported differences in elastic recovery between the volar and dorsal forearm in white subjects that were not seen in blacks and attributed the difference to protection against elasticity changes due to sun exposure in darker skin. Warrier et al.[21] reported significantly higher elastic recovery on facial skin of whites compared with blacks but no significant difference on the skin of lower legs with the same subjects. They also attributed the difference to elasticity changes due to sun exposure.

BODY SITE

There have been numerous studies demonstrating clear differences in the mechanical properties of skin from different parts of the body,[17,19,22] and it is obvious that this parameter must be controlled.

AGE

One of the major applications of skin mechanical measurements is to the study of both intrinsic and sun-induced aging of the skin,[19,23–30] so it is not surprising that this factor must be controlled and the age range of the subject should be reported.

SUBJECT FACTORS AND CONTROL SITES

When conducting studies that involve treatments, it is often feasible to overcome problems with control of subject factors by using each subject as his or her own control. Measurement made at baseline can be compared to treatment on the same subject at the same site. It is also sometimes possible to maintain a contralateral site as an untreated or placebo-treated control.

CONCLUSIONS

Mechanical measurements of skin function have progressed considerably in the past 20 years. Reliable commercial devices are now available for measuring several different aspects of skin mechanics. With adequate control of instrumental factors, it is possible to obtain valuable data relating to the effects of treatments, disease states, or the natural aging process. Because there are so many different instrumental approaches and different factors to study, it does not seem feasible to set absolute standards. Rather, general guidelines are in order and those practices that are applicable to other bioengineering methods, such as control of instrumental parameters, measurement environment, treatment site, and contact pressure, are also required to achieve good skin biomechanical measurement.

REFERENCES

1. Piérard, G.E.. Relevance, comparison and validation of techniques, in *Handbook of Non-invasive Methods and the Skin*, Serup, J. and Jemec, G.B.E., Eds., CRC Press, Boca Raton, 1995, 9–14.
2. Pinnagoda, J., Tupker, R.A., and Serup, J., Guidelines for transepidermal water loss (TEWL) measurement, *Contact Dermatitis*, 22, 164–178, 1990.
3. Berardesca, E., EEMCO guidance for the assessment of stratum corneum hydration: electrical methods, *Skin Res. Technol.*, 3, 126–132, 1997.
4. Cook, T.H., International standardization of instruments used to determine the mechanical properties of skin, in *Bioengineering and the Skin*, Marks, R. and Payne, P., Eds., Lancaster, MTP Press, U.K., 1981, 123–128.
5. Piérard, G.E., EEMCO guidance to the *in vivo* assessment of tensile functional properties of the skin. Part 1: relevance to the structures and aging of the skin and subcutaneous tissues, *Skin Pharmacol. Appl. Skin Physiol.*, 12, 352–362, 1999.
6. Rodrigues, L., EEMCO guidance to the *in vivo* assessment of tensile functional properties of the skin. Part 2: instrumentation and test modes, *Skin Pharmacol. Appl. Skin Physiol.*, 14, 52–67, 2001.
7. Auriol, F., Vaillant, L., Machet, L., Diridollou, S., and Lorette, G., Effects of short-time hydration on skin extensibility, *Acta Derm. Venereol.* (Stockholm), 73, 344–347, 1993.
8. Murray, B.C. and Wicket, R.R., Sensitivity of Cutometer data to stratum corneum hydration level, *Skin Res. Technol.*, 2, 167–172, 1996.
9. Murray, B.C. and Wicket, R.R., Correlations between Dermal Torque Meter, Cutometer and Dermal Phase Meter measurements of human skin, *Skin Res. Technol.*, 3, 101–106, 1997.
10. Wiechers, J.W. and Barlow, T., Skin moisturization and elasticity originate from at least two different mechanisms, *Int. J. Cosmet. Sci.*, 21, 425–435, 1999.
11. Wickett, R.R., Stretching the skin surface: skin elasticity measurements, *Cosmet. Toiletries*, 116(3), 47–54, 2001.
12. Piérard, G.E., Nikkels-Tassoudji, N., and Piérard-Franchimont, C., Influence of the test area on the mechanical properties of skin, *Dermatology*, 191, 9–15, 1995.
13. He, M.M., Minematsu, Y., Simion, A.S.A., and Wicket, R.R., Two-exponential rheological models that describe skin's response to the Dermal Torque Meter, in preparation, 1999 (Abstr.).

14. Boyce, S.T., Supp, A.P., Wickett, R.R., Hoath, S.B., and Warden, G.D., Assessment with the dermal torque meter of skin pliability after treatment of burns with cultured skin substitutes, *J. Burn Care Rehabil.*, 21, 55–63, 2000.
15. Diridollou, S., Berson. M., Black, D., Gregoire, J.M., Patat, F., and Gall, Y., Subcutaneous fat involvement in skin deformation following suction, presented at 12th International Society for Bioengineering and the Skin, Boston, MA, 1998 (Abstr.).
16. Diridollou, S., Berson, M., Vabre, V. et al., An *in vivo* method for measuring the mechanical properties of the skin using ultrasound, *Ultrasound Med. Biol.*, 24, 215–224, 1998.
17. Barel, A.O., Lambrecht, R., and Clarys, P., Mechanical function of the skin: state of the art, in *Skin Bioengineering Techniques and Applications in Dermatology and Cosmetology*, Elsner, P., Barel, A.O., Berardesca, E., Gabard, B., and Serup, J., Eds., Karger, Basel, 1998, 69–83.
18. Dobrev, H., Use of Cutometer to assess epidermal hydration, *Skin Res. Technol.*, 6, 239–244, 2000.
19. Cua, A.B., Wilhelm, K.P., and Maibach, H.I., Elastic properties of human skin: relation to age, sex, and anatomical region, *Arch. Dermatol. Res.*, 282, 283–288, 1990.
20. Berardesca, E., de Rigal, J., Lévêque, J.L., and Maibach, H.I., *In vivo* biophysical characterization of skin physiological differences in races, *Dermatologica*, 182, 89–93, 1991.
21. Warrier, A.G., Kligman, A.M., Harper, R.A., Bowman, J., and Wicket, R.R., A comparison of black and white skin using noninvasive methods, *J. Soc. Cosmet. Chem.*, 47, 229–240, 1996.
22. Gniadecka, M., Gniadecka, R., Serup, J., and Sondergaard, J., Skin mechanical properties present adaption to man's upright position, *Acta Derm-Venereol.* (Stockh.), 74, 188–190, 1994.
23. Adhoute, H., de Rigal, J., Marchand, J.P., Privat, Y., and Lévêque, J.L., Influence of age and sun exposure on the biophysical properties of the human skin: an *in vivo* study, *Photodermatol. Photoimmunol. Photomed.*, 9, 99–103, 1992.
24. Piérard-Franchimont, C., Letawe, C., Fumal, I., Van, C.I., and Piérard, G.E., Gravitational syndrome and tensile properties of skin in the elderly, *Dermatology*, 197, 317–320, 1998.
25. Daly, C.H. and Odland, G.F., Age-related changes in the mechanical properties of human skin, *J. Invest. Dermatol.*, 73, 84–87, 1979.
26. Escoffier, C., de Rigal, J., Rochefort, A., Vasselet, R., Lévêque, J.L., and Agache, P.G., Age-related mechanical properties of human skin: an *in vivo* study, *J. Invest. Dermatol.*, 93, 353–357, 1989.
27. Lévêque, J.L., Corcuff, P., de Rigal, J., and Agache, P., *In vivo* studies of the evolution of physical properties of the human skin with age, *Int. J. Dermatol.*, 23, 322–329, 1984.
28. Lévêque, J.L., de Rigal, J., Agache, P.G., and Monneur, C., Influence of ageing on the *in vivo* extensibility of human skin at a low stress, *Arch. Dermatol. Res.*, 269, 127–135, 1980.
29. Piérard, G.E., Henry, F., Castelli, D., and Ries, G., Ageing and rheological properties of facial skin in women, *Gerontology*, 44, 159–161, 1998.
30. Piérard, G.E., Kort, R., Letawe, C., Olemans, C., and Piérard-Franchimont, C., Biomechanical assessment of photodamage: derivation of a cutaneous extrinsic ageing score, *Skin Res. Technol.*, 1, 17–20, 1995.

16 Mapping Mechanical Properties of Human Skin

Klaus-P. Wilhelm and Howard I. Maibach

CONTENTS

Introduction 187
Regional Variability of *in Vivo* Elastic Properties 187
Conclusions 193
Acknowledgment 196
References 196

INTRODUCTION

The skin performs numerous functions vital to the maintenance of homeostasis. Faced with a variety of extrinsic factors, such as physical trauma from the sun, wind, temperature changes, and the dynamic movements of the body, as well as the intrinsic aging process, its integrity is well preserved by a complex interaction of the three layers: epidermis, dermis, and subcutaneous tissue.

The epidermis, particularly the stratum corneum layer, provides a barrier to permeability. It is tough and resilient as a result of the presence of fibrous keratin. Evidence for stratum corneum elasticity has been described;[1,2] intraepidermal elasticity has also been associated with the presence of sweat.[3] The relatively thick dermis together with the subcutaneous fat gives the skin its toughness and elasticity, an important mechanical property primarily attributable to the intra- and intermolecular cross-links of collagen, elastic, and reticulin fibers embedded in ground substance.[4,5] The extensibility of the skin, however, shows directional variation characterized as anisotropy.[6,7]

The regional variability and age dependency of the elastic properties of human skin *in vivo* has been examined utilizing a noninvasive suction device.[2]

REGIONAL VARIABILITY OF *IN VIVO* ELASTIC PROPERTIES

Measurements were performed in 33 volunteers on 11 anatomical regions: forehead (center), postauricular, upper arm (inner middle third), volar and dorsal forearm

0-8493-7521-5/02/$0.00+$1.50

(center between wrist and elbow), palm (thenar aspect), abdomen (3 cm above umbilicus), thigh (anterior, upper third), ankle (medial malleoli), upper back (scapula), and lower back (at level of L1). Except for measurements of the abdomen where volunteers were recumbent, all measurements were taken with volunteers seated.

The population was divided into four groups: Group A: 8 young females (age 25.8 ± 3.0; mean ± SD); Group B: 9 older females (age 74.7 ± 5.7); Group C: 8 young males (age 28.3 ± 1.2); and Group D: 8 older males (age 75.1 ± 4.8). All participants were Caucasians with no signs of apparent skin disease.

Skin elasticity was measured with a commercially available suction instrument (Cutometer SEM 474, Courage & Khazaka, Cologne, Germany). This instrument measures the elastic properties of skin based on the principle of suction/elongation, and is described elsewhere in this book. Because of the small test area (suction chamber ≈ 3 mm^2), skin elongation contributable from the dermo-subepidermal junction is considerably smaller than it is from instruments with larger suction chambers, e.g., the Dermaflex A (suction chamber ≈ 80 mm^2).[8]

In this study the time–strain mode was used with 5-s application of three different loads: 100, 200, and 500 mbar, followed by a 5-s relaxation period. The skin deformation was plotted as a function of time. A typical time–strain recording is illustrated in Figure 16.1. The following parameters were recorded: immediate distension (U_e), measured at 0.48 s; delayed distension (U_v); immediate retraction (U_r); and final distension (U_f).[9,10] The deformation curve consists of an initial purely elastic component, followed by a viscoelastic, and a final purely viscous component.[8]

Differences were tested for their statistical significance using Student's t-test for differences between the age groups and by an analysis of variance for regional variation. A p value less than or equal to 0.05 was considered statistically significant.

Uncorrected results of skin elasticity measurements with respect to anatomical region and age for a load of 200 mbar are summarized in Table 16.1. There are

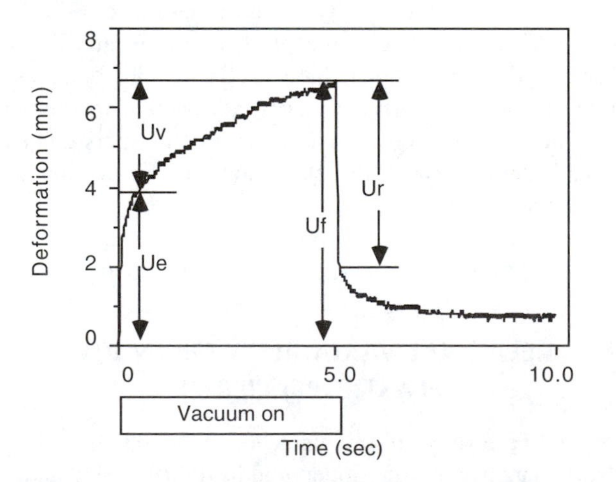

FIGURE 16.1 Typical skin deformation plotted as a function of time. The parameters are immediate distension (U_e), delayed distension (U_v), immediate retraction (U_r), and final distension (U_f).

tremendous differences between various anatomical sites for all parameters: immediate distension (U_e), delayed distension (U_e), immediate retraction (U_r), and final deformation (U_f). All parameters tended to be increased in elderly people at most anatomical regions. These direct elongation parameters are, however, influenced by skin thickness and do not provide immediate information about the elastic properties. Hence, the great difference in the young group between the amount of immediate distension (U_e) between the palm with only 0.06 mm and the postauricular region with 0.57 mm can certainly be largely ascribed to parallel differences in skin thickness, and likewise the observed age differences may be partly caused by skin atrophy in the aged group.

However, certain biologically relevant ratios of these parameters are independent of skin thickness and can be compared between subjects and anatomical regions.[6,8,9,11] For each group, two of these relative parameters independent of skin thickness were calculated: First, U_r/U_f (biological elasticity) was calculated, which measures the ability of skin to regain its initial position after deformation. A value of 1 would indicate 100% elasticity. Decreasing elasticity results in smaller ratios. Second, the viscoelastic proportion of the total distension (U_v/U_e). Increasing values of U_v/U_e indicate an increasing viscoelastic portion of the deformation process, which is controlled by the collagen fibers.

There were significant differences for biological elasticity as well as viscoelastic-to-elastic ratio between anatomical sites and between age groups. Regardless of the load applied, no significant differences (data not shown), however, were noted between the sexes for either ratio for most regions of the body.

The biological elasticity (U_r/U_f) tended to increase with increasing loads in both age groups (Table 16.2, Figure 16.2). Biological elasticity was generally higher on the back (abdomen) and proximal extremities (upper arm, thigh) than at distal parts (ankle, palm). This vertical vector of elasticity has also been confirmed by Gniadecka et al.[12] with a different suction device (Dermaflex A); they compared this decreased vertical vector of elasticity (increased stiffness) with the "congenital antigravity suit" of tall animals, e.g., giraffes. Tall animals exhibit stiffer tissue (skin, fascia, veins) in dependent areas to directly compensate for the increased hydrostatic pressure; the finding of a similar vertical vector of skin elasticity provides evidence for a similar antigravity suit in humans.

Biological elasticity (U_r/U_f) was significantly decreased in the aged group as compared with the young group at most anatomical sites (Figure 16.2). The effect of aging, i.e., photoaging, can also be seen by comparison of different anatomical sites within one age group. For example, sun-exposed areas (dorsal forearm) showed lower biologic elasticity than areas with low sun exposure (ventral forearm) in both age groups (Figure 16.3). A similar difference in biological elasticity was seen between the lower and the upper back.

The adverse effect of chronic sun exposure as well as of acute ultraviolet-irritation has been described previously by Lévêque et al.[13] and by Berardesca et al.[14]

The effect of aging on skin stiffness has not been fully elucidated. It has been proposed that diminished elasticity in aged skin is due to the damage, disintegration, or changes of the structure of elastic fibers[14–16] that appear both in intrinsic and solar

TABLE 16.1
Parameters of Skin Elasticity Measurements at 200-mbar Load

	U_e		U_v		U_r		U_f	
Region	**Young**	**Aged**	**Young**	**Aged**	**Young**	**Aged**	**Young**	**Aged**
Forehead	0.28 ± 0.05	0.43 ± 0.06*	0.12 ± 0.02	0.23 ± 0.05*	0.29 ± 0.06	0.38 ± 0.07	0.11 ± 0.02	0.28 ± 0.05*
Postauricular	0.57 ± 0.10	0.94 ± 0.17*	0.23 ± 0.03	0.42 ± 0.06*	0.63 ± 0.12	0.93 ± 0.20	0.18 ± 0.03	0.43 ± 0.08*
Upper arm	0.33 ± 0.08	0.61 ± 0.08*	0.10 ± 0.02	0.27 ± 0.03*	0.39 ± 0.09	0.73 ± 0.09*	0.04 ± 0.01	0.14 ± 0.02*
Volar forearm	0.14 ± 0.02	0.47 ± 0.08*	0.05 ± 0.01	0.24 ± 0.04*	0.16 ± 0.02	0.56 ± 0.11*	0.03 ± 0.01	0.16 ± 0.03*
Dorsal forearm	0.11 ± 0.02	0.19 ± 0.04	0.03 ± 0.01	0.11 ± 0.03*	0.11 ± 0.02	0.20 ± 0.06	0.04 ± 0.01	0.11 ± 0.02*
Palm	0.06 ± 0.01	0.08 ± 0.01	0.03 ± 0.00	0.04 ± 0.01	0.05 ± 0.01	0.06 ± 0.01	0.03 ± 0.00	0.07 ± 0.01*
Abdomen	0.68 ± 0.14	0.30 ± 0.05*	0.17 ± 0.02	0.15 ± 0.04	0.77 ± 0.15	0.39 ± 0.08*	0.07 ± 0.02	0.07 ± 0.02
Thigh	0.16 ± 0.02	0.30 ± 0.04*	0.04 ± 0.00	0.12 ± 0.02*	0.17 ± 0.02	0.32 ± 0.06*	0.03 ± 0.00	0.04 ± 0.00
Ankle	0.09 ± 0.01	0.25 ± 0.01*	0.03 ± 0.01	0.02 ± 0.00	0.09 ± 0.02	0.04 ± 0.01*	0.03 ± 0.00	0.03 ± 0.00
Upper back	0.24 ± 0.02	0.27 ± 0.04	0.09 ± 0.01	0.18 ± 0.04*	0.29 ± 0.03	0.32 ± 0.06	0.05 ± 0.01	0.13 ± 0.03*
Lower back	0.23 ± 0.04	0.35 ± 0.06	0.07 ± 0.01	0.17 ± 0.03*	0.26 ± 0.04	0.40 ± 0.07	0.03 ± 0.00	0.11 ± 0.03*

Note: A 5-s load (200 mbar) was applied followed by a 5-s relaxation period. U_e (immediate distension), U_v (delayed distension), U_r (immediate retraction), and U_f (final deformation). Generally, parameters were noted to be increased in the aged group at all anatomical regions except for the abdomen and ankle. (Mean ± S.E.M., $n = 13$ to 16 volunteers.)

* Statistically significant difference ($p < 0.05$) between the age groups (Student's t-test).

Source: Modified from Cua, A.B. et al., *Arch. Dermatol. Res.*, 282, 283–286, 1990. With permission.

TABLE 16.2
Elasticity (U_r/U_f) at Different Loads

Region	Load (mbar)	Young			Aged		
		Mean	S.D.	S.E.M.	Mean	S.D.	S.E.M.
Forehead	500	0.72	0.17	0.04	0.66	0.21	0.05
	200	0.68	0.17	0.05	0.61	0.25	0.06
	100	0.64	0.15	0.04	0.53	0.25	0.06
Postauricular	500	0.76	0.15	0.04	0.65	0.22	0.06
	200	0.73	0.14	0.03	0.66	0.22	0.06
	100	0.65	0.18	0.05	0.60	0.18	0.05
Upper arm	500	0.92	0.04	0.01	0.84	0.14	0.04
	200	0.90	0.04	0.01	0.83	0.09	0.02
	100	0.85	0.07	0.02	0.78	0.14	0.03
Volar forearm	500	0.88	0.09	0.02	0.77	0.12	0.03
	200	0.81	0.10	0.03	0.75	0.13	0.03
	100	0.72	0.13	0.04	0.67	0.12	0.03
Dorsal forearm	500	0.77	0.19	0.05	0.64	0.17	0.04
	200	0.72	0.20	0.05	0.58	0.15	0.04
	100	0.74	0.13	0.04	0.48	0.15	0.04
Palm	500	0.66	0.09	0.03	0.53	0.08	0.02
	200	0.60	0.09	0.03	0.45	0.09	0.02
	100	0.53	0.11	0.03	0.45	0.09	0.02
Abdomen	500	0.91	0.06	0.02	0.87	0.11	0.03
	200	0.91	0.05	0.01	0.83	0.11	0.03
	100	0.87	0.08	0.02	0.77	0.14	0.03
Thigh	500	0.92	0.05	0.01	0.88	0.10	0.03
	200	0.83	0.15	0.04	0.85	0.09	0.02
	100	0.82	0.06	0.02	0.79	0.10	0.03
Ankle	500	0.78	0.09	0.02	0.62	0.09	0.02
	200	0.71	0.13	0.04	0.59	0.12	0.03
	100	0.65	0.13	0.04	0.52	0.11	0.03
Upper back	500	0.91	0.07	0.02	0.81	0.08	0.02
	200	0.78	0.24	0.06	0.71	0.14	0.04
	100	0.75	0.14	0.04	0.66	0.11	0.03
Lower back	500	0.90	0.08	0.02	0.86	0.14	0.04
	200	0.88	0.05	0.01	0.80	0.19	0.05
	100	0.80	0.10	0.03	0.70	0.14	0.04

Note: U_r/U_f tended to increase with increasing load in both age groups and was generally lower in the aged than in the young group.

* Statistically significant difference ($p < 0.05$) between the age groups (Student's t-test); (n = 13 to 16 volunteers).

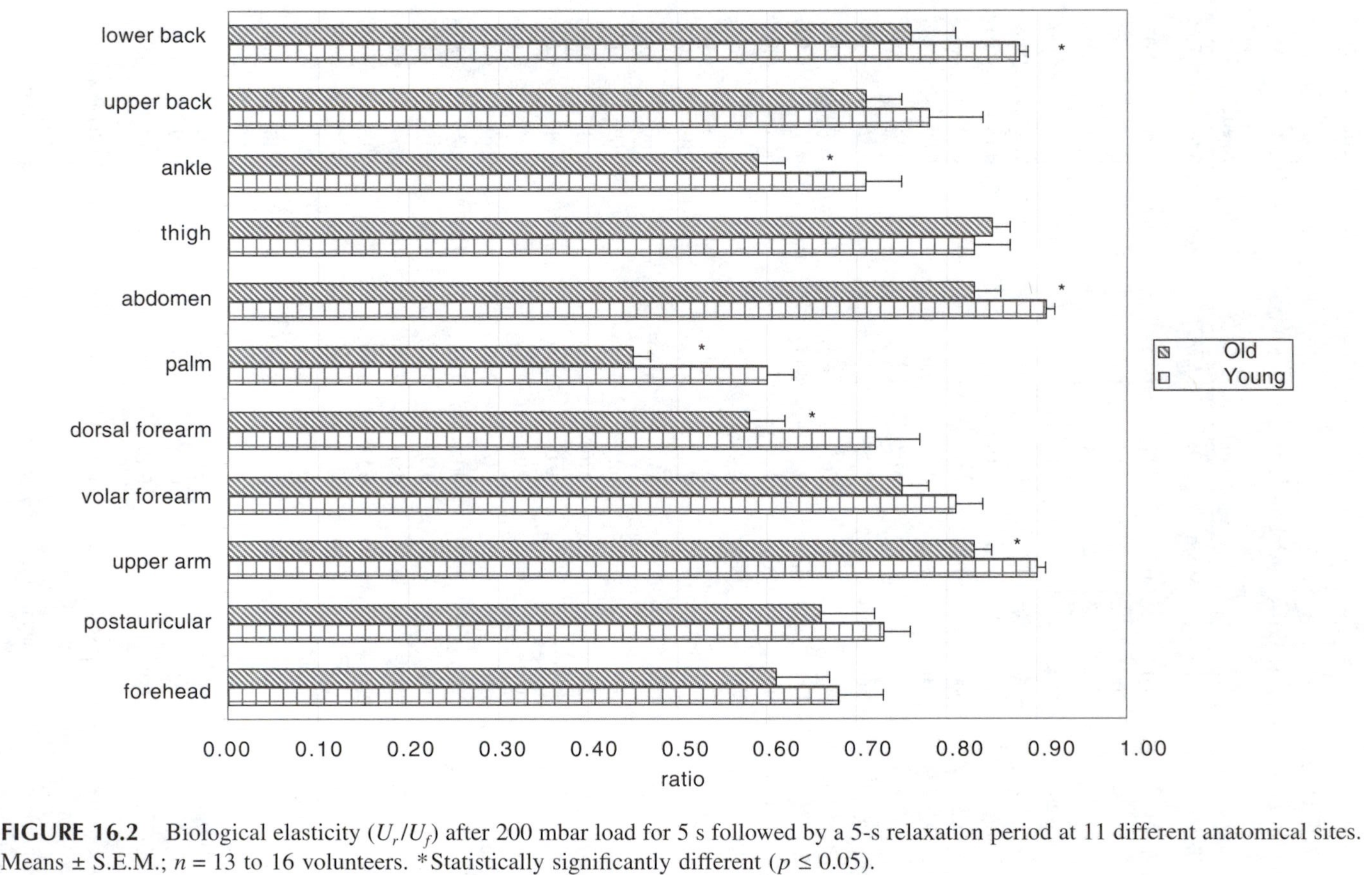

FIGURE 16.2 Biological elasticity (U_r/U_f) after 200 mbar load for 5 s followed by a 5-s relaxation period at 11 different anatomical sites. Means ± S.E.M.; $n = 13$ to 16 volunteers. *Statistically significantly different ($p \leq 0.05$).

aging. This may be caused by accumulation of glycosaminoglycans in the papillary dermis of the actinically damaged skin interfering with the elastic fiber system located in this region.

It has also been proposed that the viscoelastic properties are dependent on alterations in the equilibrium of the "ground substance" of the dermis,[17] and that the content of glycosaminoglycans or soluble collagen may be related to viscosity, decreasing with maturation and age,[18] whereas the biological elasticity (U_r/U_f) should correlate with the function of the elastic fiber network.[8]

In this investigation the viscoelastic-to-elastic ratio (U_v/U_e) was increased with increasing loads and tended to be higher in the aged group for almost all anatomical sites. Age differences were especially apparent when high loads were applied (Table 16.3, Figure 16.4). The ratios as a function of load for the dorsal and volar forearm are illustrated in Figure 16.5.

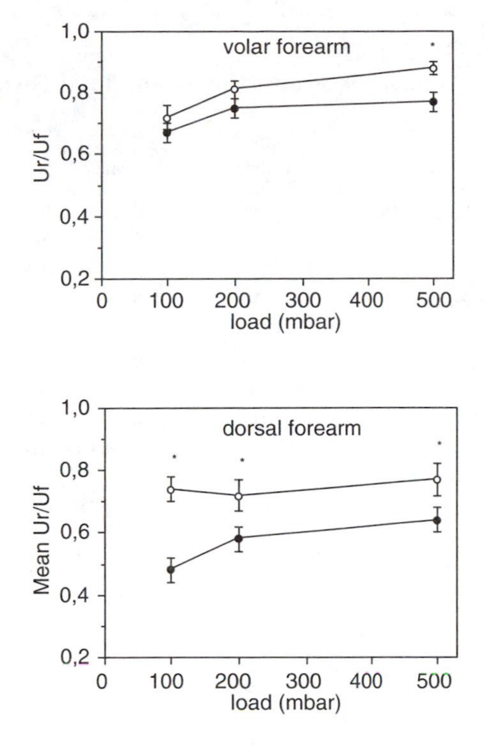

FIGURE 16.3 Influence of load on U_r/U_f on the volar and dorsal forearm in young (open circles) and aged (closed circles) volunteers. Means ± S.E.M.; $n = 13$ to 16 volunteers. *Statistically significantly different ($p \leq 0.05$).

CONCLUSIONS

To avoid such invasive techniques as biopsy, differences in mechanical properties of the skin at various regions of the human body can be evaluated using the suction device. The mechanical properties of the skin of the human body differ greatly between various anatomical sites. The largest differences are present for the direct elongation parameters, which are mostly due to differences in skin thickness. However, there are also differences in those elasticity parameters independent of skin thickness. Knowledge of these differences and expansion of the database will provide useful material for understanding human skin biology and for planning clinical studies appropriately.

TABLE 16.3
Viscoelastic Proportion of the Total Distension (U_v/U_e) at Different Loads

Region	Load (mbar)	Young			Aged		
		Mean	S.D.	S.E.M.	Mean	S.D.	S.E.M.
Forehead	500	0.56	0.26	0.07	0.61	0.28	0.07
	200	0.51	0.24	0.06	0.52	0.29	0.07
	100	0.50	0.30	0.08	0.59	0.35	0.08
Postauricular	500	0.49	0.27	0.07	0.46	0.26	0.07
	200	0.47	0.21	0.05	0.56	0.25	0.06
	100	0.49	0.13	0.03	0.60	0.20	0.05
Upper arm	500	0.43	0.18	0.05	0.42	0.18	0.05
	200	0.33	0.11	0.03	0.48	0.15	0.04
	100	0.26	0.09	0.02	0.38	0.20	0.05
Volar forearm	500	0.59	0.07	0.02	0.55	0.20	0.05
	200	0.36	0.09	0.02	0.53	0.14	0.04
	100	0.31	0.14	0.04	0.39	0.15	0.04
Dorsal forearm	500	0.46	0.17	0.04	0.64	0.18	0.05
	200	0.34	0.23	0.06	0.50	0.26	0.06
	100	0.30	0.15	0.04	0.35	0.15	0.04
Palm	500	0.49	0.19	0.05	0.53	0.14	0.04
	200	0.41	0.12	0.03	0.46	0.20	0.05
	100	0.60	0.47	0.13	0.63	0.44	0.11
Abdomen	500	0.27	0.14	0.04	0.49	0.26	0.07
	200	0.30	0.13	0.03	0.47	0.17	0.04
	100	0.26	0.08	0.02	0.33	0.15	0.04
Thigh	500	0.34	0.09	0.02	0.44	0.11	0.03
	200	0.26	0.15	0.04	0.44	0.11	0.03
	100	0.40	0.27	0.08	0.41	0.23	0.06
Ankle	500	0.42	0.15	0.04	0.52	0.18	0.05
	200	0.34	0.13	0.04	0.39	0.16	0.04
	100	0.44	0.11	0.03	0.52	0.19	0.05
Upper back	500	0.50	0.12	0.03	0.65	0.23	0.06
	200	0.37	0.11	0.03	0.61	0.16	0.04
	100	0.32	0.12	0.03	0.41	0.14	0.04
Lower back	500	0.38	0.12	0.03	0.54	0.25	0.06
	200	0.33	0.15	0.04	0.49	0.28	0.07
	100	0.35	0.38	0.11	0.38	0.35	0.09

Note: U_v/U_e increased with increasing loads and tended to be higher in the aged group. Age differences were especially apparent when high loads were applied.

* Statistically significant difference ($p < 0.05$) between the age groups (Student's t-test); (n = 13 to 16 volunteers).

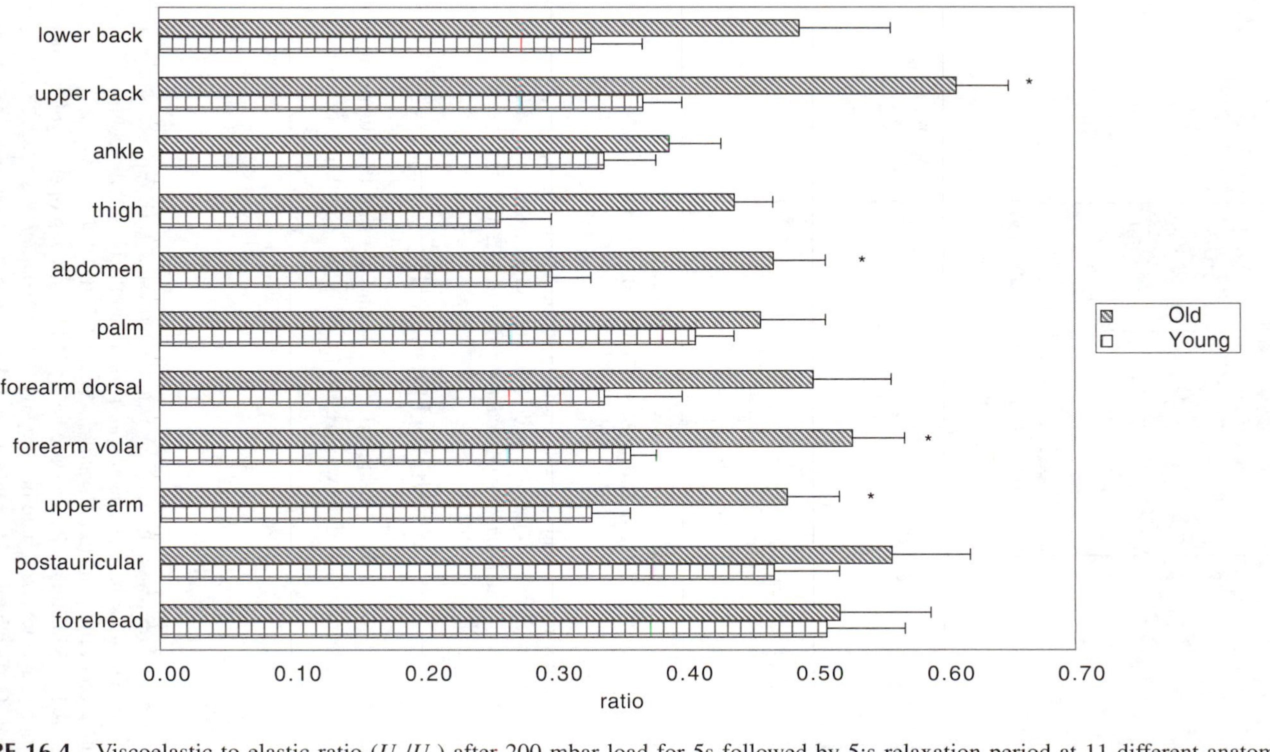

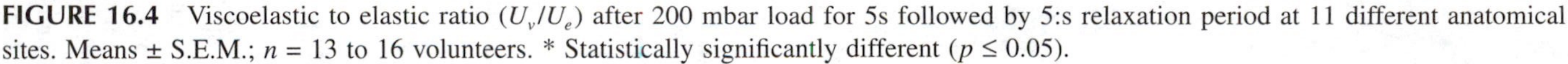
FIGURE 16.4 Viscoelastic to elastic ratio (U_v/U_e) after 200 mbar load for 5s followed by 5:s relaxation period at 11 different anatomical sites. Means ± S.E.M.; $n = 13$ to 16 volunteers. * Statistically significantly different ($p \leq 0.05$).

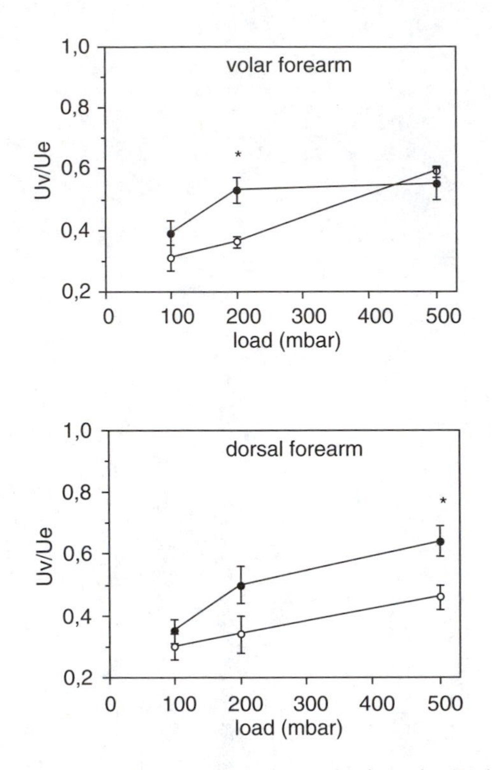

FIGURE 16.5 Influence of load on viscoelastic-to-elastic ratio (U_v/U_e) on the volar and dorsal forearm in young (open circles) and aged (closed circles) volunteers. Means ± S.E.M.; $n = 13$ to 16 volunteers. *Statistically significantly different ($p \leq 0.05$).

ACKNOWLEDGMENT

This article contains substantial material, tables, and figures that were published before by the authors[2] and that was reproduced with the kind permission of Springer-Verlag.

REFERENCES

1. Christensen, M.S., Hargens, C.W., Nacht, S., and Gans, E.H., Viscoelastic properties of intact human skin: instrumentation, hydration effects, and the contribution of the stratum corneum, *J. Invest. Dermatol.*, 69, 282–286, 1977.
2. Cua, A.B., Wilhelm, K.P., and Maibach, H.I., Elastic properties of human skin: relation to age, sex, and anatomical region, *Arch. Dermatol. Res.*, 282, 283–288, 1990.
3. Serban, G., Edelberg, R., Garcia, M., and Hambridge, A., Intraepidermal sweat levels and the viscoelastic properties of human skin, *Bioeng. Skin*, 2(2), 134, 1986.
4. Pierard, G.E.A., Critical approach to *in vivo* mechanical testing of the skin, in *Cutaneous Investigation in Health and Disease*, Lévêque, J.L., Ed., Marcel Dekker, New York, 1989, 215–240.

5. Uitto, J.J., Fazio, M.J., and Olsen, D.R., Molecular mechanisms of cutaneous aging, *J. Am. Acad. Dermatol.*, 21, 614–622, 1989.
6. Escoffier, C., de Rigal, J., Rochefort, A., Vasselet, R., Lévêque, J.D., and Agache P., Age related mechanical properties of human skin. An *in vivo* study, *J. Invest. Dermatol.*, 93, 353–357, 1989.
7. Vogel, H.G., Directional variations of mechanical parameters in rat skin depending on maturation and age, *J. Invest. Dermatol.*, 76, 493–497, 1981.
8. Elsner, P., Wilhelm, D., and Maibach, H.I., Mechanical properties of human vulvar skin, *Br. J. Dermatol.*, 122, 607–614, 1990.
9. Agache, P., Monneur, C., Lévêque, J.L., and de Rigal, J., Mechanical properties and Young's modulus of human skin *in vivo*, *Arch. Dermatol. Res.*, 269, 221–232, 1980.
10. Lévêque, J.L., de Rigal, J., Agache, P.G., and Monneur, C., Influence of aging on the *in vivo* extensibility of human skin at a low stress, *Arch. Dermatol. Res.*, 269, 127–135, 1980.
11. de Rigal, J. and Lévêque, J.L., *In vivo* measurement of the stratum corneum elasticity, *Bioeng. Skin*, 1, 13–23, 1985.
12. Gniadecka, M., Gniadecka, R., Serup, J., and Søndergaard, J., Skin mechanical properties present adaptation to man's upright position, *Acta Derm-Venereol.* (Stockh.), 74, 188, 1994.
13. Lévêque, J.L., Porte, G., de Rigal, J. et al., Influence of chronic sun exposure on some biophysical parameters of human skin, *J. Cutaneous Aging Cosmet. Dermatol.*, 1, 123–127, 1988.
14. Berardesca, E., Borroni, G., Gabba, P., Borlone, R., and Rabbiosi, G., Evidence for elastic changes in aged skin revealed in an *in vivo* extensometric study at low loads, *Bioeng. Skin*, 2, 261–270, 1985.
15. Balin, A.K. and Pratt, L.A., Physiological consequences of human skin aging, *Cutis*, 43, 431–436, 1989.
16. Bouissou, H., Pieraggi, M., Julian, M., and Savit, T., The elastic tissue of the skin: a comparison of spontaneous and actinic (solar) aging, *Int. J. Dermatol.*, 27(5), 327–335, 1988.
17. Daly, C.H. and Odland G.F., Age-related changes in the mechanical properties of human skin, *J. Invest. Dermatol.*, 73, 84–87, 1979.
18. Vogel, H.G., Age dependence of viscoelastic properties in rat skin; directional variations in relaxation experiments, *Bioeng. Skin*, 1(2), 157–173, 1985.

17 Skin Mechanics and Hydration

Tina Holst Larsen and Gregor B. E. Jemec

CONTENTS

Introduction ... 199
The Influence of Changes in Overall Skin Hydration on the Mechanical Properties of Skin ... 201
Moisturizers and Skin Plasticity ... 202
Comments ... 204
References ... 204

INTRODUCTION

The physical properties of the skin vary with location, age, and sex; but in addition to these endogenous factors, skin hydration is also thought to play a role. Skin consists of two main layers: epidermis (thickness 0.07 to 0.012 mm) and dermis (thickness 1 to 4 mm). Data suggest that both contribute to the overall mechanical properties of the skin, but only the epidermis is available for rapid modification by topical agents either directly or indirectly by, for example, reducing water loss. Topical treatment may also affect the dermis secondarily (e.g., by retinoids), although the timeframe is often much longer for this to occur. Finally, the dermis may be affected by generalized changes in the body. The mechanical effects of skin hydration can therefore involve several different structures.

The water barrier is located to the stratum corneum of the epidermis. Removal of the stratum corneum from human skin results in as much as a 50-fold increase in the rate of water loss from the skin surface.[1] This structure also allows for control of the plastic properties of the corneocytes and of the internal aqueous balance of the body. Forslind[2] describes the skin barrier as composed of two major components: a hydrophilic component, the keratin; and a hydrophobic component. The intercellular lipid material represents the lipophilic/hydrophobic constituent. The lipid phase primarily acts as a barrier against water loss from the body but also prevents free access of water and foreign substances into the body. At the same time, it allows a minute loss of water, which is thought to plasticize the stratum corneum under normal circumstances. This baseline water diffusion (*perspiratio insensibilis* or the unnoticed perspiration) amounts to 2.25 $L/m^2/s$ and is distinct and separate from

0-8493-7521-5/02/$0.00+$1.50

sweat gland secretion. The plasticity of the corneocytes is achieved by hydration of the keratin;[3] i.e., water is a plasticizer of keratin.[4] It is also a clinical impression that overall skin plasticity is dependent on sufficient hydration of the epidermal corneocytes. In skin diseases with reduced plasticity, e.g., chronic eczema or psoriasis, secondary features such as fissures may be related to a stiff, thick, and dry stratum corneum. If the corneocyte keratin is deprived of its water, the material becomes brittle, nonelastic, and breaks easily.

Lipids that can form biological membranes are characterized by a common structure: a hydrophobic component, generally fatty acid hydrocarbon chains, and a more or less hydrophilic head. For thermodynamic reasons the hydrophobic parts of the lipid molecules are segregated to form a separate region, whereas the hydrophilic head group faces the surrounding aqueous face.[5] Specific skin lipids may confer specific barrier properties and plasticity to the skin. In addition, Friberg and Osborne[6] have proposed that a liquid crystalline state of the intercellular lipids is required for optimum barrier function. It may be speculated that the state of the intercellular lipids also influence the mechanical properties of the epidermis. The balance between liquid and solid crystalline phases of the lipids is influenced by hydration, proportion of unsaturated fatty acids, and probably other unknown factors. The skin is constantly influenced by physical and chemical agents, which abrade the epidermal surface mechanically and extract or change the composition of the intercellular lipids. Therefore, a continuous renewal of the lipid structures constituting the water barrier is necessary.[2]

The dermis is traditionally divided into the papillary layer and the reticular layer. The papillary layer is adjacent to the epidermis and living cells are most abundant in this part of the skin.[7] The deeper reticular layer contains the ground substance containing glycosaminoglycans. A major component is hyaluronate, which is hydrophilic and binds most of the tissue fluid.[8] The major determining structure for the overall mechanical properties of the skin is the dermis, and in particular the deeper reticular layer. It consists of collagen, elastin, and reticulin fibers embedded in an amorphous gel-like ground substance. The fibers are thought to be responsible for the highly nonlinear behavior when the tissue is being stretched.[9] The ground substance probably plays a major role when the tissue is under compression. The large molecules of hyaluronate with the bound tissue fluid behave like a gel, which can be regarded as a solid. Some authors are of the opinion that there must be a small amount of free movable fluid in the ground substance that is not bound in the gel or the cells.[10,11] According to Oomens et al.[8] the fiber network embedded in the colloid-rich part of the ground substance can be regarded as some fiber-reinforced solid while the free fluid is considered to be a Newtonian fluid. Oomens et al.[8] compare the bulk material to a sponge. When the sponge is compressed, the fluid will be displaced and the viscosity of the fluid will create a resistance against this flow. Furthermore, the size of the pores of the sponge will decrease when the sponge is depressed. Overall, these components result in a nonlinear time-dependent behavior of the bulk material under compression. Skin hydration affects the mechanical properties of dermis as well since water is bound by both ground substance and collagen fibers.

Several structures that are all susceptible to changes in hydration are therefore potentially involved in determining the overall mechanical properties of human skin *in vivo*. Changes induced by externally applied substances may therefore occur either through a direct action on epidermal structures or through modification of the overall skin homeostasis through, e.g., changes in hydration secondary to changes in water evaporation, or through modification of dermal structures. The time–effect relationship of externally applied plasticizing substances can furthermore offer clues to the nature of involved structures, e.g., the slow action of retinoids affecting epidermal as well as dermal structures, or the near immediate effects of water affecting the stratum corneum.

THE INFLUENCE OF CHANGES IN OVERALL SKIN HYDRATION ON THE MECHANICAL PROPERTIES OF SKIN

The overall correlation between untreated skin and stratum corneum hydration has been studied using capacitance measurements (Corneometer) and the suction cup technique (Dermaflex).[12] A poor correlation was found, and it was concluded that in the absence of external modifications, such as emollients, skin capacitance was a poor single predictor of its mechanical properties. This conclusion should, however, be interpreted with caution because of the chosen method for measurement of skin hydration, which exhibits a shallow penetration of the epidermis.[13] Hydration of the deeper epidermis was therefore not measured, which may have contributed to the negative results.

A number of experiments describing the consequences of more or less selective active hydration or dehydration of various compartments of the skin, however, strongly suggest that both affect the mechanical properties of the skin. This has been shown using both shear wave propagation and suction cup methods. The viscoelasticity skin analyzer (VESA) is a fast, noninvasive, user-friendly method to evaluate skin viscoelasticity by the speed of elastic shear wave propagation (SWP). This method does not modify the fine structure of the skin.[14–16]

Vexler et al.[17] induced skin edema in rabbit ears by using croton oil as a model for changes restricted to the dermal layers of the skin, i.e., changes in hydration of the deeper dermal layers. Initially, the SWP was similar in the upper and the middle areas on both ears. After treating the upper area, the SWP was significantly ($p < 0.0001$) elevated in both horizontal and vertical directions indicating the increased stiffness of edematous skin. More-pronounced changes were observed along the long axis of the ear, indicating the increase of the skin anisotropy (directional variations in viscoelasticity) in the edematous areas by roughly 15 to 20%. The results are similar to the reduced skin viscoelasticity observed in the patients with postmastectomy lymphedema.[15] By using the Dermaflex machine, similar effects have also been noted in analogous studies of histamine-induced wheals in human skin.[18]

These experiments suggest that volume changes of the dermis affect the overall mechanical properties of the skin adversely. It is thought that this is a consequence of prestretching the fibers and reducing the possibilities for shifts of the ground substance. These changes cause the skin to lose plasticity, and are in good agreement with crude clinical assessments of overall skin elasticity in edema.

Brazelli et al.[19] made an analogous observation by using hemodialysis as a model to investigate the effects on biophysical properties of the skin induced by removal of fluids and water from the body. Removing water from one compartment to another (from skin to blood) decreases skin thickness after therapy. A significant positive correlation between changes in skin thickness and stratum corneum water content was found. The decrease in the water content of the skin implied an increase in distensibility and a decrease in elasticity. Significant linear correlations were found among stratum corneum water content, skin distensibility, and transepidermal water loss (TEWL).

MOISTURIZERS AND SKIN PLASTICITY

The mechanism of moisturization is not fully understood, although many details have been described. Querleux et al.[20] have used magnetic resonance imaging as a tool for *in vivo* quantification of water content and water behavior in living tissues.[20] Hydration profiles obtained by this method delineate two different structures of stratum corneum: an outer layer where hydration can be altered by external factors and an inner layer where hydration is not altered. Hydration may be altered by water actively penetrating into the skin, or by water accumulation due to reduced evaporation as the water barrier is improved by moisturizers.

Several *in vivo* mechanical studies of the skin have been carried out to describe the effect of moisturizers or their constituents on the mechanical properties of skin. A number of studies using different techniques have been conducted to suggest that the use of moisturizers affects the water content of the stratum corneum. The techniques use different methodological principles either to quantify skin hydration directly or to assess derived surrogate measures of skin hydration.[16,21]

Changes in skin conductance, impedance, or capacitance are frequently used to study epidermal hydration *in vivo*. These methods correlate mutually and have been shown to correlate to tissue water although the presence of other electrically active molecules may influence the measurements.[16,22] Other aspects of dry skin have also been used as surrogate measures of skin hydration. Scaling has been assessed using D-squame tapes as well as image analysis, and roughness has also been considered as a surrogate measure.[16] It has therefore been of obvious interest to study the influence of externally applied substances on the mechanical properties of skin *in vivo*.

Van Duzee[22] examined the elastic modulus of the skin as a function of water content for untreated stratum corneum and stratum corneum treated with urea and LiBr. The modulus was found to be a function of water content, not water activity. To clarify the role of water, Jemec et al.[23] conducted a simple experiment to study the effect of superficial hydration on the mechanical properties of skin *in vivo* using topically applied tap water. The water was applied to the ventral aspect of the forearm of healthy volunteers for 10 or 20 min. Afterward, the skin distensibility, resilient distensibility, elastic retraction, and hysteresis were measured.

Distensibility is the distension of the skin when negative pressure (suction) is first applied in the suction chamber, and resilient distensibility is any elevation of the skin remaining after the first suction. Elastic retraction is the ability of the skin to return to its original position after suction, and hysteresis is the distance that the skin is stretched beyond the distensibility when the suction is repeated over the same area: the creeping phenomenon. Significant increases in distensibility, resilient distensibility, and hysteresis were noted after 20 min of hydration. Most of these findings were already apparent after 10 min. The parameter that showed the greatest change was hysteresis.

The observations were supported by Auriol et al.,[24] who studied skin elasticity before and after 1, 2, 5, and 10 min of hydration by application of tap water. After hydration there was an increase in all rheological parameters linked to elasticity. These modifications occurred from the first minute and increased thereafter.

These observations were followed by studies of possible effects of some moisturizer components. Olsen and Jemec[25] studied the influence of tap water, paraffin oil, ethanol, and glycerin on skin mechanics, since these substances are common ingredients in moisturizers and emollients. Distensibility and hysteresis showed the most-pronounced changes. Water and paraffin oil application caused significant ($p < 0.03$) increases after only 10 min, persisting for at least 10 min following paraffin oil but less following water application. Glycerin had a slow onset of action, but the changes remained even after cessation of the application. Application of ethanol had a negative effect on distensibility ($p < 0.03$). The rapid onset of action suggests that the outermost layers of the epidermis play an important role in skin mechanics, and the observation that ethanol reduces distensibility suggests that the method used is sensitive to a range of changes.

These results suggested that the mechanical studies of short-term effects were viable, and in continuation of these studies Jemec and Wulff[26] therefore measured skin mechanics (distensibility and hysteresis) and capacitance when using six different types of moisturizers on healthy and atopic skin. Baseline mechanical values were similar in the two groups, but capacitance was significantly lower in atopics ($p < 0.001$). The results were that all moisturizers increased distensibility as well as hysteresis of normal and atopic skin, but that atopic skin was more influenced by moisturizers than was normal skin. Furthermore, high-lipid moisturizers caused significantly greater changes in plasticity of normal skin than low lipid moisturizers while no such effect was observed for atopic skin. A similar study was subsequently conducted in healthy skin and reproduced the findings, suggesting that the amount of lipids in the applied moisturizer was correlated with the changes in the mechanical properties of the skin.[27]

Pedersen and Jemec[4] conducted a study of the plasticizing effect of water and glycerin on hysteresis (change in maximal elevation following repeated suction, the creep phenomenon) and distensibility (maximal distension achieved) on human skin *in vivo* using the Dermaflex machine. Water or glycerin was applied to the flexor side of the forearm, regional untreated skin served as baseline, and readings were made after 3, 6, 9, 12, and 15 min. Both substances caused a significant increase in hysteresis, water after 12 and 15 min of hydration ($p < 0.01$), glycerin after 3 min ($p < 0.05$), and the effect continued to the end of the observation period.

No significant differences were seen in distensibility. Because of the rapid onset of action for both substances, the effects are thought to take place in the outermost layers of epidermis.[22–24,26]

Vexler et al.[17] also examined possible changes in viscoelasticity following local application of hydrating creams (two types) and placebo evaluated by VESA measurements in four volunteer women before and at different time intervals after cream application. The upper, middle, and lower area of the forearm were examined. Both hydrating creams significantly reduced skin stiffness (lowered SWP readings), while the placebo had no significant effect. These changes were found only along the long axis of the forearm and persisted for about 3 h.

COMMENTS

Plasticity is reduced in dry skin, and emollients have traditionally been used to ameliorate the discomfort caused by skin dryness. The water content of the stratum corneum may play a role in this reduced plasticity, and hydration has therefore been a focal point for several studies, although it is unlikely that a monocausal relationship exists between water and plasticity of the entire skin.[3] Skin hydration can nevertheless affect the mechanical properties of human skin *in vivo*, and is a factor that must be taken into consideration when measuring the mechanical properties of the skin for other purposes.[28] A body of literature exists to suggest that changes occur both in short-term studies and in studies of a longer duration.

The exact underlying mechanisms are not known but are thought primarily to involve epidermal structures because of the responsiveness of the tissue and the presence of the epidermal water barrier. Other substances may have similar effects with a longer lead time — e.g., various lipids and absorbed substances such as retinoids are known to affect a wider variety of skin structures through established biological mechanisms. Because the mechanisms are not known, and because the loss of plasticity is such a prominent feature of dry skin, mechanical measurements may be particularly pertinent to the study of dry skin.

Using the mechanical properties as a functional measure of the effects of skin hydration, however, is not an established technique, and additional methodological work is required to verify and support the methods used in the discussed studies. In particular, the predictive value — both over time and in relation to subjective measures — of the published observations remains to be described, as do the underlying biological mechanisms. Objectively measuring the mechanical properties of human skin in relation to dry skin and hydration may therefore prove even more interesting in the future.

REFERENCES

1. Montagna W., *The Anatomy and Physiology of the Skin*, Academic Press, New York, 1964.
2. Forslind, B., A domain mosaic model of the skin barrier, *Acta Derm-Venereol.* (Stockh.), 74, 1, 1994.

3. Blank, I.H., Further observations on factors which influence the water content of stratum corneum, *J. Invest. Dermatol.*, 21, 259, 1953.
4. Pedersen, L.K. and Jemec, G.B.E., Plasticising effect of water and glycerin on human skin *in vivo*, *J. Dermatol. Sci.*, 19, 48, 1999.
5. Iraelachvili, J.N., Marcelja, S., and Horn, R.G., Physical principles of membrane organization, *Q. Rev. Biophys.*, 13, 121, 1980.
6. Friberg, S.E. and Osborne, D.W., Small angle X-ray diffraction patterns of stratum corneum and a model structure for its lipids, *J. Disp. Sci. Technol.*, 6, 485, 1985.
7. Brown, I.A., Structural Aspects of the Biomechanical Properties of Human Skin, Ph.D. thesis, University of Strathclyde, Glasgow, 1971.
8. Oomens, C.W.J., Van Campen, D.H., and Grootenboer, H.J., A mixture approach to the mechanics of skin, *J. Biomech.*, 20, 877, 1987.
9. Lanir, Y., Constitutive equations for fibrous connective tissues, *J. Biomech.*, 16, 1, 1983.
10. Guyton, A.C., Granger, H.J., and Taylor, A.E., Interstitial fluid pressure, *Physiol. Rev.*, 51, 527, 1971.
11. Wiederheim, C.A., The interstitial space, in *Biomechanics; Its Foundation and Objectives*, Fung, Y.C., Perrone, N., and Anliker, M., Eds., Prentice-Hall, Englewood Cliffs, NJ, 1972.
12. Jemec, G.B.E. and Serup J., Epidermal hydration and skin mechanics, *Acta Derm-Venereol.* (Stockh.), 70, 245, 1990.
13. Blichman C. and Serup J., Assessment of skin moisture. Measurement of electrical conductance, capacitance and transepidermal water loss, *Acta Derm-Venereol.* (Stockh.), 68, 284, 1988.
14. Buras, E.M. and Dorogi, P.L., Skin, biomechanics of, in *Encyclopedia Medical Devices Instrumentation*, Webster, J.G., Ed., Wiley-Interscience, New York, 1988, 2625–2631.
15. Mridha, M., Odman, S., and Oberg, P.A., Mechanical pulse wave propagation in gel, normal and oedematous tissues, *J. Biomech.*, 25, 1213, 1992.
16. Serup, J. and Jemec, G.B.E., Eds., *Handbook of Non-Invasive Methods and the Skin*, CRC Press, Boca Raton, FL, 1995.
17. Vexler, A., Polyanski, I., and Gorodetsky, R., Evaluation of skin viscoelasticity and inisotropy by measurements of speed of shear wave propagation with viscoelasticity skin analyzer, *J. Invest. Dermatol.*, 113, 732, 1999.
18. Serup, J. and Northeved, A., Skin elasticity in psoriasis, *J. Dermatol.*, 12, 3128, 1985.
19. Brazelli, V., Borroni, G., Vignoli, G.P. et al., Effects of fluid volume changes during hemodialysis on the biophysical parameters of the skin, *Dermatology*, 1994, 113, 1994.
20. Querleux, B., Richard, S., Bittoun, J. et al., *In vivo* hydration profile in skin layers by high-resolution magnetic resonance imaging, *Skin Pharmacol. Appl. Physiol.*, 7, 210, 1994.
21. Jemec, G.B.E., Na, R., and Wulf, H.C., The inherent capacitance of moisturising creams — a source of false positive results? *Skin Pharmacol. Appl. Physiol.*, 13, 182, 2000.
22. Van Duzee, B.F., The influence of water content, chemical treatment and temperature on the rheological properties of stratum corneum, *J. Invest. Dermatol.*, 71, 140, 1978.
23. Jemec, G.B.E., Jemec, B., Jemec, B.I.E., and Serup J., The effect of superficial hydration on the mechanical properties of human skin *in vivo*: implications for plastic surgery, *Plast. Reconstr. Surg.*, 85, 100, 1990.
24. Auriol, F., Vaillant, L., Machet, L. et al., Effects of short-time hydration on skin extensibility, *Acta Derm-Venereol.* (Stockh.), 73, 344, 1993.

25. Olsen, O.L. and Jemec, G.B.E., The influence of water, glycerin, paraffin oil and ethanol on skin mechanics, *Acta Derm-Venereol.* (Stockh.), 73, 404, 1993.
26. Jemec, G.B.E. and Wulf, H.C., The plasticising effect of moisturisers on human skin *in vivo*: a measure of moisturising potency? *Skin Res. Technol.*, 1998, 88, 1998.
27. Jemec, G.B.E. and Wulf H.C., The correlation between lipidisation and the plasticising effect of moisturisers, *Acta Derm-Venerol.* (Stockh.), 79, 115, 1999.
28. Rodrigues, L., EEMCO guidance to the *in vivo* assessment of tensile functional properties of the skin, *Skin Pharmacol. Appl. Skin Physiol.*, 14, 52, 2001.

18 Skin Tensile Strength in Scleroderma

Gerald E. Piérard , T. Hermanns-Lê, and C. Piérard-Franchimont

CONTENTS

Introduction 207
Structure of the Connective Tissue 208
Clinical Assessment of Scleroderma 209
Instrumental Assessment of the Skin Tensile Strength 209
Assessment of Therapeutic Efficacy 211
References 212

INTRODUCTION

Assessment of the mechanical properties of skin is useful for evaluation of the severity and therapeutic response of a series of connective tissue disorders including scleroderma.[1] Indeed, a great contrast may exist between the tensile strength of normal and diseased skins. The whole of healthy skin exhibits both flexibility and relative resistance to deformation, thus permitting body movement and allowing temporary compression and distention of a part. Once the deformation ceases, elasticity permits the skin to return spontaneously and progressively to its initial shape. As such, progressive resistance to deformation, flexibility, and elasticity must be adequately balanced to fulfill the ideal tensile strength of the skin. In general, it is reasonable to consider that one is unlikely to be far wrong in interpreting the tensile properties on whole skin as representing the properties of the dermis and subcutaneous tissues. The behavior of the three-dimensional fibrous meshworks is the combination of the response of the components to which the deformation is applied. Such properties represent the functional expression of the structural configuration of the connective tissue from the molecular level to the microscopic level.[1–3]

In reality, the variation in mechanical properties of the skin due to body region, age, and gender outweighs the variability in biochemical composition of skin connective tissues.[1] As a result, the *in vivo* tensile strength of the skin is notoriously complex to assess precisely and may be difficult to interpret, particularly in pathological conditions.[1] This is largely due to the composite structure of the

0-8493-7521-5/02/$0.00+$1.50

dermo-hypodermal tissues. The typical mechanical response of each macromolecular component is variable within the wide range of force intensity encountered in life. This obviously adds to the difficulties in interpreting data and unraveling the tensile strength complexity of the skin.

The understanding of the skin mechanical properties in scleroderma is so complex that it is difficult to obtain anything approaching the sort of clean result one obtains from the establishment of the molecular composition of the dermis. Indeed, the correct interpretation of *in vivo* mechanical measurements requires a knowledge of the overall microstructure of the connective tissue, including the arrangement of collagen and elastic fibers, and their relation to other biological components.

STRUCTURE OF THE CONNECTIVE TISSUE

In general, there is a clear interdependence between the structure and mechanical functions of connective tissues.[1–3] Although the basic properties of the individual fibrous components are quite well identified *in vitro*, the relationship is somewhat complicated *in vivo*. The varying wavy to stretched presentation of fibers at rest governs part of the mechanical properties. Looseness or compactness of the networks is also important. The situation is further clouded by the presence of physical entanglements between the diverse fibrous structures.

The bulk of the reticular dermis in scleroderma consists of a dense network of collagen fibrils and fibers, the organization of which conditions the mechanical stability of the tissue and its resistance to deformation.[4] The collagen fibrils that are identified by electron microscopy are packed into fiber units visible in the light microscope. Primary and secondary fiber bundles denote assemblies of these fibers. How fibers are packed and held together is important with regard to the mechanical properties of the tissue. Cross-links between collagen molecules are not the primary cause for aggregation of fibrils, but rather the interfibrillar matrix plays an important role here. On a higher hierarchical level, the coarse collagen bundles are connected to each other, and most often closely packed and stretched in superimposed planes grossly parallel to the skin surface.[4] The elastic fibers present in smaller amounts normally serve to restore deformed collagen bundles to a more relaxed and wavy position. This is hardly possible in scleroderma. So-called reticulated fibers are organized in an open network around the bundles of collagen fibers. This multicomponent system of fibers is permeated by a more or less hydrated matrix composed of proteoglycans and glycoproteins in which are located cells of diverse phenotypes. These cells are responsible for maintaining and remodeling the macromolecular matrix. They may exert a contractile activity upon the fibrous networks through the intervention of adhesion molecules.[5,6]

Both the thickness and intimate structure of the dermis play prominent and independant mechanical roles. For example, a major role is sometimes ascribed to the structure, density, and location of the elastic fibers when measuring the tensile strength involved in discrete skin deformations.[3,7] In fact, although it is well acknowledged that these fibers exhibit typical elastic properties, any change in the overall elastic functional properties of the skin cannot be ascribed solely to them. Many other structures participate in elasticity, and it is practically impossible to disentangle their respective contributions in the tensile response of the skin.[1]

CLINICAL ASSESSMENT OF SCLERODERMA

Scleroderma is a heterogeneous group of diseases with distinct clinical manifestations. The main types are the localized forms (morphea, monomelic) and the systemic forms. The clinical classification of systemic scleroderma distinguishes three types. Type I is acroscleroderma including a subgroup of the Raynaud's syndrome. Type II is acroscleroderma with progression of sclerosis to proximal areas of the limbs. Type III is diffuse scleroderma usually starting on the trunk with rapid progression to other areas including the extremities.

Skin involvement in scleroderma is characterized by hardness and hidebinding as a result of massive deposition and straightening of collagen bundles. Temporary dermal edema may occur. Skin induration and tethering can be evaluated according to skin severity scoring obtained by clinical palpation. There exists an overlap between the different skin scorings such as the "skin thickness index" and the "skin tethering/hidebound index." The skin thickness/hidebinding is scored in a 4-point scale (0: normal, 1: mild, 2: moderate, 3: severe). It is generally accepted that the sclerotic skin changes correlate with the overall disease activity and prognosis of systemic sclerosis.[8–10] Unfortunately, such assessments remain subjective with low intra- and interobserver reproducibility.[8–16] In addition, they are not free of bias. According to studies, the modified Rodnan skin score has been calculated from 17, 26, or 74 body areas.[8,9,15]

INSTRUMENTAL ASSESSMENT OF THE SKIN TENSILE STRENGTH

In general, the tensile strength of scleroderma skin has been studied noninvasively by distinct methods applying forces in parallel or perpendicular fashion to the skin surface. Stretching, indentation, elevation, suction, torsion, and vibration devices can afford sound although sometimes distinct or even contradictory information.[1,17] The real value of measuring the tensile strength *in vivo* may be sometimes uncertain because of confounding factors that are not always suspected clinically. These include regional variations in the tensile strength and variability due to body posture, age, gender, and cumulative ultraviolet exposure.[1]

To add complications, tensile strength of the whole skin is normally time dependent. Indeed, when skin is stressed by a load, the overall response of the organ to alter its shape is nonlinear. A rapid elastic extension normally takes place at first to give way to a viscoelastic phase with much less extension. When the load is maintained at a particular level for a period, further extension, known as mechanical creep or viscous extension, gradually takes place. Such phenomena, representing the previous stress history, also occur when a series of stresses are consecutively applied and removed. When the strain is maintained for days and weeks, tissue remodelings progressively occur and are responsible for a biological creep corresponding to a relaxation distinct from the mechanical creep. Because of the above considerations, time-dependent tensile strength depends to some extent on the rate at which the load is applied, the duration for which it is sustained, and the previous stress history, as well as the preconditioning of the site. In scleroderma, biological creep and mechanical creep are reduced, particularly during the sclerotic stage.

The quantitative assessment of skin toughness in scleroderma may appear of interest for monitoring both the progression of the disease and the effectiveness of therapies. Surprisingly, only a few studies on the mechanical properties of skin in patients with scleroderma have been carried out until now. They were performed using different devices without any standardized protocols. Hence, results are difficult to compare, even if there is overall agreement concerning the basic and specific changes in the viscoelastic properties of skin.

Noninvasive bioengineering techniques such as high-frequency ultrasound[18–22] and possibly nuclear magnetic resonance imaging[23] are well suited to provide reliable and sensitive data. It is beyond doubt that the various tissue thicknesses and volumes affect the overall tensile strength of the skin in scleroderma. However, it is hazardous and probably misleading to use the dermal thickness measured using ultrasound methods as a single severity criterion for scleroderma. Indeed, although the dermis is often thickened, variations in time may be large. When the collagen and elastic fibers are stretched at rest during the induration phase, cutaneous edema subsides and skin thickness may decrease.[22]

Hidebinding is the most impressive and typical change in scleroderma[4,9] and is not directly related to dermal thickening. In fact, it is difficult to pinch the sclerotic skin into a fold. Only a few noninvasive bioengeneering methods can adequately assess the deep fixation and induration of the skin.[1,2,4,15,16,24–39] Among them, the suction method is a close imitation of lifting the skin between the fingers to assess hidebinding adequately. Indeed, an important aspect of the global tensile functional properties of cutaneous tissues is the tethering of skin to the fat layer. In scleroderma it limits the free mobility of the dermis.[1,2,40]

The dermis is not a homogeneous structure because it exhibits tremendous differences between its different layers, and also with the variable severity of solar elastosis. In addition, there is ample evidence that the volume of the hypodermis and of the eventual sebaceous lobules is of equal, if not greater, importance than the dermal thickness itself in its influence on the overall tensile functional properties. In fact, the fat-enriched lobules made of either sebocytes or adipocytes place the dermis under tension, limiting the mobility of skin.[1] As a consequence of these uncertainties, the calculation of ratios between *in vivo* tensile stress and tissue strain (Young's modulus, shear modulus, bulk modulus) can only be viewed as theoretical concepts with probably little relevance when applied *in vivo* to the complex composite of structures making skin and its subcutaneous tissue. This is particularly true in scleroderma.

Skin involvement in systemic scleroderma begins on the acral portion of the limbs, where it usually remains more prominent than on other body sites. It would be conceivable to choose other specific target sites for biomechanical assessments. However, the physical properties of normal skin vary greatly depending on the body site, and computation of many measurements should be performed before recommending a specific body site other than the forearm for comparative assessments of acroscleroderma.[37]

The size of the test area influences the data, which may appear conflicting in some instances. The superficial dermis of scleroderma can be atrophic, and skin distension measured on a small area (2-mm measuring probe) increases. By contrast, hidebinding the skin to underlying tissues is responsible for low distensibility when measured on a larger surface area (8-mm measuring probe). The resistance to vertical stress is essentially due to the dermis rather than the hypodermis, although the relative contribution of each is not easily distinguished. The subcutaneous fat compartment is likely more involved by high suction pressure applied to large surface areas than by lower suction pressures to smaller surface areas. Because the main alterations in scleroderma are localized in the reticular dermis and hypodermis, a large measuring probe and a high suction force might appear to yield more relevant information. At all stages of scleroderma, skin distensibility decreases. This functional characteristic is linked to the stretching of the collagen fibers and reflects the skin induration and the deep tethering. The viscoelastic ratio is highly variable among patients and even in time in a given patient. This aspect may be ascribed in part to fluctuations in dermal edema and load in proteoglycans. The immediate retraction ability following the suction release is often decreased.[36] The biological elasticity is either unmodified or increased, particularly during the terminal sclerotic phase of the disease.[2,37]

It is acknowledged that the 2- and 8-mm probes of the suction device do not measure precisely the same aspect of the tensile properties of skin.[41] Their use in combination in the same patients allows one to somewhat disentangle the contribution of the different layers of the cutaneous connective tissue.[38] Using such a combined approach, the trends in biomechanical alterations of skin appear to be alike in all forms of the disease. Skin changes are then best identified by the decrease in the value of the 8-mm/2-mm skin distensibility ratio.[37]

Skin viscoelasticity can be assessed by the suction method by determining the ratio U_v/U_e. However, such a value is arbitrarily determined by choosing the time of measurement during the suction phase. Hence, this ratio is not standardized among the different investigators. Another way to assess skin viscoelasticity consists of measuring the speed of propagation of acoustic waves. The shear wave propagation shows physiological variability and directional differences on normal skin in relation to the mechanical anisotropy responsible for the Langer's lines and the resting tension lines.[42–44] According to the authors' experience, such evaluations are sensitive and can be used to monitor the edema severity of scleroderma.

ASSESSMENT OF THERAPEUTIC EFFICACY

Objective evaluation of skin stiffening has obvious advantages over subjective clinical rating. It may be added to the range of biological and functional evaluations already in use for estimating internal organ involvement by scleroderma. The therapeutic efficacy may be markedly different on the various organs affected by scleroderma. For instance, improvement in pulmonary function is not usually correlated with recovery of skin mobility.[37]

A decrease in skin stiffness does not occur spontaneously or under various ancillary treatments. Only a few clinical studies have reported improvements of scleroderma. They are hardly confirmed when using the criteria of evidence-based medicine. Because of the natural course of acroscleroderma, any therapy resulting in stabilization of the skin involvement should be regarded as beneficial. A drug yielding a prominent regression of cutaneous sclerosis is not yet available.

REFERENCES

1. Piérard, G.E. and the EEMCO Group, EEMCO guidance to the *in vivo* assessment of tensile functional properties of the skin. Part 1: Relevance to the structures and ageing of the skin and subcutaneous tissues, *Skin Pharmacol. Appl. Skin Physiol.,* 12, 352, 1999.
2. Piérard, G.E. and Lapière, Ch.M., Physiopathological variations in the mechanical properties of skin, *Arch. Dermatol. Res.*, 260, 231, 1977.
3. Oxlung, H., Relationship between the biomechanical properties, composition and molecular structure of connective tissues, *Connective Tissue Res.*, 15, 65, 1986.
4. Piérard, G.E., Piérard-Franchimont, C., and Lapière, Ch.M., Les compartiments conjonctifs dans les sclérodermies. Etude de la structure et des propriétés biomécaniques, *Dermatologica*, 170, 105, 1985.
5. Delvoye, P., Mauch, C., Krieg, T. et al., Contraction of collagen lattices by fibroblasts obtained from patients and animals with heritable disorders of connective tissue, *Br. J. Dermatol.,* 115, 139, 1986.
6. Greiling, D. and Thieroff-Ekerdt, D., 1α, 25-Dihydroxyvitamin D3 rapidly inhibits fibroblast-induced gel contraction, *J. Invest. Dermatol.,* 106, 1236, 1996.
7. Agache, P., Monneur, C., Lévêque, J.L. et al., Mechanical properties and Young modulus of human skin *in vivo*, *Arch. Dermatol. Res.,* 269, 221, 1980.
8. Rodnan, G.P., Lipinski, E., and Luksick, J., Skin thickness and collagen content in progressive systemic sclerosis and localized scleroderma, *Arthritis Rheum.,* 22, 130, 1979.
9. Clements, P.J., Lachenbruch, P.A., Cheng, N.S. et al., Skin score: a semi-quantitative measure of cutaneous involvement that improves prediction of prognosis in systemic sclerosis, *Arthritis Rheum.,* 33, 1256, 1990.
10. Atman, R.D., Medger, T.A., Bloch, D.A. et al., Predictors of survival in systemic sclerosis (scleroderma), *Arthritis Rheum.,* 4, 403, 1991.
11. Kahaleh, M.B., Suttany, G.L., Smith, E.A. et al., A modified sclerodermal skin scoring method, *Clin. Exp. Rheumatol.*, 4, 367, 1986.
12. Clements, P.J., Lachenbruch, P.A., Seibold, J.R. et al., Skin thickness score in systemic sclerosis: an assessment of interobserver variability in 3 independent studies, *J. Rheum.,* 20, 1892, 1993.
13. Clements, P.J., Lachenbruch, P.A., Seibold, J.R. et al., Inter and intraobserver variability of total skin thickness score (modified Rodnan TSS) in systemic sclerosis, *J. Rheum.,* 22, 1281, 1995.
14. Pope, J.E., Baron, M., Bellamy, N. et al., Variability of skin scores and clinical measurements in scleroderma, *J. Rheum.,* 22, 1271, 1995.
15. Enomoto, D.N.H., Mekkes, J.R., Bossuyt, P.M.M. et al., Quantification of cutaneous sclerosis with a skin elasticity meter in patients with generalized scleroderma, *J. Am. Acad. Dermatol.,* 35, 381, 1996.

16. Seyger, M.B., van den Hoogen, F.H.J., de Boo, T. et al., Reliability of two methods to assess morphea: skin scoring and the use of a durometer, *J. Am. Acad. Dermatol.,* 37, 793, 1997.
17. Rodrigues, L. and the EEMCO group, EEMCO guidance of the *in vivo* assessment of tensile functional properties of skin. Part 2: Instrumentation and test modes, *Skin Pharmacol. Appl. Skin Physiol.,* 14, 52, 2001.
18. Cole, G.W., Handler, S.J., and Burnett, K., The ultrasonic evaluation of skin thickness in scleredema, *J. Clin. Ultrasound,* 9, 501, 1981.
19. Serup, J., Quantification of acrosclerosis, measurement of skin thickness and skin-phalanx distance in females with 15 MHz pulsed ultrasound, *Acta Derm-Venereol.* (Stockh.), 64, 35, 1984.
20. Myers, L.M., Cohen, J.S., Sheets, P.W. et al., B-mode ultrasound evaluation of skin thickness in progressive systemic sclerosis, *J. Rheum.,* 13, 577, 1986.
21. Hoffman, K., Gerbaulet, U., el-Gammal, S. et al., 20-MHz B-mode ultrasound in monitoring the course of localized scleroderma (morphea), *Acta Derm-Venereol.* (Stockh.), 164, S3, 1991.
22. Scheja, A. and Akesson, A., Comparison of high frequency (20 Mhz) ultrasound and palpation for the assessment of skin involvement in systemic sclerosis (scleroderma), *Clin. Exp. Rheum.*, 15, 283, 1997.
23. Richard, S., Querleux, B., Bittoun, J. et al., *In vivo* proton relaxation times analysis of the skin layers by magnetic resonance imaging, *J. Invest. Dermatol.,* 97, 120, 1991.
24. Bluestone, R., Grahame, R., Holoway, V. et al., Treatment of systemic sclerosis with D penicillamine. A new method of observing the effects of treatment, *Ann. Rheum. Dis.,* 29, 153, 1970.
25. Serup, J. and Northeved, A., Skin elasticity in localized scleroderma (morphea), *J. Dermatol.*, 170, 105, 1985.
26. Bjerring, P., Skin elasticity measured by dynamic admittance: a new technique for mechanical measurements in patients with scleroderma, *Acta Derm-Venereol.* (Stockh.), 120, S84, 1986.
27. Piérard, G.E., Histological and rheological grading of cutaneous sclerosis in scleroderma, *Dermatologica*, 179, 18, 1989.
28. Ballou, S.P., Mackiewicz, A., Lysiekiewicz, A. et al., Direct quantitation of skin elasticity in systemic sclerosis, *J. Rheumatol.,* 17, 790, 1990.
29. Kalis, B., de Rigal, J., Léonard, F. et al., *In vivo* study of sclerodermal by noninvasive techniques, *Br. J. Dermatol.,* 122, 785, 1990.
30. Falanga, V. and Bucalo, B., Use of a durometer to assess skin hardness, *J. Am. Acad. Dermatol.*, 29, 47, 1993.
31. Humbert, P., Dupond, J.L., Agache, P. et al., Treatment of scleroderma with oral 1,25-dihydroxyvitamin D3: evaluation of skin improvement using non-invasive techniques, *Acta Derm-Venereol.* (Stockh.), 73, 449, 1993.
32. Herrick, A.I., Gush, R.J., Tully, M. et al., A controlled trial of the effect of topical glyceryl nitrate on skin blood flow and skin elasticity in scleroderma, *Ann. Rheum. Dis.,* 53, 212, 1994.
33. Humbert, P., Delaporte, E., Dupond, J.L. et al., Treatment of localized scleroderma with oral 1,25-dihydroxyvitamin D3, *Eur. J. Dermatol.*, 4, 21, 1994.
34. Reisfeild, P.L., A hard subject: use of a durometer to assess skin hardness, *J. Am. Acad. Dermatol.,* 31, 515, 1994.
35. Aghassi, D., Monson, T., and Braverman, I., Reproducible measurements to quantify cutaneous involvement in scleroderma, *Arch. Dermatol.,* 131, 1160, 1995.

36. Ishikawa, T. and Tamura, T., Measurement of skin elastic properties with a new suction device (II): systemic sclerosis, *J. Dermatol.,* 23, 165, 1996.
37. Nikkels-Tassoudji, N., Henry, F., Piérard-Franchimont, C. et al., Computerized evaluation of skin stiffening in scleroderma, *Eur. J. Clin. Invest.,* 26, 457, 1996.
38. Piérard-Franchimont, C., Nikkels-Tassoudji, N., Lefebvre, P. et al., Subclinical skin stiffening in adults suffering from type 1 diabetes mellitus. A comparison with Raynaud's syndrome, *J. Med. Eng. Technol.,* 22, 206, 1998.
39. Drobev, H.P., *In vivo* study of skin mechanical properties in patients with systemic sclerosis, *J. Am. Acad. Dermatol.,* 40, 436, 1999.
40. Viatour, M., Henry, F., and Piérard, G.E., Computerized analysis of intrinsic forces in the skin, *Clin. Exp. Dermatol.,* 20, 308, 1995.
41. Piérard, G.E., Nikkels-Tassoudji, N., and Piérard-Franchimont, C., Influence of the test area on the mechanical properties of skin, *Dermatology,* 191, 9, 1995.
42. Vexler, A., Polyansky, I., and Gorodetsky, R., Evaluation of skin viscoelasticity and anisotropy by measurement of speed of shear wave propagation with viscoelasticity skin analyzer, *J. Invest. Dermatol.*, 113, 732, 1999.
43. Hermanns-Lê, T., Jonlet, F., Scheen, A., and Piérard, G.E., Age- and body mass index-related changes in cutaneous shear wave velocity, *Exp. Gerontol.*, 36, 363, 2001.
44. Nizet, J.L., Piérard-Franchimont, C., and Piérard, G.E., Influence of body posture and gravitational forces on shear wave propagation in the skin, *Dermatology,* 202, 177, 2001.

19 Mechanical Properties in Other Dermatological Diseases

Hristo P. Dobrev

CONTENTS

Introduction 216
General Description of the Instruments and Skin Mechanical Parameters 216
Mechanical Properties of the Skin in Dermatological Diseases 218
Ehlers–Danlos Syndrome 218
Mechanical Properties of the Skin 218
Interpretation of the Changes of Skin Mechanical Parameters 218
Correlation between Clinical Assessment and Mechanical Parameters of the Skin 219
Diagnostics by Means of Measurements of Skin Mechanics 219
Scleredema of Buschke 219
Mechanical Properties of the Skin 219
Interpretation of the Changes of Skin Mechanical Parameters 219
Correlation between Clinical Assessment and Mechanical Parameters of the Skin 220
Diagnostics by Means of Measurements of Skin Mechanics 220
Monitoring of Progress and Treatment Response of the Disease... 221
Psoriasis Vulgaris 222
Mechanical Properties of the Skin 222
Interpretation of the Changes of Skin Mechanical Parameters 222
Correlation between Clinical Assessment and Mechanical Parameters of the Skin 223
Diagnostics by Means of Measurements of Skin Mechanics 223
Monitoring and Comparison of Treatment Response of the Disease 224
Erysipelas of the Lower Legs 224
Mechanical Properties of the Skin 224
Interpretation of the Changes of Skin Mechanical Parameters 225

0-8493-7521-5/02/$0.00+$1.50

Correlation between Clinical Assessment and Mechanical Parameters of the Skin 225
Monitoring of Progress and Treatment Response of the Disease... 225
Lymphedema of the Lower Limbs 226
Mechanical Properties of the Skin 226
Interpretation of the Changes of Skin Mechanical Parameters 226
Monitoring of Progress and Treatment Response of the Disease... 226
Differential Diagnostics by Means of Skin Elasticity Measurements........ 226
Conclusions 227
Acknowledgment 227
References 228

INTRODUCTION

Human skin, as a complex multi-layered organ, has three major mechanical properties:[1,2]

- Stiffness, i.e., resistance to change of shape;
- Elasticity, i.e., ability to recover the initial shape after deformation;
- Viscoelasticity, i.e., time-dependent deformation with a "creep" phenomenon and nonlinear stress–strain properties with "hysteresis."

These properties are altered in dermatological diseases, which are accompanied with pathological induration or softening of the skin. Noninvasive bioengineering measurements allow quantifying the alterations of skin mechanics *in vivo*. On the other hand, each mechanical parameter determined is related to the corresponding structure and composition of the skin and thus gives information regarding which components are affected.

This chapter describes the mechanical properties of the skin in dermatological diseases such as Ehlers–Danlos syndrome, scleredema of Buschke, psoriasis vulgaris, erysipelas, and lymphedema of the lower limbs and discusses the possibilities for diagnostics, differential diagnostics, and disease monitoring using bioengineering measurements.

GENERAL DESCRIPTION OF THE INSTRUMENTS AND SKIN MECHANICAL PARAMETERS

Mechanical properties of the skin have been studied by Twistometer,* Dermaflex A®,** and Cutometer®† SEM 474. These instruments differ with regard

* L'Oreal, France. The instrument based on L'Oreal invention that is now commercially available is the Dermal Torque Meter®, Dia-Stron, Hampshire, U.K.

** Cortex Technology, Hadsund, Denmark.

† Courage and Khazaka, Cologne, Germany.

to measuring principle, direction of the mechanical stress, size of the test area, and mechanical parameters determined (Table 19.1).[2–5] Moreover, most studies have been performed using different measuring schemes (Table 19.2). Despite that, possibilities for comparison of general information obtained exist. Basic principles, which must be considered, are as follows:[2–6]

- The small-size measuring probes determine the mechanical properties of the epidermis and papillary dermis, whereas large measuring probes determine those of the whole skin.
- Mechanical parameters are related to the structure and composition of the skin and provide information about its three major mechanical properties. Skin distensibility parameters are linked to the stretching of collagen and elastic fibers and reflect the skin thickness and rigidity. Elastic parameters are related to the function of elastic fibers. Viscoelastic parameters are attributed to the water content of the skin and displacement of the interstitial fluid throughout the fibrous network in the dermis.
- Mechanical parameters determined are absolute and relative. Only the relative parameters are independent of the skin thickness and can be directly compared between subjects, anatomical regions, and time points.

Table 19.1 shows the comparable mechanical parameters of the skin determined by means of the three devices.

TABLE 19.1
Devices and Mechanical Parameters of the Skin

Device	Twistometer	Dermaflex	Cutometer
Measuring principle	Torsion	Suction	Suction
Test-area size	1, 3, and 5 mm	10 mm	2, 4, 6, and 8 mm
Stress direction	Plane	Vertical	Vertical
Mechanical Property	**Mechanical Parameters of the Skin**		
Distensibility	U_e	TD	U_e, U_f
Elasticity	U_r/U_e	RER	U_a/U_f, U_r/U_f, U_r/U_e, U_r
Viscoelasticity	U_v	H, RD	U_v, U_v/U_e, R8, H, R

Legend: U_e, immediate distension (elastic deformation); TD, tensile distensibility = U_f, final distension; RER, relative elastic retraction = U_a/U_f, gross elasticity; U_r/U_e, net elasticity; U_r/U_f, biological elasticity; U_r, immediate retraction; U_v, delayed distension (viscous deformation); U_v/U_e, viscoelastic to elastic ratio, H; hysteresis; R8, viscopart; RD, resilient distension = R, residual skin elevation (residual deformation) after the release of the first suction. (For details see previous book chapters.)

TABLE 19.2
Measurement Schemes

Disease	Device	Probe (test-size), mm	Mode[a]	Measuring Scheme	Ref.
Ehlers–Danlos syndrome	Twistometer	3	1	—	in 3
	Cutometer	4	1	500 mbar, 5s/5s, 3 cycles	7
Scleredema of Buschke	Cutometer	2; 8	1	400 mbar, 5s/5s, 1 cycle	9–11
Psoriasis	Dermaflex	10	2	300 mbar, 3s/3s, 9 cycles	12
	Cutometer	2; 8	1	450 mbar, 5s/3s, 5 cycles	13
Erysipelas	Cutometer	2; 8	1	450 mbar, 5s/5s, 3 cycles	14
		2; 8	2	450 mbar, 45 mbar/s, 10s/10s/5s, 3 cycles	11
Lymphedema	Cutometer	6	1	300 mbar, 3s/3s, 3 cycles	15

[a] Mode 1 = suction phase with constant negative pressure and relaxation phase at vacuum 0. Mode 2 = suction phase with linearly rising negative pressure followed by linearly decreasing negative pressure and relaxation phase at vacuum 0.

MECHANICAL PROPERTIES OF THE SKIN IN DERMATOLOGICAL DISEASES

EHLERS–DANLOS SYNDROME

Ehlers–Danlos syndrome (EDS) is a group of nine heritable connective tissue diseases of collagen metabolism characterized by clinical evidence for skin hyperextensibility.

Mechanical Properties of the Skin

Bramont et al. (cited in Reference 3) investigated the volar forearm skin of five patients with type 2 EDS by means of skin caliper and Twistometer. They found a significantly lower skinfold thickness and an increased skin extension (U_e) in patients than in controls, whereas elastic modulus and the elasticity ratio U_r/U_e were found unaltered. The authors concluded that thinning of the skin is responsible for the clinically observed hyperextensibility.

Later, Henry et al.[7] determined mechanical properties on the volar forearm skin in 17 children with type 1 to 3 EDS using a Cutometer equipped with a 4-mm probe. They found a prominent increase both in skin extensibility (U_e and U_f) and elasticity (U_a/U_f, U_r/U_f, and U_r/U_e). The viscoelastic-to-elastic ratio (U_v/U_e) was decreased due to the raised U_e and lower U_v. Children with type 1 EDS were the most affected, whereas those with type 3 EDS had nearly normal skin.

Interpretation of the Changes of Skin Mechanical Parameters

Since distensibility of the skin inversely correlates to its thickness and rigidity, the increase of U_e and U_f in EDS suggests thinning of the skin as well as some alterations in collagen tissue. On the other hand, the increased elasticity parameters suggest that pathological processes do not involve elastic tissue. This is in accord with histological examinations. Skin is atrophic in type 1 EDS; the collagen bundles are thin and rare in the dermis and hypodermic septae, whereas elastic tissue is relatively increased.[8] The alterations are much less evident in the mitis types of EDS. Data of bioengineering measurements correspond with these findings.

Correlation between Clinical Assessment and Mechanical Parameters of the Skin

Clinical evaluation of patients with EDS shows that skin hyperextensibility is severe in disease type 1, moderate in type 2, and mild in type 3. This impression is confirmed by Cutometer measurements.

Diagnostics by Means of Measurements of Skin Mechanics

Diagnosis of different types of EDS is based mostly on clinical criteria. Objective measurements of skin mechanical properties could be used for early detection and evaluation of the degree of skin involvement.

Scleredema of Buschke

Scleredema of Buschke (SB) is a rare disease of unknown etiology, which is characterized clinically by nonpitting induration of the skin.

Mechanical Properties of the Skin

The authors studied four patients with SB and one patient with SB associated with multiple myeloma.[9–11] Measurements were made by means of Cutometer equipped with 2- and 8-mm measuring probes over 9 anatomical regions. Affected skin showed lower skin distension (U_e and U_f) and higher viscoelastic parameters (U_v/U_e, R8) in comparison with the skin of healthy controls. Elastic parameters (U_a/U_f, U_r/U_f, U_r/U_e) were preserved as measured by the 2-mm probe and relatively increased as measured by the 8-mm probe (Figure 19.1). The changes were more expressive at infraclavicular regions, upper arms, and volar forearms (Figure 19.2).

Interpretation of the Changes of Skin Mechanical Parameters

The decreased skin distensibility parameters indicate an increased thickness and hardness of the skin. This is due to the thickening of the collagen bundles and enlarged volume of the subcutaneous tissue as a result of a deposition of glucosaminoglycans in the dermis. The viscoelastic-to-elastic ratio (U_v/U_e) is raised at the expense of highly decreased U_e. Values of the elasticity parameters suggest that the elastic tissue is not altered. On the other hand, the recovery of the skin is facilitated by the existing separation of collagen bundles from each other by spaces filled with mucin and reduced friction between fibers. This explains the relative increase of U_a/U_f, U_r/U_e, and U_r/U_f as measured by the 8-mm probe.

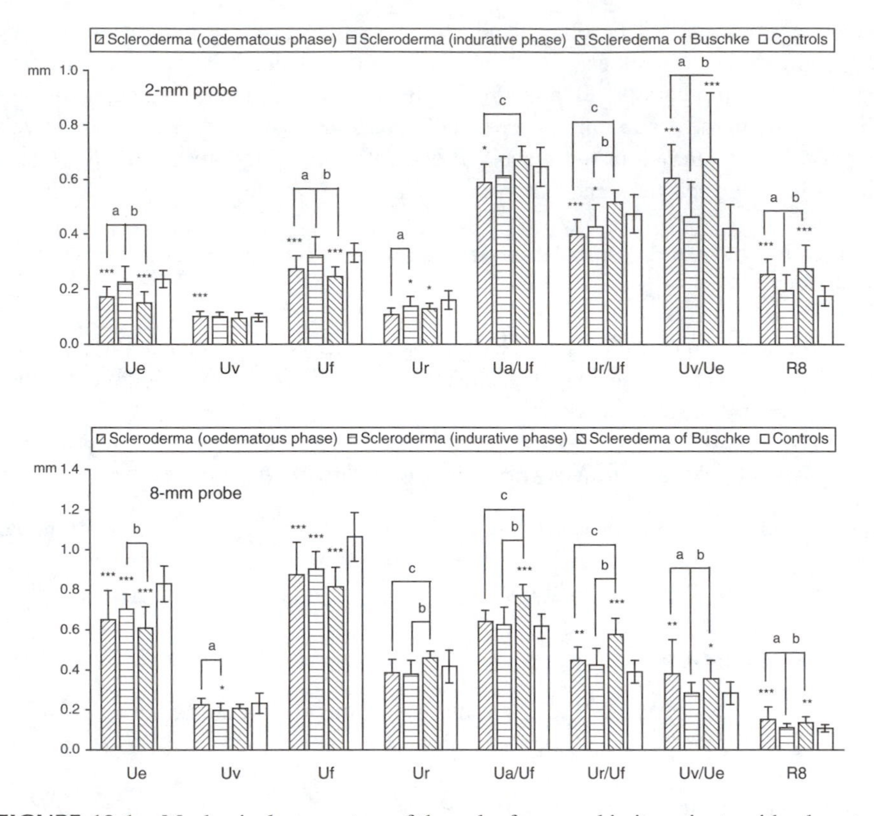

FIGURE 19.1 Mechanical parameters of the volar forearm skin in patients with edematous phase of scleroderma ($n = 14$), indurative phase of scleroderma ($n = 16$), SB ($n = 5$), and healthy controls ($n = 30$) measured by Cutometer. The average values and standard deviations are shown. t-test between patients and controls $*p < 0.05$, $**p < 0.01$, $***p < 0.001$; a, significant differences between edematous and indurative phase of scleroderma (t-test); b, significant differences between indurative phase of scleroderma and SB (t-test); c, significant differences between edematous phase of scleroderma and SB (t-test).

Correlation between Clinical Assessment and Mechanical Parameters of the Skin

A significant inverse relationship was found between the severity score of skin induration and the parameters U_e, U_v, U_f, and U_r. Low values of extensibility correlated with severe skin induration (Figure 19.3).

Diagnostics by Means of Measurements of Skin Mechanics

Bioengineering measurements of skin elasticity are very sensitive. They can detect minimal changes in skin mechanics and thus can be used for early diagnosis of SB. This is of special importance in patients with paraproteinemia. Because the pathological process involves the dermis and subcutaneous tissue, the larger measuring probes, which evaluate the viscoelastic properties of the whole skin, are more suitable

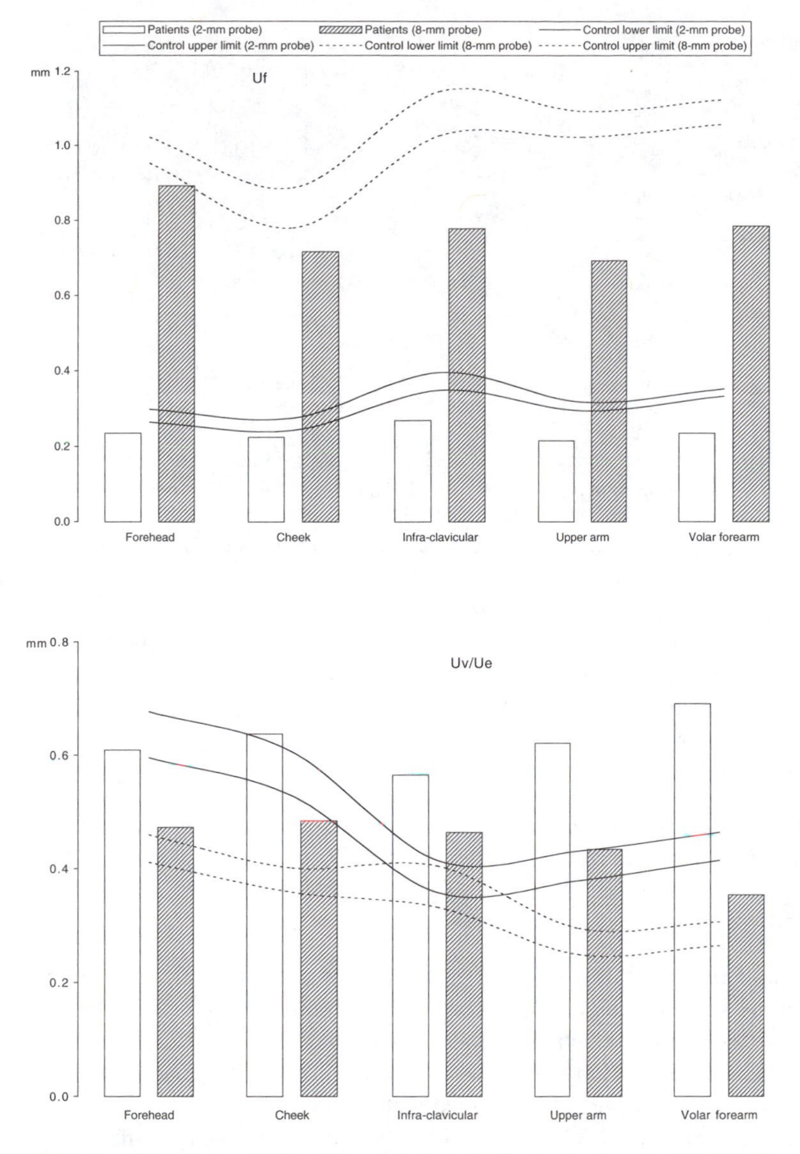

FIGURE 19.2 Skin distensibility (U_f) and viscoelastic-to-elastic ratio (U_v/U_e) in patients with SB ($n = 5$) measured by Cutometer at different anatomical regions compared with 95% confidence intervals of healthy controls ($n = 20$).

than the smaller probes. Skin distensibility (U_f) and viscoelastic-to-elastic ratio (U_v/U_e) are the most indicative mechanical parameters to skin induration in SB. Between them an inverse correlation exists.

Monitoring of Progress and Treatment Response of the Disease

The increase of skin distensibility (U_f) and the decrease of viscoelastic-to-elastic ratio (U_v/U_e) indicate a reduction of skin thickness, i.e., an improvement of the

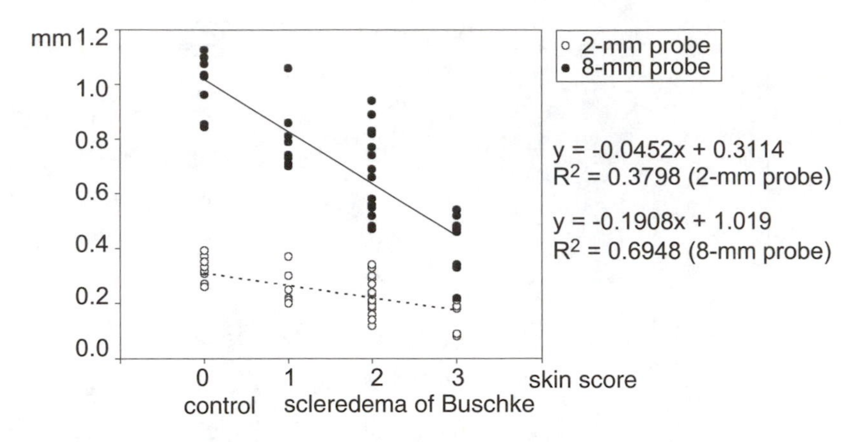

FIGURE 19.3 Correlation of final distension (U_f) with clinical skin score in SB. (From Dobrev, H., *Acta Derm-Venereol.* (Stockh.), 78, 103–106, 1998. With permission.)

disease. Repeated measurements in patients with SB suggest that the traditional treatment approaches cannot essentially influence the disease course. Mechanical parameters initially worsen and then gradually and slowly normalize.[9,11] These data show that SB is a self-limited disease. In a patient with an association of SB and multiple myeloma, Grudeva and Dobrev[10] reported an improvement of skin involvement and mechanics after six-course myeloma polychemotherapy.

Psoriasis Vulgaris

Psoriasis is a genetically determined, inflammatory, and proliferative disease of the skin, which is characterized clinically by erythematous scaly plaques.

Mechanical Properties of the Skin

Serup and Northeved[12] measured 23 plaques in ten patients suffering from psoriasis vulgaris by means of high-frequency ultrasound and Dermaflex A. They reported that skin thickness, hysteresis (H), and resilient distension (RD) were increased, whereas tensile distension (TD) was decreased in psoriatic plaques as compared with regional control sites of unaffected skin.

The authors studied 82 psoriatic plaques in 19 in-patients with a Cutometer equipped with 2- and 8-mm diameter probes.[13] Using both probes, the plaques showed lower skin distensibility (U_e and U_f) and elasticity (U_a/U_f and U_r/U_f), and higher viscoelastic-to-elastic ratio (U_v/U_e) compared with adjacent apparently normal skin. Delayed distension (U_v) and hysteresis (H) measured by the 2-mm-diameter probe were decreased, but were increased when measured by the 8-mm-diameter probe (Figure 19.4).

Interpretation of the Changes of Skin Mechanical Parameters

In respect to biomechanics, stiffness and softness simultaneously characterize the psoriatic skin. The first is due to epidermal proliferation, whereas the second is due

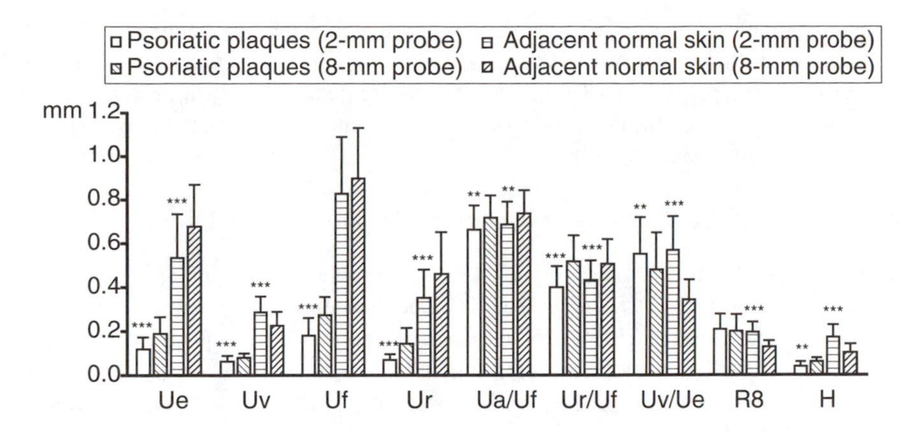

FIGURE 19.4 Mechanical parameters of psoriatic plaques ($n = 82$) and adjacent normal skin measured by Cutometer. The mean values and standard deviation are shown. t-test between psoriasis and controls **$p < 0.01$, ***$p < 0.001$. (Adapted from Dobrev, H., *Acta Derm. Venereol.* (Stockholm), 80, 263–266, 2000. With permission.)

to vasodilatation and edema in the papillary dermis. The rigidity of the epidermis and the stretched dermal fiber network at rest reduce immediate distension (U_e) and final distension (U_f) of the skin. Additional degenerative changes in elastic fibers are most likely responsible for the decrease in skin elasticity parameters. The viscoelastic-to-elastic ratio (U_v/U_e) is increased at the expense of the decreased U_e. Delayed distension (U_v) and hysteresis (H) are attributed to the displacement of the interstitial fluid throughout the fibrous network. These parameters as measured by the 2-mm-diameter probe are decreased, whereas they are increased as measured by the 8-mm-diameter probe. This discrepancy can be attributed to the different aspects of skin mechanics that the two probes measure. The increased epidermal thickness of the psoriatic plaques restricts the depth of measurement with the small-diameter probe. The values of U_v and H measured by the 8-mm-diameter probe reflect solely the availability of dermal edema in the psoriatic skin.

Correlation between Clinical Assessment and Mechanical Parameters of the Skin

The relationships between Cutometer mechanical parameters and clinically assessed induration of the psoriatic plaques were influenced by the diameter of the measuring probe.[7] Using both probes, an inverse correlation between the induration and skin distensibility of the psoriatic plaques was established. Severe induration significantly correlated with higher U_v/U_e (8-mm probe).

Diagnostics by Means of Measurements of Skin Mechanics

Measurements performed by means of Cutometer (2-mm probe) established statistically significant lower values of skin distensibility (U_e and U_f) and higher values of viscoelasticity parameters (U_v/U_e and H) of the uninvolved volar forearm skin in psoriatic patients compared with the skin of age-matched healthy controls.[13] Besides the suggestion that the normal-appearing psoriatic skin is not entirely normal, this demonstrates the sensitivity of Cutometer measurements. More studies are needed to elucidate the value of noninvasive skin elasticity measurements for diagnosis and prediction of psoriasis.

Monitoring and Comparison of Treatment Response of the Disease

After treatment with dithranol for 11 to 27 days, the mechanical parameters of psoriatic plaques altered toward those of adjacent control skin.[13] The changes in U_f and U_v/U_e were most indicative for the improvement of psoriasis. Between them, an inverse correlation exists. An increase of U_f and a decrease of U_v/U_e accompany the thinning of the plaques. These parameters can be used for comparison of different treatment methods.

Using the Cutometer, Dobrev[11] compared two methods for occlusive therapy of plaque psoriasis. The first method included an application of 0.05% betamethasone dipropionate cream under hydrocolloid occlusive dressing for 7 days with three changes of the dressing. The second method included a 12-h daily application of ointment containing 1.5% coal tar, 0.02% flumethasone pivalate, and 1% salicylic acid under plastic occlusion for 21 days. The first method produced a significant increase in skin distensibility (U_f) and a decrease in viscoelastic-to-elastic ratio (U_v/U_e) of the plaques at the first week, whereas the second method provoked a gradual improvement with a maximum at day 21.

Erysipelas of the Lower Legs

Erysipelas is a streptococcal infection of the dermis and upper subcutaneous tissue with typical clinical features. Inflammatory dermal edema results in thickening of the skin and an alteration of its mechanical properties.

Mechanical Properties of the Skin

The authors investigated 25 patients with erysipelas of the lower legs by means of a Cutometer equipped with 2- and 8-mm-diameter probes.[11,14] Two measuring modes were applied. Mode 1 included a suction phase with an application of constant negative pressure and a relaxation phase at vacuum 0. Mode 2 included a suction phase with an application of linearly rising negative pressure followed by a linearly decreasing negative pressure and relaxation phase at vacuum 0. Using measuring mode 1, affected skin was characterized by increased parameters U_v, U_f, R, U_v/U_e, H and decreased U_e, U_r, U_a/U_f, U_r/U_f (Figure 19.5). Using measuring mode 2, all parameters determined by Cutometer software were higher on the indurated skin except for R8. The parameters of final skin distension and hysteresis were universal for the two measurement modes. They showed identical alterations. When the Cutometer is used, the application of measuring mode 2 does not provide any advantage and only burdens the vacuum pump of the device. The changes in skin mechanical parameters were more pronounced with the large than with the small measuring probe.

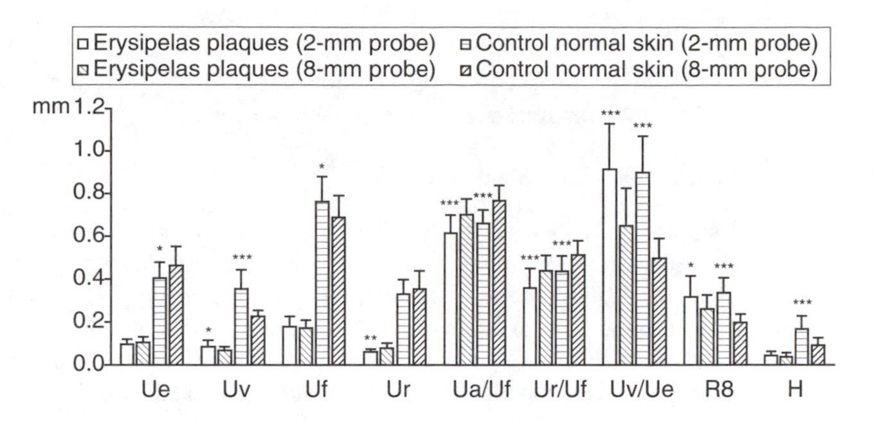

FIGURE 19.5 Mechanical parameters of erysipelas plaques ($n = 25$) and control normal skin measured by Cutometer. Measuring mode 1 is applied. The mean values and standard deviation are shown. t-test between erysipelas and controls $*p < 0.05$, $**p < 0.01$, $***p < 0.001$. (Adapted from Dobrev, H., *Skin Res. Technol.*, 4, 155–159, 1998. With permission.)

Interpretation of the Changes of Skin Mechanical Parameters

Erysipelas plaques are characterized by intense edema and vascular dilatation in the dermis. Therefore, the large-diameter measuring probe is more appropriate for evaluation of the disease. The collagen and elastic fibers are stretched at rest and the immediate distension (U_e) is reduced. The great increase in delayed distension (U_v), which indicates a decrease in the viscoelasticity of the interstitial fluid as a result of the increased water content, results in raised final distension (U_f) of the skin. These changes of Cutometer parameters are indicative of the increased thickness but softness of the affected skin in erysipelas. The increase in parameters U_v, U_v/U_e, H, and R provide evidence of dermal edema. The decrease of relative elastic parameters can be attributed to the increased skin thickness and some alterations of the elastic fibers.

Correlation between Clinical Assessment and Mechanical Parameters of the Skin

The visual score of edema and leg circumference significantly correlated with the increase in viscoelastic parameters U_v, U_v/U_e, H and the decrease in elastic parameters U_a/U_f, U_r/U_f.

Monitoring of Progress and Treatment Response of the Disease

After a 20-day treatment, a reduction of U_v, U_f, R, U_v/U_e, H and an increase of U_a/U_f, U_r/U_f were observed. The changes in relative parameters U_v/U_e and U_a/U_f measured by the large, 8-mm-diameter probe are most suitable for evaluation of erysipelas. The decrease in U_v/U_e indicates a reduction of inflammatory edema, whereas the increase in U_a/U_f indicates an improvement of skin mechanics of the affected limb. Thus, the degree of restoration of skin mechanical properties could be used for prediction of the development of lymphovenous insufficiency.

Lymphedema of the Lower Limbs

Lymphedema is due to disturbance of lymphatic drainage. It produces a firm, nonpitting edema with epidermal thickness that alters skin mechanics.

Mechanical Properties of the Skin

Auriol et al.[15] studied 11 patients with lymphedema of the lower limbs using a Cutometer equipped with a 6-mm-diameter probe. They applied three consecutive measuring cycles, but, instead of the first cycle, they considered the last cycle for calculation of the skin mechanical parameters. In the authors' opinion, this reduces measurement sensibility. Moreover, the authors reported only values of absolute mechanical parameters. From data published, the authors were able additionally to calculate the values of the relative parameters. Decreased U_e and U_r, increased U_v and U_v/U_e, and lower biological elasticity (U_r/U_f) characterized affected skin. The changes in residual deformation (R) were not significant.

Interpretation of the Changes of Skin Mechanical Parameters

Decreased extensibility of the skin is mainly due to the increased volume of the subcutaneous tissue. The fiber network is stretched at rest and immediate skin distension (U_e) is reduced. Some degenerative changes of the collagen and elastic fibers are probably additional factors. The interpretation of the parameter residual deformation (R) is difficult because it is a mixed parameter; i.e., it results from both the elastic and viscoelastic properties of the skin. On the one hand, the increased skin thickness and alterations in the elastic fibers raise the value of R. On the other hand, existing edema diminishes the friction between fibrils and tends to decrease R.

Monitoring of Progress and Treatment Response of the Disease

The increase in immediate extensibility (U_e) after treatment significantly correlates with the decrease in limb volume and diameter, and decrease in delayed distension (U_v). According to Auriol et al.,[15] U_e is the parameter most indicative of the evolution of lymphedema and can be used as an objective indirect indicator for disease monitoring.

DIFFERENTIAL DIAGNOSTICS BY MEANS OF SKIN ELASTICITY MEASUREMENTS

This section discusses the meaning of Cutometer measurements for differentiation of diseases that are characterized by induration of the skin. Table 19.3 presents the change tendency of skin mechanical parameters determined by a large 8-mm-diameter suction probe in four dermatoses.[9–11,13,14,16]

Generally, immediate distension (U_e) and final distension (U_f) of the indurated skin are simultaneously decreased, except in erysipelas. Thickening of the skin in the edematous phase of scleroderma and incipient SB produces identical alterations of skin mechanical parameters. Therefore, Cutometer measurements are not able to distinguish the two diseases.

TABLE 19.3
Mechanical Parameters of the Indurated Skin Determined by Cutometer Equipped with an 8-mm-diameter Suction Probe

Disease/Parameter	U_e	U_v	U_f	U_r	U_a/U_f	U_r/U_f	U_v/U_e	R8	Ref.
Scleroderma (edematous phase)	↓	→	↓	↓	↑	↑	↑	↑	16
Scleroderma (indurative phase)	↓	↓	↓	↓	→	↑	→	→	16
Scleredema of Buschke	↓	→	↓	→	↑	↑	↑	↑	9–11
Psoriasis	↓	↑	↓	↓	↓	↓	↑	↑	13
Erysipelas	↓	↑	↑	↓	↓	↓	↑	↑	14

Note: The change direction toward control skin is presented. ↑, increase; ↓, decrease; and →, no alteration of the values.

The changes of the viscoelastic-to-elastic ratio (U_v/U_e) are helpful to differentiate the edematous from the indurative phase of scleroderma as well as indurative phase of scleroderma from SB. The increase of U_v/U_e indicates the prevalence of the viscoelastic over elastic component of skin deformation. It is increased in all analyzed diseases except for the indurative phase of scleroderma. However, in the edematous phase of scleroderma and SB this is predominantly due to the decrease in U_e, whereas in psoriasis and erysipelas this is predominantly due to the increase in U_v. Thus, Cutometer measurements are able to distinguish the firm nonpitting edema in SB from the soft pitting edema in erysipelas.

The final distension (U_f) is the skin mechanical parameter that can help to differentiate psoriatic plaques from erysipelas plaques. The decreased U_f in psoriasis is related to the increased epidermal thickness and stiffness, whereas U_f is increased in erysipelas at the expense of greatly increased delayed distension (U_v).

CONCLUSIONS

Skin diseases characterized by thickening, induration, or softness of the skin can be successfully studied and monitoring by means of bioengineering measurement of skin mechanics. The use of measuring probes with different apertures allows one to obtain information on the mechanical properties of epidermis, dermis, and subcutaneous tissue. Regardless of the device used, all skin mechanical parameters determined must be simultaneously considered. Noninvasive measurements correlate well with clinical skin scoring systems and are more sensitive than human eyes and hands. They provide objective and quantitative information and can be used for evaluation of the therapy effect and for comparison of different treatment approaches.

ACKNOWLEDGMENT

The author would like to thank Prof. P. Elsner, Department of Dermatology and Allergology, University of Jena, for his expert review of the manuscript.

REFERENCES

1. Piérard, G.E., A critical approach to *in vivo* testing of the skin, in *Cutaneous Investigation in Health and Disease,* Lévêque, J-L., Ed., Marcel Dekker, New York, 1989, 215–240.
2. Gniadecka, M. and Serup, J., Suction chamber method for measurement of skin mechanical properties: the Dermaflex®, in *Handbook of Non-Invasive Methods and the Skin,* Serup, J. and Jemec, G.B.E., Eds., CRC Press, Boca Raton, FL, 1995, 329–334.
3. Agache, P.G., Twistometry measurement of skin elasticity, in *Handbook of Non-Invasive Methods and the Skin,*, Serup, J. and Jemec, G.B.E., Eds., CRC Press, Boca Raton, FL, 1995, 319–328.
4. Barel, A.O., Courage, W., and Clarys, P., Suction method for measurement of skin mechanical properties: the Cutometer®, in *Handbook of Non-Invasive Methods and the Skin,* Serup, J. and Jemec, G.B.E., Eds., CRC Press, Boca Raton, FL, 1995, 335–340.
5. Elsner, P., Skin elasticity, in *Bioengineering of the Skin: Methods and Instrumentation,* Berardesca, E., Elsner, P., Wilhelm, K-P., and Maibach, H.I., Eds., CRC Press, Boca Raton, FL, 1995, 53–64.
6. Dobrev, H., Use of Cutometer to assess epidermal hydration, *Skin Res. Technol.*, 6, 239–244, 2000.
7. Henry, F., Goffin, V., Piérard-Franchimont, C., and Piérard, G.E., Mechanical properties of skin in Ehlers–Danlos syndrome, types I, II, and III, *Pediatr. Dermatol.*, 13, 464–467, 1996.
8. Piérard, G.E., Piérard-Franchimont, C., and Lapiere, C.M., Histopathological aid at the diagnosis of the Ehlers–Danlos syndrome, gravis and mitis types, *Int. J. Dermatol.*, 22, 300–304, 1983.
9. Dobrev, H., *In vivo* study of skin mechanical properties in scleredema of Buschke, *Acta Derm. Venereol.* (Stockholm), 78, 103–106, 1998.
10. Grudeva-Popova, J. and Dobrev, H., Biomechanical measurement of skin distensibility in scleredema of Buschke associated with multiple myeloma, *Clin. Exp. Dermatol.*, 25, 247–249, 2000.
11. Dobrev, H., Value of Non-Invasive Bioengineering Investigation of the Skin Mechanical Properties *in Vivo*, dissertation, Higher Medical Institute, Plovdiv, Bulgaria, 2000.
12. Serup, J. and Northeved, A., Skin elasticity in psoriasis. *In vivo* measurement of tensile distensibility, hysteresis and resilient distension with a new method. Comparison with skin thickness as measured with high-frequency ultrasound, *J. Dermatol.*, 12, 318–324, 1985.
13. Dobrev, H., *In vivo* study of skin mechanical properties in psoriasis vulgaris, *Acta Derm. Venereol.* (Stockholm), 80, 263–266, 2000.
14. Dobrev, H., Use of Cutometer to assess dermal oedema in erysipelas of the lower legs, *Skin Res. Technol.*, 4, 155–159, 1998.
15. Auriol, F., Vaillant, L., Pelucio-Lopes, C., Machet, L., Diridollou, S., Berson, M., and Lorette, G., Study of cutaneous extensibility in lymphoedema of the lower limbs, *Br. J. Dermatol.*, 131, 265–269, 1994.
16. Dobrev, H., *In vivo* study of skin mechanical properties in patients with systemic sclerosis, *J. Am. Acad. Dermatol.*, 40, 436–442, 1999.

Part 2

Product Testing

20 Skin Biomechanics: Antiaging Products

Claudia Rona and Enzo Berardesca

CONTENTS

Introduction 231
Skin Aging and Biomechanical Properties[5,7] 231
Skin Biomechanical Property Assessment by Noninvasive Methods 232
Biomechanical Properties and Skin Intrinsic Aging 232
Biomechanical Properties and Photoaging 233
Biomechanical Properties and Antiaging Products 233
AHAs 233
Retinoids 234
Fish Polysaccharides 235
Metalloproteinase Inhibitors 236
Ascorbic Acid 237
References 237

INTRODUCTION

Antiaging products are used to ameliorate various signs generally attributed to senescence; these signs develop from two different events,[1–6] characterized by specific peculiarities known as chronoaging, or intrinsic aging, and photoaging, or extrinsic aging.[5] Young people, who are often very concerned about aging, seek preventive approaches to reverse or slow down its manifestations, especially in view of the likelihood of a prolonged lifetime and an intensification of social relations. These preventive approaches are available mainly for the extrinsic aging process, which in fact seems to be responsible for 80% of facial aging.[1,2]

This chapter focuses on the relationship between antiaging treatment and skin biomechanical properties, since the major changes induced by aging (i.e., wrinkles) are localized at the dermal level and can be assessed and monitored via biomechanical parameters.

SKIN AGING AND BIOMECHANICAL PROPERTIES[5,7]

Normal human dermis consists primarily of a ground substance, or extracellular matrix, and a fibrous component formed by collagen and elastic fibers. In particular,

0-8493-7521-5/02/$0.00+$1.50

the network of collagen fibers, which represents the most important element accounting for the 80% of the dry weight of the skin, provides resistance to deformation; elastic fibers, which account for 2 to 4% of the dry weight of the skin, provide elasticity, i.e., the ability to recover to the original position after a deformation; finally, glycosaminoglycans (GAGs–complex sugars, including hyaluronic acid) and proteoglycans (GAGs linked to a proteic core) function to retain large amounts of water and provide the viscous component of the deformation itself.

A general reduction of the synthesis of collagen fibers and of the ground substance is an event that contributes to the appearance of wrinkles and sagging skin. Elastic fibers are also primarily responsible for the biomechanical properties of the skin,[8] and their degeneration leads to a loss of elasticity. There is general agreement in the literature that a decrease in biological elasticity can be detected (independent of the method used) both in chronoaging and photoaging, due to the presence of elastotic material as well as to a rarefaction of the elastic network leading to a loss of functionality.[9–14]

SKIN BIOMECHANICAL PROPERTY ASSESSMENT BY NONINVASIVE METHODS

Antiaging assessments aimed at measuring the recovery of skin mechanical function are performed by applying different mechanical stresses (suctional, torsional, tensional) that cause deformation to the skin; different mathematical models, depending on test, have been proposed. Deformation of the epidermis or dermis depends on the applied load and the probe size, and different structures are tested with different loads and probe sizes. When assessing skin aging, probes and stresses must be applied to all the skin components whose integrity is to be assessed. Low loads allow evaluation of the effects of cosmetic products, which are thought to exert superficial effects, whereas higher loads permit assessment of deeper structures, which should be the main target of an effective antiaging treatment.

Biomechanical Properties and Skin Intrinsic Aging

Uniassial and biassial tests exhibit a strong age-dependence because the magnitude of initial elastic deformation decreases with age in accord with the degenerative changes induced in the dermal elastic network. Compression tests provide similar results, with a progressive loss of elastic recovery of the skin and thus a progressive increase in the time required for viscoelastic recovery after a deformation as a result of changes in the ground substance.[19] Torsional techniques allow the quantification of several biomechanical parameters, such as distensibility, elasticity, viscoelastic deformation, and creep relaxation time. While elasticity decreases from early lifetime with the decades, distensibility remains constant until approximately 70 years of age[15,19] and then starts to change progressively, depending on the measuring technique. By using torsional techniques, distensibility appears to decrease if the area under study is well defined, due to an increase of Young's modulus:[21] with aging the structure becomes more rigid, allowing a lower deformation magnitude, perhaps due to an increased interaction between collagen bundles and the size of the molecules. In contrast, when the measuring area is not well demarcated, for a given torque the deformation is larger for older people than for young; similar results have been reported with an identometric technique.[16,17,19] This demonstrates that distensibility is not a good marker for the quantification of skin aging effects.

According to Daly and Odland,[13] the viscous part in total cutaneous deformation is due to displacement of the interstitial fluid,[12,20] which could explain the decrease in the creep relaxation time recorded among older subjects.

In conclusion,[15] biomechanical parameters vary with age in two different ways: some may change continuously from early age (elasticity, relaxation time following strong deformation), whereas others may remain somewhat constant until around 70 years of age and then rapidly change (distensibility). Among all parameters, skin elasticity shows a clear clinical biological relevance and represents the parameter of choice when quantifying the effect of the aging process on the dermis.

Biomechanical Properties and Photoaging

Lévêque et al.[18] first described the effects of chronic sun exposure on the mechanical properties of the skin. The study reported striking differences in all the parameters investigated between chronically exposed and protected areas. These differences were not detectable in a control group of matched age. Long-term solar irradiation led to a decrease in skin extensibility and elastic recovery. The skin appears thicker, more rigid, and less susceptible to deformation. The authors suggest that changes in skin extensibility could be related to the increase in epidermal thickness.[22] In photoaging, distensibilty decreases in parallel with the thickening of the skin as a result of actinic damage. Berardesca and Maibach[19] confirmed these results during ultraviolet A (UVA) treatment. In particular, impairment of skin elasticity after low UVA and UVB exposure can occur both in the short term, due to the acute effects of rays on skin (inflammation and edema), and in the long term, when elastosis develops after prolonged exposure.

BIOMECHANICAL PROPERTIES AND ANTIAGING PRODUCTS

Many product categories for topical or oral use can be helpful in restoring and repairing the mechanical properties of the skin to treat or prevent skin aging. Among these, most of the data available in the literature refers to topical retinoids, AHAs in various combinations with vitamin C, and systemic antioxidants associated with fish polysaccharides.[24–26]

AHAs

Together with retinoids, AHAs today play a major role in skin care and dermatological therapy; they include glycolic, citric, and lactic acids, the so-called fruit acids. AHAs act both on epidermis and dermis,[6,15,28–35] exerting opposite effects. They have a thinning effect on stratum corneum, inducing desquamation and promoting cell renewal as well as stimulating collagen synthesis at the dermal level, increasing dermal thickness.

At the epidermal level, AHAs weaken intercorneocitary cohesion and stimulate stratum corneum desquamation by interfering with the formation of ionic bonds and desmosomes by inhibiting transferases and kinases. This leads to a reduction of the polarity of sulfate and phosphate groups on the corneocyte membrane; application of AHAs reduces pH even in the deeper layers of stratum corneum, contributing to dissolution of the desmosomes. Furthermore, AHAs stimulate epidermal turnover. These effects induce an increase in the flexibility of the stratum corneum.[36]

On the dermis, which is the primary contributor to skin biomechanical properties, AHAs increase glycosaminoglycan deposition in the ground substance and replace collagen and elastic degradation by promoting new collagen synthesis, without any sign of inflammation.[15] This effect is responsible for wrinkle improvement and recovery of such dermal mechanical properties as elasticity and "tone," as shown by Smith and co-workers[37] using a ballistometer to assess and confirm changes in skin firmness.

Retinoids

Originally used for acne treatment, today retinoids are a "standard" for the cosmetic treatment of aging signs. Initial studies of topical tretinoin were focused on photo-aging: L. Kligman[38] showed in hairless mice irradiated with UVA and UVB that topical tretinoin was able to revert the histologic changes induced by UV irradiation and to restore the dermal structure by increasing fibroblast activity and collagen deposition. Later, A. Kligman et al.[39] published the first paper on human volunteers, which suggested that tretinoin can have the same effects on human photoaged skin. He described histological and physiological changes in skin treated with tretinoin 0.05% cream: in particular, the skin showed replacement of atrophic epidermis by hyperplastic tissue, elimination of dysplastic and atypical keratinocytes, disappearance of microscopic actinic keratoses, and uniform dispersion of melanin granules. Tretinoin-treated skin also showed new collagen formation in the papillary dermis, new vessels, and exfoliation of retained follicular horns.

Ellis et al.[23] performed a double-blind, randomized, vehicle-controlled trial to assess the efficacy of 0.1% tretinoin cream in the treatment of photoaged skin: Patients were clinically graded on face and forearm skin for fine and coarse wrinkles, teleangectasia, tactile roughness, pinkness, and dermal edema. In 30 patients who completed the treatment, a significant improvement was found for all parameters investigated in treated sites, whereas no changes were seen in vehicle-treated sites.

From a mechanical point of view, Berardesca et al.[19] found no changes in skin distensibility, but a significant increase in skin elasticity after a 4-month treatment with 0.05% topical tretinoin (Figure 20.1). The results were detected using high deformation loads, confirming that retinoids were useful in improving dermal mechanical properties. Indeed, low vacuum forces monitor epidermal and upper dermal physical properties.

Hysteresis (i.e., the viscoelastic component of the deformation) remained unchanged as did distensibility during the study. The authors hypothesize that

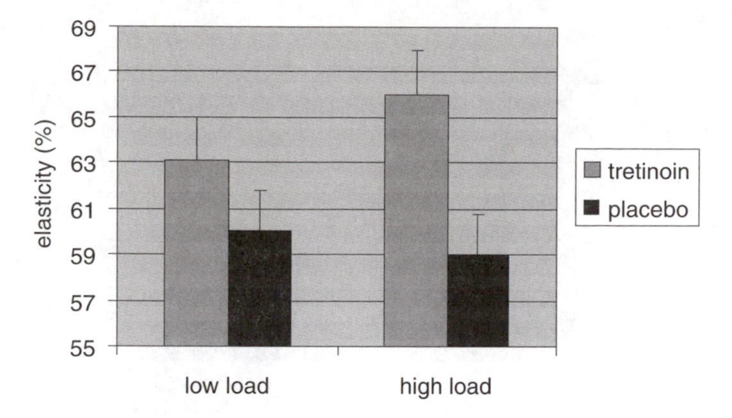

FIGURE 20.1 Skin elasticity (U_e/U_e) measured at low (100 mbars) and high (300 mbars) loads. Significant differences ($p < 0.01$) are detected after 4 months of topical tretinoin treatment.

changes in resilient distension of the skin that determine the elastic variation consequent to an increase of newly formed collagen in the dermis are responsible for this effect. This confirms data reported on the dermal effects of retinoids after long-term application.

Fish Polysaccharides

When taken orally, fish polysaccharides are reported to be effective in treating photoaging (with particular application to elastosis and wrinkles) and chronological aging (especially thinning of the skin). Their mechanism of action is still basically unknown. Previously reported data suggest a possible role in promoting collagen synthesis, in particular, procollagen and type III collagen, which stimulate repair processes within the dermis. The high content of mucopolysaccharides can improve dermal ground substance, thus improving skin mechanical behavior and skin thickness.[25–27] A study[26] reported clinical improvement of skin aging and some biophysical parameters such as thickness, microcirculation, transepidermal water loss (TEWL), and wrinkling. Other studies reported both improvement of skin mechanical properties and skin thickness.[40] In particular, when taken orally at a dose of at least 500 mg in combination with flavonoids and antioxidants, fish polysaccharides seem to exert a thickening effect at the dermal level, with significant reduction of deep lines. From a mechanical viewpoint, a significant increase in viscoelasticity is detectable (Figure 20.2) as compared with placebo treatment. This seems to be a peculiar marker of their effect on the dermis.[41] As a consequence, other mechanical parameters such as elasticity are also improved (Figure 20.3).

These results confirm that fish polysaccharides taken orally can be helpful in ameliorating some aspects of skin aging. Because of their peculiar properties, combination treatments in association with topical retinoids/AHAs would be interesting to investigate.

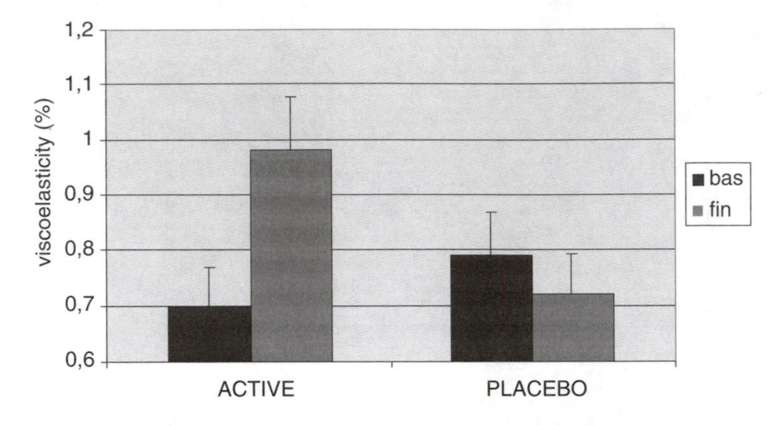

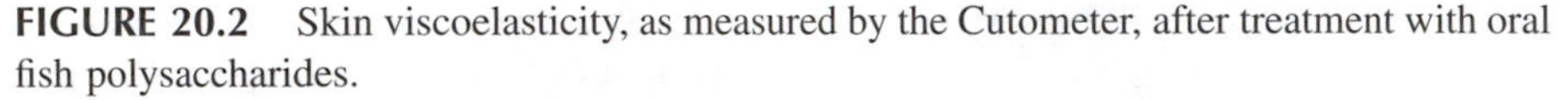

FIGURE 20.2 Skin viscoelasticity, as measured by the Cutometer, after treatment with oral fish polysaccharides.

Metalloproteinase Inhibitors

The matrix metalloproteinases (MMP)[42] are enzymes present in the skin that are responsible for the degradation of extracellular components. The extracellular matrix (EMC) establishes the three-dimensional integrity and functionality of the skin, so that it plays an essential omeostatic role. Damage to the EMC causes important alterations in the different biological functions of the skin. In healthy skin, the balance of the omeosthasis of collagen fibers depends on several factors: the synthesis of the collagen fibers themselves; the slow basal MMP action necessary to proceed to the natural and planned degradation of old or deranged fibers; the action of tissue inhibitors to control the MMP degradative activity.

The enzymatic balance of the metalloproteinases is in fact naturally controlled and ensured by tissue inhibitors (TIMPs). Aging and environmental insults, such as long-term exposure to UV rays, alter the physiological balance, increasing enzyme activity and decreasing TIMP expression and collagen fiber synthesis. These events

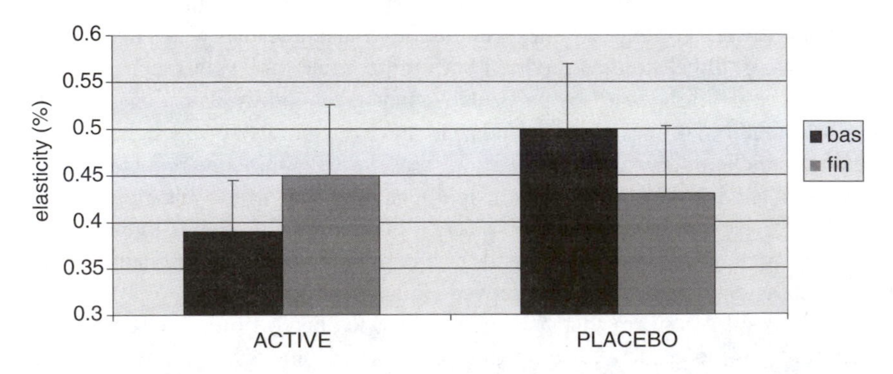

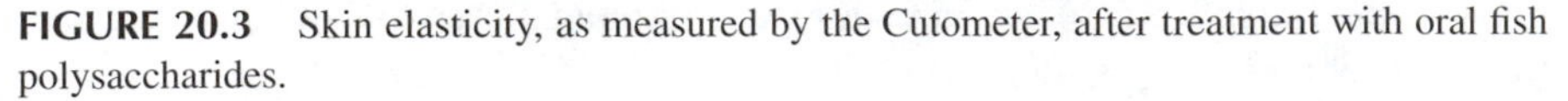

FIGURE 20.3 Skin elasticity, as measured by the Cutometer, after treatment with oral fish polysaccharides.

give rise to the collapse of the meshwork in the ground substance and contribute to the visible effects of UV damage: wrinkling, loss of elasticity and tone, teleangectasias. MMP inhibitors represent new and interesting tools that can be used by formulators in addressing different cosmetic issues and protection of the skin from external injury. In particular, the improvement in EMC properties could help in promoting skin tone, thereby reducing wrinkles and fine lines; cosmetics aimed at improving skin firmness and elasticity can, in turn, benefit from the addition of such active ingredients in their formulation. However, although preliminary findings have reported the usefulness of these molecules in improving some parameters such as skin microcirculation and skin smoothness, they have failed to show effects on mechanical properties.[43]

Ascorbic Acid

Among topical antiaging treatments, ascorbic acid has been introduced only recently because of its instability. It is an essential nutrient involved in many physiological processess and functions, by virtue of its excellent reducing capacity, which provides its antioxidant efficacy.[44] Thus, ascorbic acid can protect living tissues and cells from oxidative damage due to free radicals and oxygen-derived species. In particular, it works synergistically with vitamin E,[45] by serving as a donor antioxidant to restore the tocoferoxil radical; this reaction represents the major way to export oxidative free radicals from cell membranes. Thus, vitamin E protects membranes from free radical insult, while ascorbic acid in turn continuously restores the vitamin E.[46] This capability offers an important tool to prevent skin aging, with particular regard to tone and elasticity. In fact, ascorbic acid can preserve dermis components from sun damage by inhibiting free radical injuries charged to collagen and elastin. Regarding its mechanism of action in the treatment of skin aging, it seems to stimulate dermal deposition of new collagen.[47]

REFERENCES

1. Kligman, A.M. and Lavker, R.M., Cutaneous aging: the difference between intrinsic aging and photoaging, *J. Cutaneous Aging Cosmet. Dermatol.*, 1, 5, 1988.
2. Gilchrest, B.A., Skin aging and photoaging: an overview, *J. Am. Acad. Dermatol.*, 21, 610, 1985.
3. Gilchrest, B.A., *Photodamage*, Blackwell Science, Cambridge, MA, 1995, 1.
4. Gniadecka, M. et al., Quantitative evaluation of chronologic aging and photoaging *in vivo*, studies on skin ecogenicity and thickness, *Br. J. Dermatol.*, 139(5), 815, 1998.
5. Uitto, J. et al., *J. Invest. Dermatol. Symp. Proc.*, 3, 41, 1988.
6. Bernstein, F. and Uitto, J., Connective tissue alterations in photoaged skin and the effects of alfa hydroxy acids, *Geriatr. Dermatol.*, Suppl. A3, 7A, 1995.
7. Uitto, J., Molecular pathology of collagen in cutaneous diseases, *Advances in Dermatology*, Mosby Year Book, St. Louis, MO, 1990, 313.
8. Uitto, J., Biochemistry of the elastic fibres in normal connective tissues and its alterations in disease, *J. Invest. Dermatol.*, 72, 1, 1979.
9. Bravermann, I.M. and Fonferko, E.J., Studies in cutaneous aging: I. The elastic fiber network, *Invest. Dermatol.*, 78, 434, 1982.

10. Cua, A.B., Wilhelm, K.P., and Maibach, H.I., Elastic properties of human skin: relation to age, sex, and anatomical region, *Arch. Dermatol. Res.*, 282, 283, 1990.
11. Ellis, C.N., Weiss, J.S., and Voorhees, J.J., Tretinoin: its use in repair of photodamage, *J. Cutaneous Aging Cosmet. Dermatol.*, 1, 33, 1988.
12. Uitto, J., Fazio, M.J., and Olsen, D., Cutaneous aging: molecular alterations in elastic fibers, *J. Cutaneous Aging Cosmet. Dermatol.*, 1, 13, 1988.
13. Dali, C. and Odland, G.F., Age related changes in the mechanical properties of human skin, *J. Invest. Dermatol.*, 73, 84, 1979.
14. Piérard, G.E. and Lapiere, M., Physiopathological variations in the mechanical properties of skin, *Arch. Dermatol. Res.*, 260, 231, 1977.
15. Escoffier, C., de Rigal, J., Rochefort, A., Vasselet, R., Lévêque, J.L., and Agache, P., Age-related mechanical properties of human skin: an *in vivo* study, *J. Invest. Dermatol.*, 93(3), 353, 1989.
16. Lanir, Y., Manny, V., Zlotogorski, A., Shafran, A., and Dikstein, S., Influence of aging on the *in vivo* mechanics of the skin, *Skin Pharmacol.*, 6, 223, 1993.
17. Salter, D.C., McArthur, H.C., Crosse, J.E., and Dickens, A.D., Skin mechanics measured *in vivo* using torsion: a new and accurate more sensitive to age, sex, and moisturizing treatment, *Int. J. Cosmet. Sci.*, 15, 200, 1993.
18. Lévêque, J.L., Porte, G., de Rigal, J., Corcuff, P., Francois, A.M., and Saint Leger, D., Influence of chronic sun exposure on some biophysical parameters of the human skin: an *in vivo* study. *J. Cutaneous Aging Cosmet. Dermatol.*, 1, 123, 1988.
19. Berardesca, E. and Maibach, H.I., Mechanical properties and photoaging, *Aging Skin — Properties and Functional Changes*, Lévêque, J.L. and Agache, P., Eds., Marcel Dekker, New York, 29, 1993.
20. Fleishmajer, R., Perlish, J.S., and Bashey, L.R., Human dermalGAG and aging, *Biochim. Biophys. Acta*, 279, 265, 1972.
21. Agache, P., Monneur, C., Leveque, J.L., and de Rigal, J., Mechanical properties and Young's modulus of human skin *in vivo*, *Arch. Dermatol. Res.*, 269, 221, 1980.
22. De Rigal, J. and Lévêque, J.L., *Bioeng. Skin*, 1, 13, 1985.
23. Ellis, C.N., Weiss, J., and Voorhees, J.J., Topical tretinoin in the treatment of skin aging, *J. Cutaneous Aging Cosmet. Dermatol.*, 1, 33, 1988.
24. Elsner, P., Skin elasticity, in *Bioengineering and the Skin: Methods and Instrumentation*, Berardesca, E., Elsner, P., Wilhelm, K., and Maibach, H.I., Eds., CRC Press, Boca Raton, FL, 1995.
25. Lassus, A., Jeskanen, L., Happonen, H.P. et al., Imedeen for the treatment of degenerated skin in females, *J. Int. Med. Res.*, 19, 147, 1991.
26. Eskelinin, A. and Santalahti, J., Special natural cartilage polysaccharides for the treatment of the sun damaged skin in females, *J. Int. Med. Res.*, 20, 99, 1992.
27. Kieffer, M.E. and Efsen, J., *J. Eur. Acad. Dermatol.*, 11, 129, 1998.
28. Berardesca, E. and Maibach, H.I., AHA mechanisms of action, *Cosmet. & Toiletries*, 110, 30, 1995.
29. Van Scott, E. and Yu, R.J., Hyperkeratinization, corneocyte cohesion and alpha hydroxy acids, *J. Am. Acad. Dermatol.*, 11, 867, 1984.
30. Smith, W.P., *Cosmet. & Toiletries*, 109(9), 41, 1994.
31. Van Scott, E. and Yu, R.J., Alpha hydroxy acids, procedures for use in clinical practice, *Cutis*, 43, 222, 1989.
32. Takahashi, M. and Machida, Y., The influence of AHA on the rheological properties of the stratum corneum, *J. Soc. Cosmet. Chem.*, 36, 177, 1985.
33. Van Scott, E. and Yu, R.J., Control of keratinization with alpha hydroxy acids and related compound, *Arch. Dermatol.*, 110, 586, 1974.

34. Van Scott, E. and Yu, R.J., Alpha hydroxy acids, terapeutic potentials, *Can. J. Dermatol.*, 1, 108, 1989.
35. Ridge, J.M., Siegle, R.J., and Zuckerman, J., Use of AHA in therapy for photoaged skin, *J. Am. Acad. Dermatol.*, 23, 932, 1990.
36. Darbyshire, J., Alfa idrossi acidi: gli acidi da frutta per ringiovanire, *Cosmet. News*, 91/93, 253, 1993.
37. Fthenakis, C., Maes, D.H., and Smith, W.P., *J. Soc. Cosmet. Chem.*, 42, 222, 1991.
38. Kligman, L.H., Effect of all-trans-retinoic acid on the dermis of hairless mice, *J. Am. Acad. Dermatol.,* 15, 779, 1986.
39. Kligman, A., Grove, G.L., Hirose, R., and Leyden, J.J., Topical tretinoin for photoaged skin, *J. Am. Acad. Dermatol.*, 15, 836, 1986.
40. Rona, C., Zuang, V., and Berardesca, E., I polisaccaridi di pesce marino nel trattamento del fotoinvecchiamento, *It. J. Derm. Venereol.*, 136, 85, 2001.
41. Distante, F. et al., Oral fish cartilage polysaccharides in the treatment of photoaging: biophysical findings, *Int. J. Cosmet. Sci.*, in press.
42. Thibodeau, A., Metalloproteinase inhibitors: matrix metalloproteinase inhibitors can reduce the visible sign of aging, *Cosmet. & Toiletries,* 115, 75, 2000.
43. Meloni, M., Rona, C., and Berardesca, E., *In vivo* indirect assessment of inhibition of MMP-1 activity by noninvasive techniques, *Int. J. Cosmet. Sci.*, in press.
44. Buettner, G.R. et al., *Handbook of Antioxidants*, Cadenas, E., Ed., Marcel Dekker, New York, 1996, 91.
45. Rousseau-Richard, C. et al., *New J. Chem.,* 283, 1991.
46. Frei, E. et al., *Antioxidants in Therapy and Preventive Medicine*, Emerit, I., Ed., Plenum, New York, 1990.
47. Pinnel, S.R., Induction of collagen synthesis by ascorbic acid, a possible mechanism, *Arch. Dermatol.,* 123, 1684, 1987.

21 Product Testing: Moisturizers

André O. Barel

CONTENTS

Hydration State of the Skin 241
Mechanical Properties of the Skin and Experimental Systems Developed to Assess Biomechanical Properties 243
Torsional Deformation of the Skin: Twistometer and the Dermal Torque Meter 244
Suction Deformation of the Skin: Cutometer SEM 474/575. 244
Suction Deformation: Dermaflex A 245
Experimental Devices to Assess *in Vivo* the Biomechanical Properties of the Skin 245
Description of Recorded Viscoelastic Parameters 245
Strain–Time Mode 246
Stress–Strain Mode 248
Relation between Skin Mechanics and Hydration and the Applicability Potential of Biomechanical Measurements to Cosmetics Claiming Support of Hydration of the Skin 249
Results with the Torsional Device 249
Results with the Suction Devices 250
Results with Experimental Devices 252
Conclusions 252
References 254

HYDRATION STATE OF THE SKIN

The presence of an adequate amount of water in the stratum corneum is important for maintaining the following properties of the skin: general appearance of a soft, smooth, flexible, and healthy-looking skin; and an intact barrier function allowing a slow rate of transepidermal water loss (TEWL) under dry external conditions, which are frequently encountered.[1–4] Clinically, dry skin is defined as a dry, less-flexible, scaly, and rough aspect of the upper layers of the epidermis.[1–4] Itching may accompany these symptoms. In more severe cases of dryness (a condition described by dermatologists as xerosis) and in pathological conditions, fissuring, scaling, and cracking occur.[5–8] There is no universally accepted definition of *dry skin*.[9] Some

0-8493-7521-5/02/$0.00+$1.50

authors consider that dry skin is related to disorders of corneocyte adhesion and desquamation (rough and scaly surface), to modifications in the composition of certain epidermal lipids (ceramides), or to disorders of the water-retaining properties of the horny layer (presence of natural mosturizing factors). The perception of a dry skin state by the patients themselves may differ from the diagnosis of the clinician.

There are currently very few data confirming that a situation of dry skin is linked solely to a diminution of the water content of the horny layer. However, the positive pharmacological effect of application of water and moisturizers to the skin surface to relieve the condition of dry skin has been repeatedly confirmed. Nevertheless, given the fact that the presence of an adequate amount of water is an essential prerequisite for the maintenance of the normal structure and function of the stratum corneum, research has been directed at evaluation of the water content of this tissue. The skin surface is covered with the stratum corneum, which acts as an efficient barrier protecting the whole skin from loss of water. The structure of the horny layer *in vivo* is very complex. The upper part of the horny layer, which is exposed to air, contains relatively much less water (about 10%), whereas the lower part of the stratum corneum is in direct contact with the water-saturated viable epidermis.

The water content of the stratum corneum is influenced by several factors:[3,4,10]

- *Water diffusing from the deeper viable layers of the epidermis.* This water is normally retained by the barrier function of the lipid bilayers in the stratum corneum.
- *Water present in the horny layer.* This water is fixed, among other elements, by the natural moisturizing factors (NMF), which constitute amino acids, lactic acid, pyrrolidone carboxylic acid, urea, and other salts. These substances present a high affinity for holding water (hygroscopy). The amount of water entrapped between the lipid bilayers may also play an important role. Stratum corneum lipids, especially ceramides, contribute to the water-retaining properties through the lamellar lipid structures.[11]
- *Equilibrium between water in the upper layers of the horny layer and the external ambient humidity of air.* Depending on the value of the external relative humidity, the upper layers of the stratum corneum may take on or release water. Dry skin is generally more prevalent in the winter (winter xerosis).
- *Treatment of the skin surface with moisturizers and/or emollients.* Moisturizers are hydrating products used to combat the symptoms of dry skin.[3,4] They may hydrate the skin in different ways either by reducing the loss of water due to the occlusive effects on the skin surface of oils and fats contained in the preparations or by adding water to the stratum corneum and fixing it with the help of humectants, or by a combination of both.[3,4] As a consequence, the *in vivo* determination of the degree of hydration of the horny layer is important for characterization of normal and dry skin, for assessment of pathological skin conditions with extreme dry skin, and, finally, for assessment of the efficacy of various moisturizing products. The use of such dermato-cosmetic hydrating products to restore softness, smoothness, and moisture to dry skin is widely practiced.

This chapter is concerned with evaluation of the hydration state of the skin and assessment of the efficacy of moisturizers with the use of noninvasive bioengineering measurements of the elastic and viscoelastic properties of the skin. The chapter provides a critical overview of the different biomechanical methods (experimental devices and commercially available instruments) available to carry out mechanical measurements on normal skin, on well-hydrated skin, and on dry skin, respectively, and to assess quantitatively the efficacy of moisturizers. This chapter devotes a discussion to the significance in general of the quantitative variables obtained from *in vivo* biomechanical measurements. It is also important to evaluate the participation of the horny layer and epidermis in the global mechanical behavior of the skin *in vivo*. Some data suggest clearly an effective contribution of the horny layer to the overall mechanical properties of the skin. This is an important argument for the claim support related to biomechanical improvement of the epidermis horny layer after a topical cosmetic treatment, such as a moisturizing cream.

Several pertinent articles and reviews concerning the applicability of biomechanical measurements for cosmetic claim support have been published.[12–15]

MECHANICAL PROPERTIES OF THE SKIN AND EXPERIMENTAL SYSTEMS DEVELOPED TO ASSESS BIOMECHANICAL PROPERTIES

Assessment of the mechanical properties of the skin is not simple, since the skin is a stratified composite material and the relationship between the various layers is complex. It is clear that the mechanical characteristics of human skin result from the global contribution of connective tissue, dermis, and hypodermis and, at least to some degree, of the epidermis and horny layer.[13] The properties of the fiber constituents, their orientation, network, and their relation with the interstitial medium constituted of proteoglycans are fundamental factors determining the overall properties of the entire organ.[13] Flexibility is characteristic of human skin, fundamental to allow movement without cracking or fissuring. Flexibility contributes to overall cutaneous cohesion and to the barrier function of the skin, depending on elastic, plastic, and viscous components, which are responsible for the extremely complex mechanical behavior (viscoelastic properties of the skin). The stratum corneum is a heterogeneous structure composed of protein-enriched corneocytes embedded in a lipid matrix. These lipids are arranged in lamellar structures and contribute considerably to the barrier function of the stratum corneum.

The different methods that have been developed for testing skin mechanics are based on forces and deformations working either in the skin surface or perpendicular to the skin surface. Methods based on forces working in the plane of the skin are stretching, torsion, and wave propagation. Methods based on forces perpendicular to the skin surface are suction, indentation, elevation, and ballistometry. Many very interesting experimental devices are not commercially available but are used only in research laboratories. Several instruments using the suction method are commercially available at the present time: the Dermaflex A from Cortex Technology (Denmark), the Cutometer SEM 475 (more recently replaced by the SEM 575 version) of Courage–Khazaka Electronic (Germany), the Dermalab from Cortex (Denmark), and the BTC-2000 from SRLI (United States). One commercial instrument using the torsion method is commercially available: the Dermal Torque Meter based on the Twistometer from the L'Oréal group.

Among these commercially available instruments, most biomechanical measurements have been carried out with the Torque Meter (Dias-Stron, Andover, U.K.) and two suction devices, the Dermaflex A (Cortex Technology, Hadsund, Denmark) and the Cutometer SEM 474 and more recently SEM575 (Courage–Khazaka Electronic, Cologne, Germany). Biomechanical measurements for evaluation of the hydration state of the skin surface using the two very recent commercial devices (Dermalab and BTC-2000) have not yet been published.

Torsional Deformation of the Skin: Twistometer and the Dermal Torque Meter

Torsional testing of the mechanical properties of the skin has been used and developed by several research groups over the past 20 years, mainly by the L'Oréal group as described by Agache.[12] This work has led to the commercial version of the Dermal Torque Meter. The measurement technique consists of applying a disk to the skin surface (the disk is firmly adhered to the skin surface with double-sided adhesive tape). The Dermal Torque Meter applies a constant torque to the skin and measures the resulting angular distortion. This angular distortion induces an elongation of the skin, which can be limited by the introduction of a concentric guard ring (also firmly adhered to the skin surface), which delimits an annular area of skin and limits the sliding of the skin over the hypodermis. Thus, the angular deformation of the skin is confined to a narrow annulus (generally of 1 mm). According to the L'Oréal group, with low values of torque and with a ring gap of 1 mm, the obtained mechanical measurements result from the most superficial structures of the skin (horny layer and superficial part of the epidermis). Thus, the torsion method allows mechanical measurements pertaining to the efficacy of topical cosmetic applications. The instrument generates two types of experimental curves: strain–time and stress–strain.

Suction Deformation of the Skin: Cutometer SEM 474/575

With the Cutometer, a variable vacuum ranging from 50 to 500 mbar is applied to the skin surface through the opening of a probe. Probes with apertures 2, 4, 6, and 8 mm in diameter are available. In the procedure, the skin surface is pulled vertically upward into the probe opening by the vacuum, and the vertical deformation of the skin is measured via an optical system (diminution of light intensity of an infrared light beam). The skin adjacent to the opening of the probe is maintained in position with an external guard ring. The measuring probe is manually applied vertically onto the surface of the skin, and a constant pressure is assured by a built-in spring. The instrument is linked to a personal computer using a standard software program provided by the company. The Cutometer generates two types of experimental curves: strain–time and stress–strain. According to some authors,[14–17] when the small, 2-mm probe is used, the Cutometer is capable of more selective study of the mechanical properties of the upper epidermis.

Suction Deformation: Dermaflex A

With the Dermaflex A, a variable vacuum ranging from 0 to 500 mbar is applied on the skin surface through the opening of a probe. The only probe aperture available is 10 mm diameter. In the procedure, the skin surface is pulled vertically upward into the probe opening by the vacuum. Then, the vertical deformation of the skin is measured electronically by measuring the electric capacitance between skin surface and the electrode placed in the top of the suction chamber. The skin adjacent to the opening of the probe is maintained in position with an external guard ring, and the double-sided adhesive tape on the external steel ring prevents skin creeping during suction. The measuring probe is manually applied vertically onto the surface of the skin. The instrument delivers only one experimental curve: strain–time.

Experimental Devices to Assess *in Vivo* the Biomechanical Properties of the Skin

Several experimental systems have been developed in research laboratories to assess the biomechanical properties of skin *in vivo*. The experimental devices are grouped in two major categories according to the orientation of the load toward the skin.

1. Methods using forces parallel to the skin surface:
   Uniaxial stretching devices[18]
   Gas Bearing Electrodynamometer[19,20]
2. Methods using forces perpendicular to the skin surface:
   Levarometry[21]
   Indentometry[22]
   Ballistometry[23]
   Resonance frequency[24]

A detailed description of each system can be found in other chapters of this book. Therefore, only those experiments dealing with the effect of moisturizers on biomechanical properties as measured with these experimental devices are discussed in this chapter.

DESCRIPTION OF RECORDED VISCOELASTIC PARAMETERS

As already mentioned, the skin is a complex, five-layered organ,[25] which, similar to many other biological materials, presents the typical properties of elastic solids and various liquids in a combined way known as viscoelastic properties.[26–28]

Deformation of the skin as a function of time consists of an immediate, incomplete elastic deformation and a creeping viscoelastic deformation. These are followed by an immediate elastic recovery, a creeping recovery, and, always, a residual deformation (Figure 21.1).

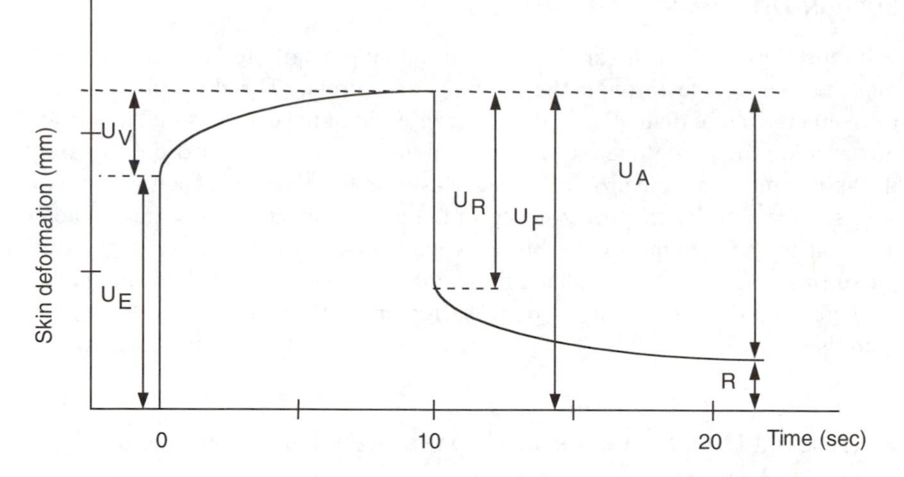

FIGURE 21.1 Graphical representation of a strain–time curve obtained for forearm skin. Vacuum of 500 mbar, application and recovery time of 10 s, 2-mm probe.

Typical mechanical properties of viscoelastic material are nonlinear stress–strain properties with hysteresis (the stress–strain curve obtained during loading is superposed by the curve obtained during unloading (Figure 21.2).

Strain–Time Mode

The vertical deformation of the skin is measured as a function of time (see Figure 21.1). In this mode, for a given aperture of the probe, the pressure of the vacuum (from 50 to 500 mbar), the total duration of suction time (stress-on) and relaxation time (stress-off), and the number of measurement cycles can be selected. The parameters used in this measuring mode describe the elastic deformation and recovery of the skin, the viscoelastic creep after the initial deformation and the initial recovery, and the residual deformation.[2] Parameters of interest are as follows:

- U_e is defined as the immediate elasticity, which corresponds with the steep linear section of the curve and which must be computed at a very short interval after application of the suction.
- U_f is defined as the maximal deformation; it is equal to the sum of the elastic deformation U_e and the viscoelastic deformation U_v.
- U_v is the delayed viscoelastic part of the skin deformation (creep).
- U_r is defined as the immediate elastic recovery of the skin after removal of the suction.
- U_r is measured in the steep linear part of the recovery.
- U_a is equal to the total recovery deformation of the skin at the end of the recovery phase.

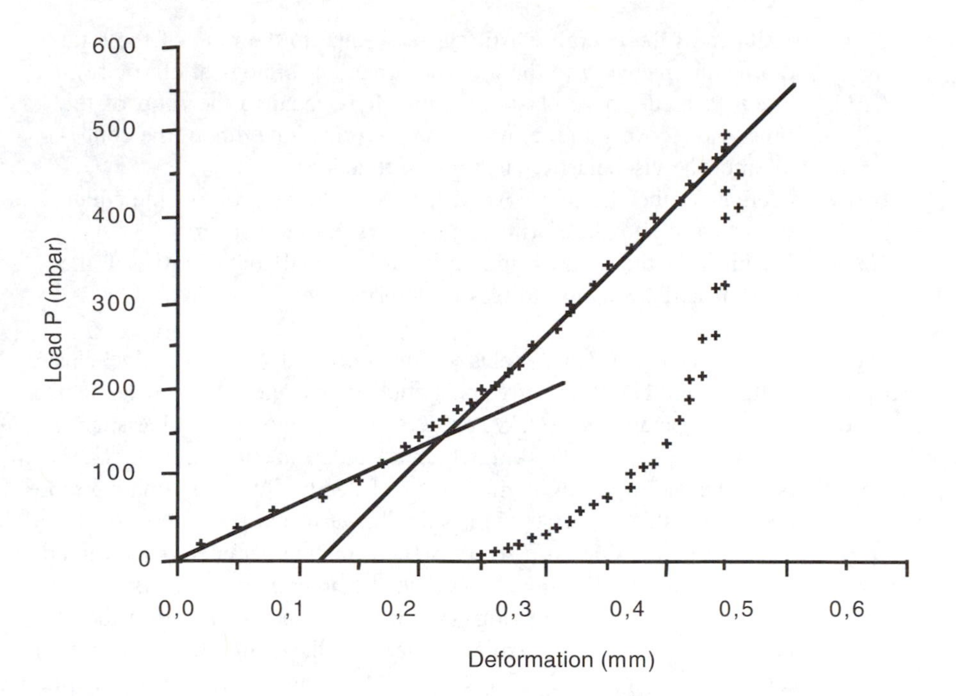

FIGURE 21.2 Graphical representation of a stress–strain curve obtained for forearm skin. Linear increase and decrease of vacuum, 2-mm probe, total stress-on and stress-off time of 20 s. Maximum value of vacuum is 500 mbar.

R is the residual deformation of the skin, which has not recovered after completion of the stress-off measuring time. When working with repetitive measuring cycles, the deformation–time curves obtained for the second, third, and subsequent deformation cycles are similar to the first, but progressively shift upward as a consequence of the residual deformation.

Hysteresis is defined as the difference between the maximal deformation U_f of the last curve and that of the first curve.

The deformation parameters are extrinsic parameters influenced by skin thickness.[27] To take into account potential differences in skin thickness when comparing men and women, young and old, different anatomical skin sites, and so forth, intrinsic skin deformation parameters must be computed (deformation × skin thickness).

As pointed out by many authors,[29–31] the problem of standardized definitions of the different skin deformation parameters is important. For example, Gniadecka and Serup[25] propose the terms *skin stiffness* or *distensibility* for U_f and *resilient distensibility* for U_a, whereas Couturaud et al.[32] propose the terms *extensibility* for U_e, *plasticity* for U_v, and *tonicity* for U_r. From these extrinsic deformation parameters (see Figure 21.1), different elastic and viscoelastic ratios have been proposed to characterize the mechanical properties of the skin. Escoffier et al.[27] propose:

U_a/U_f is defined as the overall elasticity. It is equal to the ratio of the total deformation recovery to the total deformation (biological elasticity).
U_r/U_e is defined as the pure elasticity ratio. It is equal to the ratio of the immediate recovery to the immediate deformation without the contribution of the viscoelastic part (elastic function).
U_r/U_f is defined as the elastic recovery. It is equal to the ratio of immediate recovery to the total deformation (relative elastic recovery).
U_v/U_e is defined as the viscoelastic ratio. It is equal to the ratio of the viscoelastic deformation to elastic deformation.

The computed values of U_r/U_e (elastic function) and U_r/U_f (relative elastic recovery) parallel each other closely for most studied skin sites. As a consequence, either one of the two parameters (U_r/U_e or U_r/U_f) is computed for characterizing the intrinsic elastic properties of the skin, which are independent of skin thickness.[27] Some authors use the ratio U_a/U_f as a criterion for elasticity.[33] When evaluating more specifically the viscoelastic properties of the skin, the intrinsic parameter viscoelastic ratio, U_v/U_e is generally considered.[27] All these deformation parameters are directly computed from the Cutometer using the provided standard software. By using a standard software program, the following parameters can be computed in the Dermaflex: U_f distensibility; $R = U_f - U_a$ resilient distension; the relative elastic retraction also described as the elasticity index RER; $R = R/U_f$; and the hysteresis creeping phenomenon, which corresponds to the increase of U_f after four cycles $(U_{f4} - U_{f1})$.[14,15,25]

All the deformation parameters described here can also be directly computed from the Dermal Torque Meter using the provided standard software.

Stress–Strain Mode

In this mode (see Figure 21.2), the total vertical skin deformation U_f is measured as a function of vacuum (mbar) during a 10-s linear increase (from 0 to 500 mbar) of vacuum followed by a similar linear decrease of suction during a relaxation period of 10 s. For almost all anatomical skin sites studied, nonlinear curves were obtained with hysteresis. The stress–strain curve on loading is not superposed by the relaxation curve. During the relaxation period, the values of strain do not return to zero since there is an intercept of the curve on the strain axis. Theoretically, considering the nonlinearity of the curve for human skin, Young's modulus E is not applicable.[12,34] As proposed by Piérard,[28] one can always define a pseudo-Young's modulus E as a type of coefficient of elasticity that corresponds to the derivative of the stress to the strain for a given stress value. However, most experimental ascending curves present a clear linear section between 150 and 500 mbar. From that linear section of the stress–strain curve, the slope can be calculated, corresponding to the pseudo-Young's modulus E or the coefficient of elasticity.[28,30]

For calculation of E, a theoretical simple model for the deformation of the skin in the circular aperture was proposed by Agache and co-workers.[15,34]

RELATION BETWEEN SKIN MECHANICS AND HYDRATION AND THE APPLICABILITY POTENTIAL OF BIOMECHANICAL MEASUREMENTS TO COSMETICS CLAIMING SUPPORT OF HYDRATION OF THE SKIN

From a mechanical point of view the skin is a stratified composite material whose most superficial layer, the horny layer, is the stiffest but represents only 1/100 of the total thickness. As a consequence, its role in whole-skin mechanical properties is often overlooked.[15]

Many *in vivo* studies using noninvasive bioengineering techniques have been published to evaluate the relationship between the mechanical properties of the skin and the water content of the upper layers of the epidermis (for recent reviews, see Bernengo and de Rigal[14] and Agache and Varchon).[15] Determination of the hydration is based on electrical measurements of the conductance, capacitance, and impedance of the skin (for a recent review see Barel et al.[10]). The immediate hydration effects in the stratum corneum after a single application of water or moisturizers (short-term effect) and the hydration effect in the stratum corneum after repeated application of moisturizers (long-term effect) are considered here.

RESULTS WITH THE TORSIONAL DEVICE

Torsional experiments with the Twistometer enabled Lévêque and de Rigal[36] to investigate the mechanical behavior of the horny layer *in vivo*. Using a low torque value and a small area of twisted skin (1-mm ring opening), they investigated the immediate effect of a simple application of water and the occlusion of the stratum corneum with a plastic sheet. Significant increases in the U_e parameter were observed. The variations in U_e were inversely related to the value of Young's modulus.

Application of simple vehicles such as O/W and W/O emulsions or petrolatum also induces a significant short-term increase in the U_e parameter. Addition of moisturizers such as 10% lactic acid, sodium pyrrolidone carboxylate, urea, and, particularly, glycerol provokes a significant increase in the U_e parameter.[36]

In a long-term study,[37] twice daily application for 3 weeks of an O/W vehicle containing 10% urea or glycerol as a moisturizer on the legs caused a significant rise in the U_e of the treated skin, whereas the skin treated with the vehicle alone did not differ from the control site. This study also found that total immersion of the forearm skin for 3 min in hot water (30°C) provokes a significant increase of the elasticity parameters U_e, U_r, and U_r/U_e.[37] Significant modifications of skin extensibility and elasticity were also observed after removal of the stratum corneum by adhesive tape stripping (10-15-20 strippings on the forearm skin). During stripping, U_e, U_r, and U_r/U_e increased steadily and significantly in comparison with the untouched baseline.[37]

Correlations between visual clinical assessments of symptoms of dry skin on the forehead and cheeks (using a 0 to 3 numerical scale) and torsional mechanical properties were investigated by Lévêque et al.[38] Clearly, significant inverse relations were established between clinical dryness and the U_e mechanical parameter (r varies from –0.41 to –0.66).

Results with the Suction Devices

Many *in vivo* studies conducted with the Cutometer and Dermaflex suction devices have demonstrated that the elastic and viscoelastic properties of the skin are modified after application of water or moisturizers to the skin.

It was shown by Jemec and Serup[39] that application of tap water for 20 min under occlusion changes the forearm skin mechanical properties substantially as measured with a Dermaflex suction instrument (10-mm probe aperture, repetitive cycles of 4 s stress-on and stress-off times and 300 mbar suction). The increase in hydration (capacitance method) is accompanied by the following changes in viscoelastic properties of the skin: an increase in distensibility and resilient distensibility and a decrease in elasticity. Subsequently, Olsen and Jemec[40] compared the short-term (10-min application) influence of tap water, paraffin oil, ethanol, and glycerine to the forearm skin on the mechanical properties of the skin. Significant increases in resilient distensibility, distensibility, and hysteresis were observed after application of a gauze pad soaked with tap water and paraffin oil. The results of Jemec and Serup[39] show a correlation between hydration measurements and some mechanical properties of the skin.

Murray and Wickett[33] conducted studies both on acute hydration, on the forearm skin, and on the long-term moisturizing effects on the outer calf region of the legs. Hydration was assessed using the DPM Nova instrument and the Cutometer for determination of the biomechanics of the skin (2-mm probe aperture, 2-s stress-on and stress-off times, and 100 and 500 mbar suction). The acute hydration treatment involved occlusive application of a warm, wet paper towel covered with Saran wrap for 10 min to the forearm skin. Pre- and post-treatment measurements were taken with both instruments. In the moisturizing study, a hydration lotion containing 10% glycerine was applied twice a day for 2 weeks on the calf region of the legs. Baseline, post-treatment, and post-regression measurements were made with the same instruments.

U_v and U_v/U_e were the mechanical parameters most sensitive to hydration changes in the acute hydration test (10-min application of the wet paper towel). U_e, U_v, U_f, U_a, and R were the most sensitive parameters in the moisturizing study. That the Cutometer data parameters did not consistently correlate with the DPM Nova hydration data was interpreted by the authors as the result of the Nova instrument measuring much shallower skin depth than the Cutometer.

Subsequently, Murray and Wickett[41] performed long-term hydration studies on dry leg skin to determine the sensitivity of the Cutometer and Dermal Torque Meter parameters to changes in hydration of the stratum corneum, as indicated by electrical capacitance measurements with the NOVA Dermal Phase Meter. The subjects used a moisturizung lotion twice daily for 2 weeks. The experimental conditions of the Cutometer were a 2-mm probe, 10-s stress-on and stress-off times, and 100 and 500 mbar suctions. For the Dermal Torque Meter, they were 1-mm annulus between disk and guard ring and 10-s stress-on and stress-off times. They found significant changes in most mechanical parameters: U_e, U_v, U_f, U_r, and U_a for the Dermal Torque Meter and U_e, U_v, U_f, and U_a for the Cutometer. The U_e and U_a mechanical parameters as obtained with the Dermal Torque Meter showed the highest sensitivity to the effect of moisturizing lotions (compared with other mechanical parameters of the Dermal Torque Meter and with all the parameters of the Cutometer). Correlations between the torsional and suction instruments existed, but were weak ($r = 0.41$ for U_r, $r = 0.54$ for U_e, and $r = 0.62$ for U_f). Again, no correlations were found between hydration and mechanical properties.

Determination of the biomechanical properties of the skin after topical application of glycerine and paraffin oil for 15 min on the volar forearm skin was carried out by Barel et al.[31] Cutometer measurements (2-mm probe aperture, 10-s stress-on and stress-off times, and 500 mbar suction) indicated no significant differences in the elasticity parameter U_r/U_f after application of glycerine and/or paraffin oil on the forearm skin were observed. On the contrary, the viscoelastic deformation ratio U_v/U_e was significantly increased after these topical treatments.

The same study also investigated stripping of the horny layer with adhesive tape. Intensive treatment of the skin with 40 strippings, which corresponds to the removal of the horny layer, provoked a small decrease in elasticity (U_r/U_f), but a large increase in the viscoelastic parameter (U_v/U_e). Because the study used a 6-mm probe aperture and a vacuum of 400 mbar, it is thought that, in fact, the mechanical properties of the whole dermis and perhaps of the subcutis were investigated. However, these preliminary data would indicate that changes in the viscoelastic properties of more superficial layers of the skin can also be detected with the suction method.

Other workers investigated the biomechanical properties of the skin after short-term topical application of O/W emulsions containing 5% glycerine or urea or both on the volar forearm skin with Corneometer hydration measurements.[42] Cutometer measurements were carried out with a 2-mm probe, 3- and 10-s stress-on and stress-off times, and 500 mbar suction. No significant changes in the elasticity parameters U_r/U_e and U_r/U_f after application of the three moisturizers were observed. On the contrary, the viscoelastic deformation ratio U_v/U_e was significantly increased after these topical treatments. There were no correlations between the capacitance hydration measurements and the three mechanical parameters U_r/U_e, U_r/U_f, and U_v/U_e.

In a very recent study by Dobrev,[43] the topical effects of five different moisturizers and emollients (petrolatum, paraffin oil, glycerine, W/O and O/W emulsions) on the volar forearm skin were examined by hydration (Corneometer) and by mechanical properties (Cutometer, 2-mm probe, 5-s stress-on and stress-off times, and 400 mbar suction). Delayed distension U_v and viscoelastic ratio U_v/U_e were the most sensitive parameters to indicate changes in the hydration of the skin. No correlations between skin capacitance hydration measurements and mechanical parameters were found in this study.

The evaluation of the long-term moisturizer effect of a lotion containing 5% glycerin on the legs was recently carried out by Wickett[17] using the Cutometer and the Nova instrument. The Cutometer was used with the 2-mm probe at 200 and 500 mbar suction. This study indicated that using a lower vacuum level may improve sensitivity to moisturizer-induced changes in mechanical parameters. The increase of U_e and U_r appeared to be more sensitive to the moisturizer effect than that of U_v.

Results with Experimental Devices

Experimental data obtained with noncommercially available laboratory devices have also clearly shown the effect of hydration of the skin on its biomechanical properties.

Very interesting results were obtained with the Gas Bearing Extensometer.[15,20,44,45] With this apparatus an ellipsoidal stress–strain curve is obtained for a viscoelastic material such as the skin. The slope of the major axis of the ellipse, which is called the dynamic spring rate, is an analogue of Young's modulus and reflects the softness of the skin. The openness of the elliptical loop reflects the viscoelastic behavior of the skin.[20]

Application of an emollient such as glycerol on the skin modifies the slope and the surface of the ellipse in a stress–strain diagram obtained by the Gas Bearing Extensometer.[45] Topical application of glycerol induces a reduction in the slope, indicating an increase in flexibility of the skin and a broadening of the ellipse, which is compatible with a diminution of the viscosity of the stratum corneum.[15,45]

Similarly, Maes et al.[19] demonstrated an improvement in skin softness as measured with the slope of the strain–stress curves obtained with the Gas Bearing Extensometer after application of water and O/W and W/O emulsions. Water had a clear short-term effect on the elliptical loops, whereas the O/W and W/O emulsions displayed a distinctly longer lasting effect.

The tactile sensor,[24] which determines the changes in resonance frequency that occur when a vibrating probe comes in contact with the skin surface, can be used to assess the stiffness/softness of the skin. Using this experimental device, Sakai et al.[24] have recently shown a correlation between the shift in the resonance frequency of the sensor and the hydration of the horny layer (as measured by conductance) after topical application of water and a moisturizer. The softness of the skin as determined with the tactile sensor was compared with the skin elasticity parameter U_r/U_f obtained with the Cutometer. At a low pressure of application of the sensor, highly significant negative correlations were observed ($r = -0.73$) between the shift in resonance frequency and the conductance hydration data. Moderately significant positive correlations ($r = +0.49$) between shift in resonance frequency and the Cutometer elasticity parameter U_r/U_f were seen.

CONCLUSIONS

Results of mechanical measurements clearly show that upon hydration of the horny layer by topical addition either of simple water or of moisturizers (glycerine, urea, etc.) or emollients (petrolatum), some elastic mechanical parameters and viscoelastic parameters are significantly modified. These changes in the hydration state of the stratum corneum, which were assessed by electrical hydration measurements and by mechanical properties, occurred during both short-term and long-term hydration treatments.

The mechanical devices that were capable of detecting in a significant way changes in elastic and viscoelastic properties of the skin were the Twistometer–Dermal Torque Meter, the Cutometer, the Dermaflex, and some experimental systems, such as the Gas Bearing Extensometer and the tactile sensor.

When considering only the commercially available devices, it is surprising to note that, although there are great differences in the methods (torsion vs. suction), in the surface of the skin that is mechanically modified (1 mm annulus between disk and guard ring in the torsion device, from 2- to 10-mm probe aperture in the suction device), and in the amount of stress (low torque and from 100 to 500 mbar suction), the three instruments are sensitive enough to measure significant changes in the biomechanical properties of the skin when the hydration state of the horny layer is modified.

Which instrument is the most sensitive device for measuring most specifically the mechanical properties of the horny layer? Considering the fact that in the torsion method a small skin surface between rotating disk and guard ring is submitted to small values of stress (torque), the Dermal Torque Meter has a significant sensitivity to measure more specifically the mechanical properties of the horny layer.

Because of the size of the Cutometer probe (2 to 8 mm) and Dermaflex probe (10 mm), and the deformation magnitudes involved, it is obvious that their mechanical measurements involve not only the entire stratum corneum but also other layers, such as the viable epidermis, dermis, and possibly the subcutis.[41] However, by using the Cutometer with the small, 2-mm probe and lower values of suction (200 mbar), a better sensitivity to moisturization effect is observed and the instrument measures mainly the contribution of the mechanical properties of the horny layer.[17] It is clear that measurements with the Dermaflex involve the mechanical properties of the whole skin with probably a contribution of the subcutis. However, Jemec and Serup[39] found correlations between some mechanical properties and capacitance measurements with the Corneometer.

Which mechanical parameters are the most sensitive to changes in hydration of the horny layer? Very contradictory results are found in the literature. Surprisingly, pure elastic parameters as well as viscoelastic parameters are sensitive to hydration of the horny layer and to the effect of moisturizers. Some elastic and/or viscoelastic parameters are more sensitive than others to moisturization effects and some moderate correlations are observed between the mechanical parameters as measured with the Dermal Torque Meter and with the Cutometer.

Concerning the correlations between electrical hydration measurements (capacitance and conductance) and mechanical properties, very contradictory results are observed. Some studies have clearly shown a correlation between the electrical hydration measurements and elastic and viscoelastic parameters (although the correlations were generally moderate), and in other studies no correlations at all were found.

In conclusion, hydration of the horny layer as produced by various moisturizers influences the elastic and viscoelastic properties of the skin. Noninvasive skin elasticity measurements carried out with the Dermal Torque Meter and with the Cutometer (2-mm probe at low vacuum) are appropriate for quantitative evaluation of the complex effects of various cosmetic hydrating products on the mechanical properties of the skin.

Ideally, the effects of moisturizers on the skin should be first assessed by electrical hydration measurements (Corneometer, Nova, DermaLab, or Skicon) and these assessments complemented by study of the effect of the moisturizers on epidermal elastic and viscoelastic properties.

REFERENCES

1. Tagami, H., Impedance measurements for evaluation of the hydration state of the skin surface, in *Cutaneous Investigation in Health and Disease, Non-Invasive Methods and Instrumentation*, Lévêque J.L., Ed., Marcel Dekker, New York, 1989, 79–111.
2. Tagami, H., Measurement of electrical conductance and impedance, in *Handbook of Non-Invasive Methods and the Skin*, Serup, J. and Jemec, G.B.E., Eds., CRC Press, Boca Raton, FL, 1995, 159–164.
3. Loden, M. and Lindeberg, M., Product testing of moisturizers, in *Bioengineering of the Skin: Water and Stratum Corneum*, Elsner, P., Berardesca, E., and Maibach, H.I., Eds., CRC Press, Boca Raton, FL, 1994, 275–289.
4. Loden, M., Biophysical methods of providing objective documentation of the effects of moisturizing creams, *Skin Res. Technol.*, 1, 101–108, 1995.
5. Edwards, C. and Marks, R., Hydration and atopic dermatitis, in *Bioengineering of the Skin: Water and Stratum Corneum*, Elsner, P., Berardesca, E., and Maibach, H.I., Eds., CRC Press, Boca Raton, FL, 1994, 235–242.
6. Serup, J., Hydration in psoriasis and eczema: the dry surface–high evaporative water loss paradox, in *Bioengineering of the Skin: Water and Stratum Corneum*, Elsner, P., Berardesca, E., and Maibach, H.I., Eds., CRC Press, Boca Raton, FL, 1994, 243–249.
7. Serup, J., EEMCO guidance for the assessment of dry skin (xerosis) and ichtyosis: clinical scoring systems, *Skin Res. Technol.*, 1, 109–114, 1995.
8. Berardesca, E., Fideli, D., Borroni, G. Rabiosi, G., and Maibach, H.I., *In vivo* hydration and water retention capacity of stratum corneum in clinically uninvolved skin atopic and psoriatic patients, *Acta Derm-Venereol.* (Stockh.), 70, 400–404, 1990.
9. Piérard, G., What does "dry skin" mean? *Int. J. Dermatol.*, 26, 167–168, 1987.
10. Barel, A.O., Clarys, P., and Gabard, B., *In vivo* evaluation of the hydration state of the skin: measurement and methods for claim support, in *Cosmetics: Controlled Efficacy Studies and Regulation*, Elsner, P., Merk, H.F., and Maibach, H.I., Eds., Springer-Verlag, Berlin, 1999, 57–80.
11. Moss, J., The effect of 3 moisturizers on skin surface hydration. Electrical conductance, capacitance and trans epidermal water loss, *Skin Res. Technol.*, 2, 32–36, 1996.
12. Agache P., Twistometry measurement of elasticity, in *Handbook of Non-Invasive Methods and the Skin*, Serup, J. and Jemec, G.B.E., Eds., CRC Press, Boca Raton, FL, 1995, 319–328.
13. Rodrigues, L., The *in vivo* biomechanical testing of the skin and the cosmetological efficacy claim support: a critical overview, in *Cosmetics: Controlled Efficacy Studies and Regulation*, Elsner, P., Merk, H.F., and Maibach, H.I., Eds., Springer-Verlag, Berlin, 1999, 197–208.
14. Bernengo, J.C. and de Rigal, J., Techniques physiques de mesure de l'hydratation du stratum corneum *in vivo*, in *Physiologie de la peau et explorations fonctionnelles cutanées*, Agache, P., Ed., Editions Médicales Internationales, Cachan, 2000, 117–162.

15. Agache, P. and Varchon, D., Exploration fonctionnelle mécanique, in *Physiologie de la peau et explorations fonctionnelles cutanées*, Agache, P., Ed., Editions Médicales Internationales, Cachan, 2000, 423–443.
16. Barel, A.O., Lambrecht, R., and Clarys, P., Mechanical function of the skin: state of the art, in *Skin Bioengineering Techniques and Applications in Dermatology and Cosmetology*, Elsner, P., Barel, A.O., Berardesca, E., Gabard, B., and Serup, J., Eds., Karger, Basel, 1998, 69–83.
17. Wickett, R., Stretching the skin surface: skin elasticity, *Cosmet. & Toiletries*, in press, 2001.
18. Wijn P., The Alinear Viscoelastic Properties of Human Skin *in Vivo* for Small Deformations, Ph.D. thesis, Catholic University of Nijmegen, Nijmegen, Holland, 1980.
19. Maes, D., Short, J., Turek, B.A., and Reinstein, J.A., *In vivo* measuring of skin softness using the Gas Bearing Electrodynamometer, *Int. J. Cosmet. Sci.*, 5, 189–200, 1983.
20. Hargens, C.W., The Gas Bearing Electrodynamometer, in *Handbook of Non-Invasive Methods and the Skin*, Serup, J. and Jemec, G.B.E., Eds., CRC Press, Boca Raton, FL, 1995, 353–357.
21. Manny-Aframian, V. and Dikstein, S., Levarometry, in *Handbook of Non-Invasive Methods and the Skin*, Serup, J. and Jemec, G.B.E., Eds., CRC Press, Boca Raton, FL, 1995, 345–347.
22. Manny-Aframian, V. and Dikstein, S., Indentometry, in *Handbook of Non-Invasive Methods and the Skin*, Serup, J. and Jemec, G.B.E., Eds., CRC Press, Boca Raton, FL, 1995, 349–352.
23. Hargens, C.W., Ballistometry, in *Handbook of Non-Invasive Methods and the Skin*, Serup, J. and Jemec, G.B.E., Eds., CRC Press, Boca Raton, FL, 1995, 359–364.
24. Sakai, S., Sasai, S., Endo, Y., Matue, K., Tagami, H., and Inoue, S., Characterization of the physical properties of the stratum corneum by a new tactile sensor, *Skin Res. Technol.*, 6, 128–134, 2000.
25. Gniadecka., M. and Serup, J., Suction chamber method for measurement of skin mechanical properties: the Dermaflex, in *Handbook of Non-Invasive Methods and the Skin*, Serup, J. and Jemec, G.B.E., Eds., CRC Press, Boca Raton, FL, 1995, 329–334.
26. Lévêque, J.L., *In vivo* methods for measuring the visoelastic properties of the skin, *Bioeng. Skin*, 3, 375–382, 1987.
27. Escoffier, C., de Rigal, J., Rochefort, A.D., Vasselet, R., Lévêque, J.L., and Agache, P.G., Age-related mechanical properties of human skin: an *in vivo* study, *J. Invest. Dermatol.*, 93, 353–357, 1989.
28. Piérard, G., A critical approach to *in vivo* mechanical testing of the skin, in *Cutaneous Investigation in Health and Disease*, Lévêque, J.L., Ed., Marcel Dekker, New York, 1989, 215–240.
29. Piérard, G.E., Relevance, comparison and validation of technique, in *Handbook of Non-Invasive Methods and the Skin*, Serup, J. and Jemec, G.B.E., Eds., CRC Press, Boca Raton, FL, 1995, 9–14.
30. Barel, A.O., Courage, W., and Clarys, P., Suction method for measurement of skin mechanical properties: the Cutometer, in *Handbook of Non-Invasive Methods and the Skin*, Serup, J. and Jemec, G.B.E., Eds., CRC Press, Boca Raton, FL, 1995, 335–340.
31. Barel, A.O., Lambrecht, R., and Clarys, P., Mechanical function of the skin: state of the art, in *Skin Bioengineering. Techniques and Applications in Dermatology and Cosmetology*, Elsner, P., Barel, A.O., Berardesca, E., Gabard, B., and Serup, J., Eds., Karger, Basel, 1998, 69–83.

32. Couturaud, V., Coutable, J., and Khaiat, A., Skin biomechanical properties: *in vivo* evaluation of influence of age and body site by a non-invasive method, *Skin Res. Technol.*, 1, 68–73, 1995.
33. Murray, B.C. and Wickett, R., Sensitivity of Cutometer data to stratum corneum hydration level. A preliminary study, *Skin Res. Technol.*, 2, 167–172, 1996.
34. Agache, P.G., Monneur, C., Lévêque, J.L., and de Rigal, J., Mechanical properties of Young's modulus of human skin *in vivo*, *Arch Dermatol. Res.*, 269, 221–232, 1980.
35. Panisset, F., Le stratum corneum: sa place dans la fonction mécanique de la peau humaine *in vivo*, thesis, l'Université de Franche-Comté, Faculté de Médecine et de Pharmacie, Besançon, France, 1992.
36. Lévêque, J.L. and de Rigal, J., *In vivo* measurement of the stratum corneum elasticity, *Bioeng. Skin*, 1, 13–23, 1985.
37. Aubert, L., Anthoine, P., de Rigal, J., and Lévêque, J.L., An *in vivo* assessment of the biomechanical properties of human skin modifications under the influence of cosmetic products, *Int. J. Cosmet. Sci.*, 7, 51–59, 1985.
38. Lévêque, J.L., Grove, G., de Rigal, J., Corcuff, P.A.M., and Saint-Leger, D., Biophysical characterization of dry facial skin, *J. Soc. Cosmet. Chem.*, 82, 171–177, 1987.
39. Jemec, G.B.E. and Serup, J., Epidermal hydration and skin mechanics, *Acta Derm. Venereol.* (Stockholm), 70, 245–247, 1990.
40. Olsen, L.O. and Jemec, G.B.E., The influence of water, glycerine, paraffin oil and ethanol on skin mechanics, *Acta Derm. Venereol.* (Stockholm), 73, 404–406, 1993.
41. Murray, B.C. and Wickett, R.R., Correlations between Dermal Torque Meter, Cutometer and Dermal Phase meter measurements of human skin, *Skin Res. Technol.*, 3, 101–106, 1997.
42. Barel, A.O. and Gabard, B., Unpublished results, 1999.
43. Dobrev, H., Use of Cutometer to assess epidermal hydration, *Skin Res. Technol.*, 6, 239–244, 2000.
44. Christensen, M.S., Hargens, C.W., Nacht, S., and Gaus, E.H., Viscoelastic properties of intact human skin, instrumentation, hydration effect and the contribution of the stratum corneum, *J. Invest. Dermatol.*, 69, 282–286, 1977.
45. Cooper, E.P., Missel, P.J., Hannon, D.P., and Albright, G.B., Mechanical properties of dry, normal and glycerol-treated skin as measured by Gas Bearing Electrodynamometer, *J. Soc. Cosmet. Chem.*, 36, 335–347, 1985.

22 Antikeloidal Products

Wolf-Ingo Worret

CONTENTS

Historical Overview 257
Pathophysiology 258
Antikeloidal Products with Proven Effects 258
Intralesional Corticosteroids 259
Silicone Gel Sheets 260
Comparison of Corticoid Injections and Silicone Gel Sheets 262
Results with the Cutometer 262
Results with Ultrasonography 265
Extactum Cepae Gel 266
Conclusion 266
References 267

HISTORICAL OVERVIEW

Since ancient times, scars have caused cosmetic and psychic impairment. As early as 3000 to 2500 B.C., keloids were described in the "Smith papyrus."[1] The English term *scar* is derived from the Greek word *eschara*, i.e., fireplace. The Saxons at this time had a term for this phenomenon, as well: *scaur*. In France, scars were called *eschare*. The present scientific Latin term *cicatrix* was used in the Middle Ages for white spots of the skin as a sign of imperfection.

In dermatology, keloids and hypertrophic scars are separate entities. Keloids are scars in which the wound healing tissue grows out over the actual wound edges. In contrast, hypertrophic scars respect the wound edges. Such symptoms as itching, burning, and pain are more often observed in keloids than in hypertrophic scars

Jean Louis Alibert (1768–1837), the French dermatologist, first described the clinical characteristics of the condition. He used the Greek term *chele* (which means claw of a crayfish) and changed it into *cancroide* (i.e., similar to cancer) in 1806. Later he changed the term into *cheloid* to avoid confusion with malignant tumors, in particular with cancer.[2] In 1854, Addison suggested the terms *true* and *false keloids*. As true keloids he described ones that appeared spontaneously, whereas false keloids are ones that appear post-traumatically.[3]

0-8493-7521-5/02/$0.00+$1.50

PATHOPHYSIOLOGY

Scar formation and the pathophysiology of the origin of scars is not fully understood. However, there are some extraordinary phenomena in the genesis of keloids and hypertrophic scars. In early wound healing, in the early inflammation phase,[4] virtually no differences between normal tissue restoration and keloid formation can be observed. This phase is followed by fibroplasia, consisting of increased vessel formation and moderate perivascular cellular infiltration with mast cells, plasma cells, and lymphocytes. These also begin the formation of collagen clusters, fibers, and proteoglycans (components of the matrix). Contrary to normal wound healing, the massive fibroplasia of keloids originates in the third week.

Typically, keloids form nodular, vascular proliferations with an avascular, knotty accumulation of collagen and coating proteoglycans.[5] This mass mainly comprises fibroblasts. Fibroblasts are arranged in a storiform pattern and continue to proliferate in a hyalinated collagen cluster, which is typical for disturbed wound healing. One unique cell type in keloids — the myofibroblast — is a mixture of muscle cells and fibroblasts.[4] Its exact function is not understood, and it diminishes in fully developed scars. In hypertrophic scars and keloids, there is an increased cellular activity from the elevation of the DNA concentration with increased production of type I procollagen mRNA.[6] The fibroblasts in keloids are larger than those in hypertrophic scars. Also, the activity of these cells is significantly greater than that of hypertrophic scars.[7]

The raised metabolic activity results from an increase in various enzymes, for example, glycolytic enzymes, glycoproteins, and the fibronectin deposits. Not only the quantity of fibroblasts, but also their density increases.[7] The expectation that collagen degradation is enhanced is not correct; the opposite is true: degradation is decreased.[8] The local extent of keloids is mainly caused by an increase of ground substance.[4]

A further difference between keloids and hypertrophic scars is water content: the higher proportion of water in keloids can be explained by the increased production of hyaluronic acid, which has a strong water-binding capacity. This fact is important for the mode of action of glucosteroid treatment.[9]

ANTIKELOIDAL PRODUCTS WITH PROVEN EFFECTS

Keloid treatments include surgical excision, pressure bandaging, connective tissue massages (now with a new hydraulic device: LPG, Sophia Antipolis, France), cryotherapy, low-dose radiotherapy, topical silicone gel sheet, a topical onion extract gel (Contractubex®, Merz&Co, Frankfurt, Germany), interferon, and intralesional injection of corticosteroids, mostly triamcinolone acetonide. Today, in most clinics a polypragmatic procedure is accepted: excision, then cryotherapy, then triamcinolone injections, rarely X-ray radiation, and after that pressure bandages. However, there remain keloids unresponsive to this management.

Earlier, primarily subjective gradings (for example "good/poor results") or subjective clinical scores were used for assessing efficacy. For quantitative assessment, however, noninvasive bioengineering methods, i.e., skin thickness measurements by means of high-frequency ultrasound and measurements of mechanical properties or of skin color, are used. Because keloids diminish with time and are virtually not seen in elderly people, it is mandatory to test antikeloidal products by objective methods over a short period of time, i.e., some months.

This chapter discusses antikeloidal substances that have been proved to be efficient by objective bioengineering methods: corticoid injections, silicone gel sheet, and Contractubex. Nevertheless, not all measurements were reliable, and this is discussed below.

Intralesional Corticosteroids

The mechanical properties of 17 keloids in nine patients before and during treatment with intralesional triamcinolone acetonide were studied using a noninvasive suction device (Cutometer®, Courage and Khazaka, Cologne, Germany) for measuring skin elasticity and viscosity *in vivo*.[10] This method provides quantitative assessment of the mechanical properties of scars. It is well suited for comparative studies on the efficacy of various scar therapies. Each keloid was treated with intralesional injections of triamcinolone acetonide 10 mg/ml without local anesthetic at intervals of 3 weeks. Four measurements per keloid were performed before treatment and 3 weeks after the first, second, and third treatments (Figure 22.1).

The parameters used were immediate distension (U_e), delayed distension (U_v), immediate retraction (U_r), and final distension (U_f). Relative parameters independent of skin thickness were calculated: U_v/U_e, the ratio between the viscous and the elastic deformation of the skin, and U_r/U_f, representing the ability of the skin to return to its initial position after deformation (biologic elasticity).

After three injections of triamcinolone acetonide a marked decrease of U_v/U_e and a less-pronounced increase of U_r/U_f compared to baseline values was observed

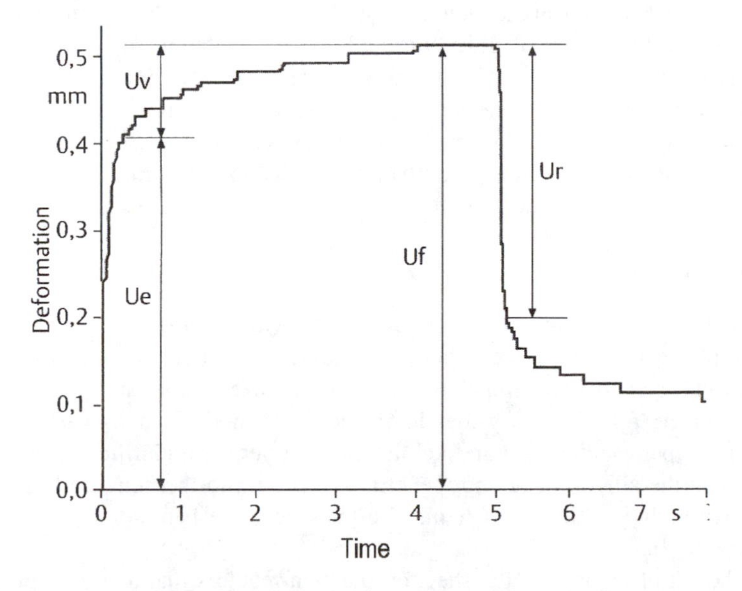

FIGURE 22.1 Graph from physiological skin as released from the Cutometer.

(Figure 22.2). These findings indicate that the primary effect of intralesional corticoids on the connective tissue of keloids is decreased viscosity as a result of a loss of ground substance. A comparison of the results of Cua et al.[11] for normal skin and the baseline values of keloids in the present study shows that U_v/U_e is higher and U_r/U_e is considerably lower in keloids, indicating that the viscous component in keloid tissue is increased whereas elasticity is decreased. The results of the investigation described above suggest that the decrease of viscosity, represented by the ratio U_v/U_e, is the main parameter that marks the beginning of the triamcinolone action on the connective tissue of keloids. Although U_r/U_f, the biological elasticity, was enhanced significantly after therapy, it seems untrue that this is the hallmark of the improvement of scars.

What alterations in the structure of keloid tissue might correspond to these changes in mechanical properties? The viscous component of the mechanical properties is thought to be dependent on the ground substance in the dermis.[11] The ground substance contains water and glycosaminoglycans, which are linked to peptide chains to form macromolecules called proteoglycans. Vogel[12] showed that the viscosity of rat skin correlated to the content of glycosaminoglycans and soluble collagen in the ground substance.

Glucocorticoids inhibit the synthesis of glycosaminoglycans at low concentrations in human skin fibroblast culture,[13] and this inhibition may be important for mediating the rapid thinning of the dermis after glucocorticoid treatment. This may be due to the ability of the glycosaminoglycans to retain water, and thus a reduced amount of glycosaminoglycans in the dermis may lead to a decrease in water content and to smaller interfibrillar spaces between the collagen and elastic fibers.[14] After 6 weeks of occlusive application of clobetasol propionate to normal skin, a loss of ground substance was observed by electron microscopy,[15] resulting in a rearrangement of the dermal fibrous network, while the fibers themselves were not altered.

Although the effects of glucocorticoids on the metabolism of the glycosaminoglycans can be observed within a few weeks, the degradation of collagen and elastin occurs much later. The apparent resistance of dermal fibers to short-term corticoids may reflect the longer half-life of these components.[14] Thus in this study, which was completed 9 weeks after the first application of intralesional steroids, no considerable effect on the collagen and elastic fibers can be expected.

Silicone Gel Sheets

The results in the reduction and improvement with the treatment of keloids and hypertrophic scars with silicone gel sheets are undisputed.[16–18] The several similar silicone sheets available differ only minimally in size, shape, and price. The composition of the foils is nearly identical. The exact mode of action of the silicone sheets on hypertrophic scars and keloids has not yet been clarified. However, it is certain that the silicone does not penetrate directly into the skin, and that there is no foreign body reaction.[18,19] Chemical effects have not been totally excluded, but are unlikely.[20]

Although placement of the sheet on the skin requires that a certain pressure be applied, this pressure is not sufficiently powerful to provoke an improvement of the

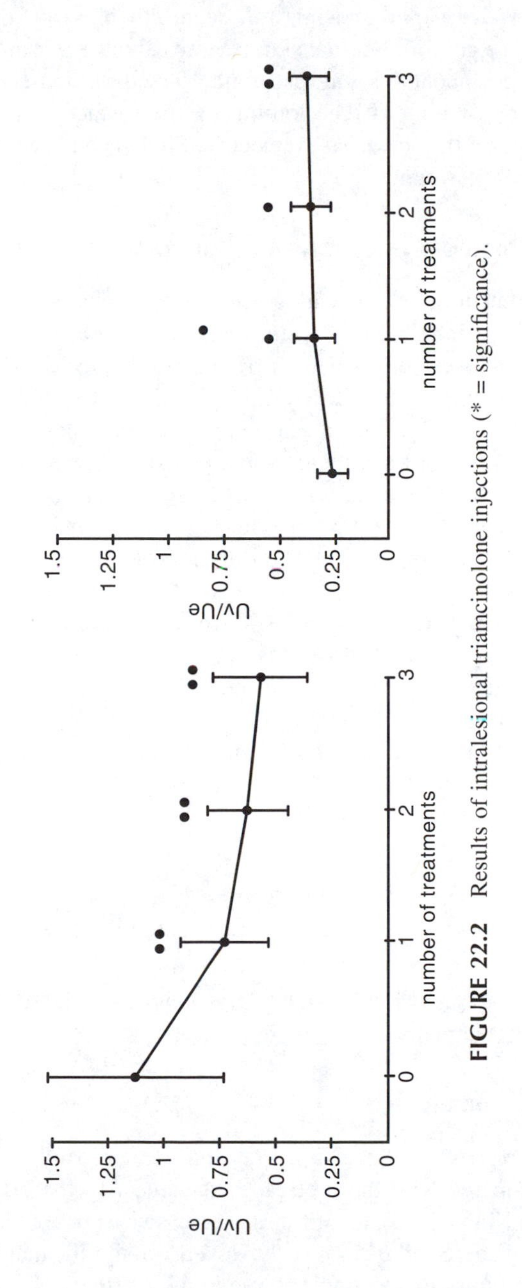

FIGURE 22.2 Results of intralesional triamcinolone injections (* = significance).

scar. To be successful with bandage therapy it is necessary to reach pressures between 15 and 40 mm Hg[21] over at least 4 to 6 months. Under the silicone sheet only pressure values between 1 and 12.8 mm Hg[20] were recorded, which is not enough to affect the scar. Water vapor transmission seems to play an important role in treatment with silicone gel foil. Normal skin releases about 8.5 g/m^2 water per hour, hypertrophic scars only about 4.5 g/m^2 per hour.[22] The development of an electrical field, for example, by rubbing of the clothing on the surface of the foil, probably has a positive impact on the scars. These electrical fields can reduce the number of the mast cells in healing wounds.[16]

Comparison of Corticoid Injections and Silicone Gel Sheets

To compare the treatment of keloids with corticoid injections vs. silicone gel sheets, a study was carried out in the author's clinic.[23] The study examined 25 keloids in 11 patients: 11 keloids were injected with triamcinolone acetate (Volon A 10®); 8 were treated with silicone gel foil (EPI-DERM™ Silicon Gel Folie; Fa. INAMED GmbH, Düsseldorf, Germany); and 6 scars served as the control group and were not treated. The mean age of the patients was 28 years (between 21 and 54 years). The sex distribution male to female was 7:4. The scars were mostly located above the sternum,[5] on the back,[4] and in the scapular region.[4] Other keloids were situated on buttocks, nape, and the upper arm. The duration of the scars varied between 3 months and 12 years.

Measurements were taken before and 4 weeks after the individual first and second treatment, respectively. The keloids were photographed, first, to trace the specific spot of measurement, especially in multiple scars, and, second, to demonstrate the clinical results of the treatment.

For determination of elasticity and viscosity, the Cutometer was used as above. For our calculations we used:

$$R_0 = U_f$$
$$R_6 = U_v/U_e \text{ (viscosity)}$$
$$R_7 = U_r/U_f \text{ (biological elasticity)}$$

Skin thickness was measured by 20 MHz ultrasonography (DUB 20, Taberna pro Medicum, Lüneburg, Germany).

Results with the Cutometer

Both the maximal expansion of the skin (R_0) and the viscosity (R_6) did not show any changes as compared with the pretreatment values. In the elasticity modulus (R_7) (Figure 22.5 and Table 22.3), there was an increase in time, but it was not related to therapy. By using the Friedmann test, however, no significance was seen. This observation is in contrast to the findings regarding corticoid injections alone as mentioned above.

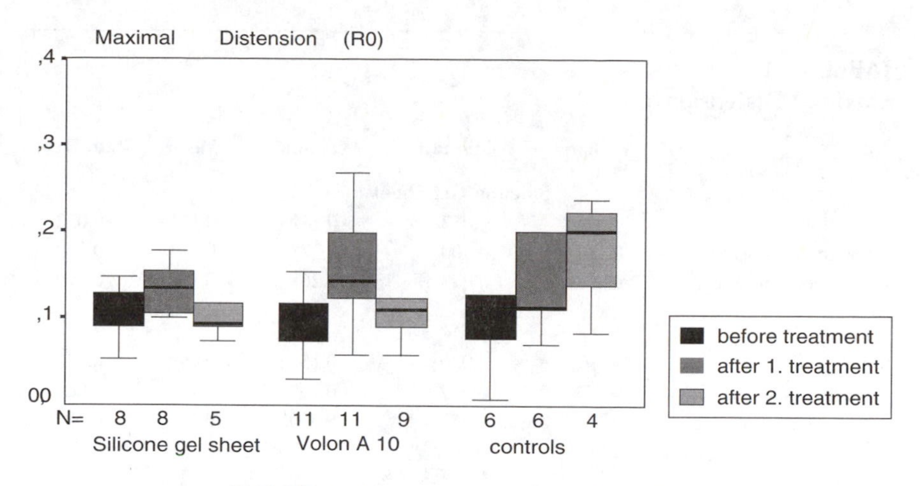

FIGURE 22.3 Results of maximal distension.

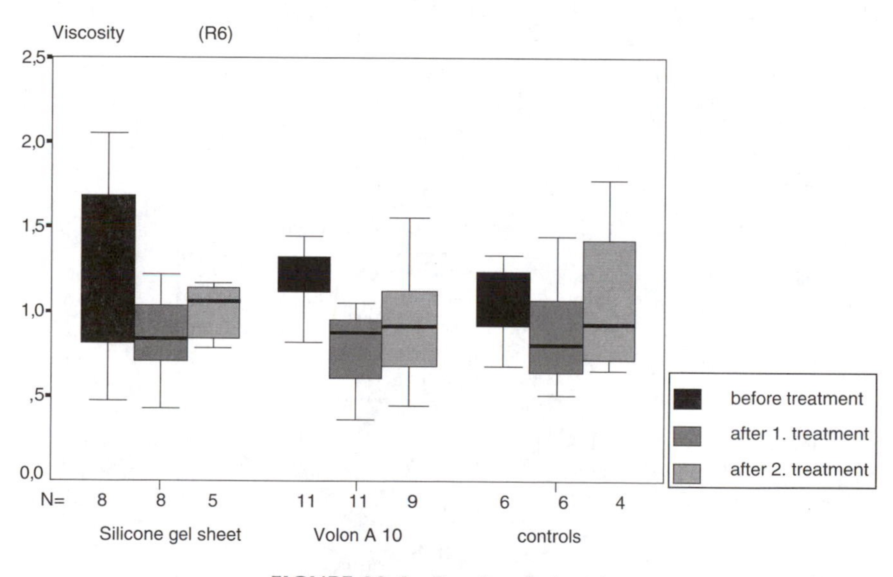

FIGURE 22.4 Results of viscosity.

In discussing the divergent results obtained using an objective skin bioengineering device, one has to reflect upon the possibilities and the limits of the Cutometer. The measurement principle of this device is a probe with an aperture 2 mm in diameter. Through this opening about 3 mm^2 of skin is drawn by a defined negative pressure of 500 mbar. Relaxation to zero occurs 5 s later, and this relaxation of the skin is calculated during this time via photoelements.

To obtain precisely reproducible results with such a small aperture, it is mandatory to perform the measurements on the same point of the keloid. Also, the forces of the negative pressure into the deeper tissues are not known and may vary.[24] In

TABLE 22.1
Maximal Distension

	Median	Minimum	Maximum	Mean	Std. Dev.
		Silicone Gel Sheet			
Before treatment	0.125	0.053	0.147	0.111	0.032
After first treatment	0.134	0.100	0.177	0.133	0.029
After second treatment	0.093	0.073	0.267	0.128	0.079
		Volon A 10			
Before treatment	0.107	0.030	0.153	0.099	0.037
After first treatment	0.143	0.057	0.267	0.158	0.068
After second treatment	0.110	0.057	0.390	0.130	0.101
		Controls			
Before treatment	0.117	0.007	0.207	0.109	0.066
After first treatment	0.113	0.070	0.373	0.163	0.111
After second treatment	0.200	0.083	0.237	0.180	0.067

TABLE 22.2
Viscosity

	Median	Minimum	Maximum	Mean	Std. Dev.
		Silicone Gel Sheet			
Before treatment	1.088	0.467	2.056	1.212	0.568
After first treatment	0.835	0.429	1.214	0.850	0.254
After second treatment	1.056	0.783	1.167	0.997	0.175
		Volon A 10			
Before treatment	1.194	0.782	2.0	1.228	0.326
After first treatment	0.876	0.363	1.833	0.863	0.389
After second treatment	0.916	0.445	1.556	0.923	0.370
		Controls			
Before treatment	1.089	0.678	1.333	1.057	0.234
After first treatment	0.802	0.508	1.444	0.878	0.339
After second treatment	0.924	0.652	1.776	1.069	0.502

addition, one must take the pretension of the tissue into account. This must be absolutely similar in each measurement.[25] The areas near joints must be measured always in the same angle of the joint to calculate with the same pretension. From all these points, it becomes evident that the Cutometer is not the correct device to obtain exact reproducible data of the improvement of keloids and hypertrophic scars *in vivo*.

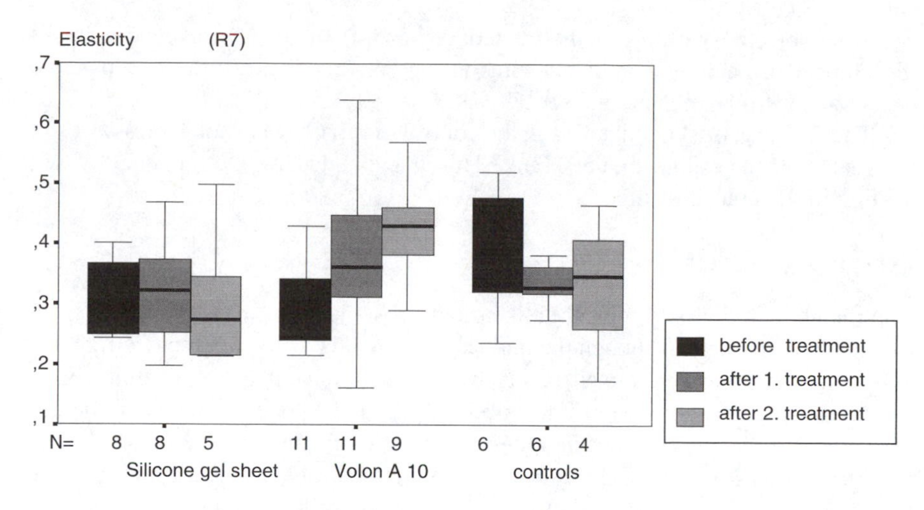

FIGURE 22.5 Results of elasticity.

TABLE 22.3
Elasticity

	Median	Minimum	Maximum	Mean	Std. Dev.
		Silicone Gel Sheet			
Before treatment	0.294	0.244	0.400	0.308	0.063
After first treatment	0.322	0.198	0.467	0.320	0.089
After second treatment	0.274	0.213	0.498	0.309	0.119
		Volon A 10			
Before treatment	0.274	0.215	0.429	0.298	0.073
After first treatment	0.362	0.161	0.640	0.381	0.130
After second treatment	0.430	0.289	0.568	0.417	0.084
		Controls			
Before treatment	0.367	0.236	0.520	0.381	0.106
After first treatment	0.327	0.274	0.382	0.331	0.038
After second treatment	0.347	0.173	0.465	0.333	0.120

Results with Ultrasonography

In the group treated with corticosteroids, the thickness of the scars was reduced as much as 60%. Also the skin thickness regressed quickly for 29% after the first treatment. In the silicone sheet–treated group this regression was only about 5%. In the course of the treatment, echogenity of the dermis and the input echo increased. A disadvantage of the ultrasonograph is that the maximal input depth of the device is only about 7 mm. The thickness of keloids, of course, could be larger. In these cases the objectivity of this method is reduced. With both, the ultrasonography and the clinical inspections combined with interviews, treatment with corticosteroids was superior to the silicone gel sheet.

The decrease of skin thickness in the control group (-7%) is noteworthy. Despite no treatment, the keloid tissue shrank. This is explained by the tendency of self-healing in the course of time.

EXTACTUM CEPAE GEL

Contractubex is the only topical antikeloidal substance with proven effects on the international markets. Although the main active agent is a phytoextract from onions, it has not yet been determined precisely which ingredient is responsible for the improvement of scars. Several effects on keloids and hypertrophic scars could be demonstrated that act together in softening and diminishing such lesions.

Antiproliferative effect: In cultures of human fibroblasts of different sources (scars, keloids, embryonal tissue) the inhibition of skin fibroblasts was maximal between 43 and 46%; in fibroblasts of keloids it was more pronounced between 38 and 53%.[26] An inhibition of proliferation of keloid fibroblasts and fibroblasts of other tissues was established for all three components of Contractubex: heparin, allantoin, and the onion extract.[27]

Inhibition of proteoglycan and collagen formation: The effect of the formulation on the formation of such extracellular matrix components as proteoglycans and collagen was studied in cultures of human skin.[28] The results showed a clear and dose-related reduction of proteoglycans and collagen. In an animal experiment it was established that pathologically enhanced synthesis of collagen was inhibited, which was true in scar tissue but not in normal skin.[29] Based on these results, it is assumed that Contractubex in scars with pathologically increased concentration of collagen and proteoglycans — as is the case in keloids and hypertrophic scars — can lead to improvement of scar structures.

Breaking up the arrangement of scar tissue: In specimens of human scars it has been histologically proved that the combination in Contractubex inhibits the polymerization of collagen and stimulates the decartelization of collagen.[30] In this manner, a certain breakup of the scar tissue is achieved. In fresh scars, excessive collagen polymerization is prevented.

Improvement of scar elasticity: The effect of Contractubex on the elasticity of scars was studied in animal experiments on the basis of viscoelastic characteristics.[31] At the end of the treatment the "stiffness index" was measured and calculated against the untreated scars of the control group. The index was markedly diminished in treated areas. It also became evident that normal skin treated with the onion extract was not affected by the therapy. The improvement of scar elasticity is of clinical importance in scars that cause functional impairment.

CONCLUSION

Noninvasive bioengineering methods seem to be well suited for studies on the efficacy of various therapies in keloids and hypertrophic scars. As the mechanical properties are only one aspect of keloid tissue, elasticity measurements should be combined with other bioengineering methods such as skin thickness assessments by means of high-frequency ultrasound or quantitative measurement of skin color with a colorimeter. In the author's experience, the Cutometer is probably not the optimal device for the evaluation of keloid tissue. Other methods should be defined to obtain more precise measurements. A more versatile instrument may be the uniaxial extensometer described by Edwards (Cardiff) at the 13th Congress of the International Society of Bioengineering and Skin in March 2000. At the same meeting, Marshall (Glasgow) presented a new method of picture analysis applied to scar tissue. The disadvantage is that it is an invasive histological technique. The method is based on calculating the alignment of the nuclei of fibrocytes. In hypertrophic scars, the bridges of collagen and the fibrocytic nuclei are oriented in a parallel manner to the skin surface, whereas in a keloid, most nuclei are located within a 30° angle. Further, by measuring these angles it is possible to obtain an accurate notion of the severity of a scar.

REFERENCES

1. Murray, J.C., Pollack, S., and Pinnell, S.R., Keloids: a review, *J. Am. Acad. Dermatol.*, 4, 461–470, 1981.
2. Kelly, A.P., Keloids, *Dermatol. Clin.*, 6, 413–424, 1988.
3. Rockwell, W.B., Cohen, I.K., and Ehrlich, H.P., Keloids and hypertrophic scars: a comprehensive review, *Plast. Reconstr. Surg.*, 84, 827–837, 1989.
4. Murray, J.C., Scars and keloids, *Dermatol. Clin.,* 11, 697–708, 1993.
5. Linares, H.A. and Larson, D.L., Proteoglycanes and collagenase in hypertrophic scar formation, *Plast. Reconstr. Surg.,* 52, 589, 1978.
6. Kikuchi, K., Kadono, T., and Takehara, K., Effects of various growth factors and histamine on cultured keloid fibroblasts, *Dermatology,* 190, 4–8, 1995.
7. Nakaoka, H., Miyauchi, S., and Miki, Y., Proliferating activity of dermal fibroblasts in keloids and hypertrophic scars, *Acta Derm. Venereol.* (Stockholm), 75, 102–104, 1995.
8. Craig, R.D.P., Schofield, J.D., and Jackson, D.S., Collagen biosynthesis in normal and hypertrophic scars and keloids: a function of the duration of the scar, *Br. J. Surg.*, 62, 741–744, 1975.
9. Asboe-Hansen, G., Hypertrophic scars and keloids: etiology, pathogenesis and dermatologic therapy, *Dermatologica,* 120, 178–180, 1960.
10. Krusche, T. and Worret, W.-I., Mechanical properties of keloids *in vivo* during treatment with intralesional triamcinolone acetonide, *Arch. Dermatol. Res.*, 287, 289–293, 1995.
11. Cua, A.B., Wilhelm, K.-P., and Maibach, H.I., Elastic properties of human skin: relation to age, sex, and anatomical region, *Arch. Dermatol. Res.,* 282, 283–288, 1990.
12. Vogel, H.G., Age dependence of viscoelastic properties in rat skin; directional variations in relaxation experiments, *Bioeng. Skin.* 1, 157–174, 1985.
13. Saarni, H. and Hopsu Havu, V.K., The decrease of hyaluronate synthesis by anti-inflammatory steroids *in vitro*, *Br. J. Dermatol.*, 98, 445–449, 1978.
14. Oikarinen, A., Dermal connective tissue modulated by pharmacologic agents, *Int. J. Dermatol.*, 31, 149–156, 1992.

15. Lehmann, P.L., Zheng, P., Lavker, R.M., and Kligman, A.M., Corticosteroid atrophy in human skin. A study by light, scanning, and transmission electron microscopy, *J. Invest. Dermatol.*, 81, 169–176, 1983.
16. Hirshowitz, B., Ullmann, Y. et al., Silicone occlusive sheeting (SOS) in the management of hypertrophic and keloid scarring, including the possible mode of action of silicone, by static electricity, *Eur. J. Plast. Surg.*, 16, 5–9, 1993.
17. Quinn, K.J., Silicone gel in scar treatment, *Burns,* 13, 33–40, 1987.
18. Wong, T.W., Chiu, H.C. et al., Symptomatic keloids in two children. Dramatic improvement with silicone cream occlusive dressing, *Arch. Dermatol.*, 131, 775–777, 1995.
19. Clugston, P.A., Vistnes, M.D. et al., Evaluation of silicone-gel sheeting on early wound healing of linear incisions, *Ann. Plast. Surg.*, 34, 12–15, 1995.
20. Ahn, S.T., Monafo, W.W., and Mustoe, T.A., Topical silicone gel: a new treatment for hypertrophic scars, *Surgery,* 106, 781–787, 1989.
21. Quinn, K.J., Evans et al., Non-pressure treatment of hypertrophic scars, *Burns*, 12, 102–108, 1985.
22. Ohmori, S., Effectiveness of Silastic sheet coverage in the treatment of scar keloid (hypertrophic scar), *Aesth. Plast. Surg.*, 12, 95–99, 1988.
23. Schuster, K., Die vergleichende Treatment von pathologischen Narben mit Silikonfolien und Steroidinjektionen und deren Evaluierung mittels hautphysiologischen Meßmethoden, Thesis, Technische Universität München, Germany, 2001.
24. Clark, J.A., Cheng, J.C., Leung, K.S., and Leung, P.C., Mechanical characterisation of human postburn hypertrophic skin during pressure therapy, *J. Biomech.*, 20, 397–402, 1987.
25. Wilkes, G., Brown, I., and Wildnauer, R., The biomechanical properties of skin, *Crit. Rev. Bioeng.*, 453–495, 1973.
26. Wülfroth, P., Investigation of the anti-proliferative effect of the Contractubex ingredients on human fibroblasts, *Merz Forschungsber.,* 1989.
27. Majewski, S. and Chadzynska, M., Effects of heparin, allantoin and cepae extract on the proliferation of keloid fibroblasts and other cells *in vitro*, poster presentation MS-No 330/P16, 17th World Congress of Dermatology, 24–29 May 1987, Berlin, Germany.
28. Wiesmann, P., Effect of Contractubex, an anti-keloid-medication, and its individual components on the synthesis and secretion of glycosaminoglycan, sulfated glycosaminoglycans and collagen in cultured skin fibroblasts, *Merz Forschungsber.,* 1991.
29. Janicki, S. and Sznitowska, M., Effect of ointments for treating scars and keloids on metabolism of collagen in scar and healthy skin, *Eur. J. Pharm. Biopharm.*, 37, 188–191, 1991.
30. Heine, H., Narbentreatment durch transepidermale Heparinisierung, *Therapeutikon,* 6, 369–375, 1989.
31. Janicki, S., Sznitowska, M., Maceluch, J., and Jaskowski, J., Studies on the viscous–elastic properties of the skin in evaluating ointments to improve the elasticity of scars, *Przegl. Dermatol.*, 75, 352–357, 1988.

Index

A

Abdomen
  human *in vivo* Cutometer studies, 187–195
  human *in vivo* Dermagraph studies, 131–134
Acetate, cortisol, 19
Acetate, prednisolone
  animal *in vitro* (*ex vivo*) isorheological point studies, 24
  animal *in vitro* (*ex vivo*) stress–strain studies, 19
  animal *in vivo* recovery studies, 30
Acne treatment, 234
Acoustic wave testing, *see* Shear wave propagation (SWP) testing
Acquired skin diseases, 6
Acral skin vs. truncal skin, 44
Acroscleroderma. *see also* Scleroderma
  inhibition of joint motion, 45
  Type I systemic scleroderma, 209, 211, 212
ACTH, 20
Actinic elastosis
  collagen, appearance of, 10, 11
  elastic fibers in, 11
  elastin, abnormal accumulation of, 10
  and exposure to UV radiation, 11, 233
  human *in vivo* Cutometer studies, 94
  and MMPs, 11–12
  papillar dermis, appearance of, 193
  skin thickening in, 233
  Ultraviolet (UV) radiation, 11, 233
Adhesion model of friction, 56
Adipocytes, 210
Adipose panniculus, 43, 64
Adrenalectomized animals, 20
Age of skin, impact on
  animal *in vitro* (*ex vivo*) anisotropy studies, 21
  animal *in vitro* (*ex vivo*) creep(ing) studies, 24
  animal *in vitro* (*ex vivo*) hysteresis studies, 22
  animal *in vitro* (*ex vivo*) isorheological point studies, 24
  animal *in vitro* (*ex vivo*) relaxation studies, 22
  animal *in vitro* (*ex vivo*) repeated strain studies, 24
  animal *in vitro* (*ex vivo*) stress–strain studies, 19, 20
  animal *in vitro* (*ex vivo*) thermocontraction studies, 26–27
  animal *in vivo* recovery studies, 29–30
  animal *in vivo* repeated strain studies, 28–29
  atrophy, skin, 6, 7
  biassial tests, 232
  biological elasticity, 189, 232
  collagen, quantity of, 6, 10
  compression test, 232
  creep(ing), 24, 232–233
  deformation, 232–237
  elastic fibers, 189, 193
  elasticity, 11, 232–233
  by glycosaminoglycans, 193
  human *in vivo* AMP studies, 155–159
  human *in vivo* CAC studies, 157–159
  human *in vivo* COR studies, 157–158
  human *in vivo* creep(ing) studies, 24
  human *in vivo* Cutometer studies, 94, 188–194
  human *in vivo* Dermagraph studies, 133
  human *in vivo* DTM studies, 64, 74–75
  human *in vivo* Durometer studies, 142
  human *in vivo* elasticity studies, 44, 183, 188–194
  human *in vivo* IDRA studies, 155–159
  human *in vivo* skin friction studies, 49
  human *in vivo* stiffness studies, 155–159
  human *in vivo* stress–strain studies, 19
  human *in vivo* tensile strength studies, 207, 209
  human *in vivo* Twistometer studies, 64
  human *in vivo* viscoelasticity studies, 94
  keloid(s), 262
  plasticity, 22
  recovery, 11, 232
  relaxation, 232–233
  skin thickness, 10, 11, 19
  strain rate, 20
  uniassial tests, 232
  viscoelastic-to-elastic ratio, 189, 193–196
  viscoelasticity, 232
  of wounds, healing of, 45
  Young's modulus, 232
AHAs, *see* Alpha Hydroxy Acids (AHAs)
Alanines, 8
Aldehydes, 6
Algodystrophy, reflex, 139

Allantoin, 266
Alpha Hydroxy Acids (AHAs), 233–234
Amino–acetonitrile, 20
Amino acids
  alanines, 8
  aldehydes, formation of, 6
  ε-amino groups, oxidation of, 6
  glycine, 4, 5
  4–hydroxyproline, 4, 5
  lysine, 4, 8
  natural moisturizing factor, 242
  proline, 4, 5
ε-amino groups, oxidation of, 6
Aminoterminal propeptides of Type I procollagens (PINP), 5
Aminoterminal propeptides of Type III procollagens (PIIINP), 5
AMP, *see* Displacement, amplitude or angular (AMP)
Anchoring fibrils, 4
Andrade's constant, 68–69
Androgens, 20
Anetoderma, 9, 10
Angular deformation, 65
Angular distortion, *see* Torsion, test method
Animal *in vitro* (*ex vivo*) anisotropy studies, 21
Animal *in vitro* (*ex vivo*) creep(ing) studies, 24
Animal *in vitro* (*ex vivo*) hysteresis studies, 21–23
Animal *in vitro* (*ex vivo*) isorheological point studies, 23–24
Animal *in vitro* (*ex vivo*) relaxation studies, 22–23, 25
Animal *in vitro* (*ex vivo*) repeated strain studies, 24
Animal *in vitro* (*ex vivo*) stress–strain studies, 18–21
Animal *in vitro* (*ex vivo*) thermocontraction studies, 26–27
Animal *in vitro* (*ex vivo*) tissue expansion studies, 20–21
Animal *in vitro* (*ex vivo*) wound studies, 25–26
Animal *in vivo* extensometer studies, 83
Animal *in vivo* hydration studies, 201–202
Animal *in vivo* recovery studies, 29–30
Animal *in vivo* stiffness studies, 83
Animal *in vivo* stress–strain studies, 27–28
Animal SWP studies, 201–202
Anisotropic properties of skin
  animal hydration studies, 201
  animal *in vitro* (*ex vivo*) studies, 21
  animal *in vivo* studies, 27–30
  animal SWP studies, 201–202
  DermaLab, 120
  and extensibility, 187
  human *in vivo* extensometer studies, 83
  human *in vivo* skin friction studies, 50
  human *in vivo* studies, 21, 187
  Langer's lines, *see* Langer's lines
  measurement of, 79
  modeling of, 64, 78, 87, 120
  muscular fibers, impact on, 21
  Young's modulus, 47, 120
Ankle
  human *in vivo* Cutometer studies, 187–195
  human *in vivo* diabetes studies, 143–145
  human *in vivo* elasticity studies, 187–195
Anti–inflammatory drugs, nonsteroidal, 20
Antiaging treatments, 231–237
Antikeloidal treatments, 258–267
Antioxidants, 235, 237
Apocrine glands, 3
Arms. *see also* Forearm
  human *in vitro* indentation studies, 163
  human *in vivo* Cutometer studies, 187–195
  human *in vivo* Dermagraph studies, 131–134
  human *in vivo* Durometer studies, 142
  human *in vivo* elasticity studies, 187–195
  human *in vivo* extensometer studies, 83
  human *in vivo* SB studies, 219–220
  keloid treatment assessment, 262
Ascorbate, 5
Ascorbic acid, 237
Asparagine residues, 5
Astronauts, 46
ATCH, 20
Athletes, 73
Atrophic morphea, Pasini–Pierini, 44
Atrophogenicity, 20
Atrophy, skin
  age–related, 6, 7
  animal *in vitro* (*ex vivo*) stress–strain studies, 19, 20
  collagen, appearance of, 10
  collagen, reduced amounts of, 6, 7
  corticosteroids, impact on, 45
  human *in vivo* extensometer studies, 82
  steroid–induced, 6, 7
  thinning of the skin, 7
  Type I collagen in, 7
  wrinkle formation, 10
Axis of skin under test, *see* Test axis
Axons, unmyelinated, 3

B

Back
  human *in vivo* Cutometer studies, 187–195
  human *in vivo* Dermagraph studies, 132–134
  human *in vivo* DTM studies, 73

human *in vivo* elasticity studies, 187–195
keloid treatment assessment, 262
Baker, Bader, and Hopewell extensometer, 83
Ball–elastomer test method, 50
Ballistometry, test method
human *in vivo* stress studies, 64
IDRA, 147–159
of perpendicular skin mechanics, 243
of photoaging, 234
Basal lamina, *see* Basement membrane (BM)
Basal plasma membrane of epithelial cells, 12
Basement membrane (BM), 4, 12, 41
Betamethasone dipropionate cream, 224
Bi–valent cross–links of polypeptide chains, 6
Biassial tests, 232
Biaxial elongation test method
elasticity studies, 43
INSTRON, 18–24, 26–27
tensile tests, 79
Biological elasticity
age of skin, 189, 232
and elastic fibers, 193
of the horny membrane, 251
human *in vivo* Cutometer studies, 94, 189, 191–192, 251, 259–260
human *in vivo* EDS studies, 218
human *in vivo* elasticity studies, 189, 191–192
human *in vivo* erysipelas studies, 224–225, 227
human *in vivo* hydration studies, 251
human *in vivo* keloid treatment studies, 259–260, 262, 265
human *in vivo* lymphedema studies, 226
human *in vivo* psoriasis studies, 222–223, 227
human *in vivo* SB studies, 219, 220, 227
human *in vivo* scleroderma studies, 211, 220, 227
parameter, Cutometer, 94, 180, 217
parameter, DTM, 180
and preconditioning, 182
stripping away of horny membrane, impact on, 251
Biosynthetic human growth hormone, 20
Biothesiometer, 144
Blanching, during tests, 167
Blistering, from epidermolysis bullosa, 12
BM, *see* Basement membrane (BM)
Bonding, skin to tester, 85
Bones and Type I collagen, 8
BPAG1, *see* Bullous pemphigoid antigen 1 (BPAG1)
Breast cancer, 114, 201
Bruising, 13
BTC–2000, 244
Bulk modulus, 210
Bullous pemphigoid antigen 1 (BPAG1), 12
Burger's rheological model, 68–69, 71
Burns, treatment of, 181
Buttocks, location of test site, 262

C

C–terminal disulfide bonds, 4
CAC, *see* Cutaneous Absorption Coefficient (CAC)
Calipers, skin thickness measurement, 18, 27, 28, 29
Calorimetry, differential scanning, 27
Calves, 131–134
Capacitance, skin
Corneometer, 201
Dermaflex A, 245
and hydration, 202–203, 249–251
Carboxyterminal propeptides of procollagens, 5
Cartilage, 11
Cells
endothelial, 12
epithelial, 12
Langerhans, 3
mast, 258, 262
Merkel, 3
plasma, 258
Cellular senescence, 11
Ceramides, 242
α chains, 4, 5
Cheeks, site of test, 249–250
Chele, 257
Chemical peelings, 96
Chondroitine sulfates, 10
Chromosome 7, 8
Chronoaging, 10, 231–237, *see also* Age of skin, impact on
Cicatrix, 257
Circadian variations of tensile strength, 20
Citric acid, 233
Clobetasol propionate, 260
Coal tar, 224
Coefficient of elasticity, 248
Coefficient of Restitution (COR), 152–158
Collagen
in actinic elastosis affected skin, 10, 11
age of skin, 6, 10
alterations in synthesis, 10–11, 234–237
animal *in vitro* (*ex vivo*) wound studies, 26
atrophy, 6, 7, 10
in basement membrane, 12
corticosteroids, impact on, 10
degradation of, 11–12
in diabetes affected skin, 6, 7

disease characterization, 13
in EDS affected skin, 7, 8, 9, 218–219
in erysipelas affected skin, 225
in fibroblasts, 4, 157–158
glucocorticoids, effect of, 6
and hydration, 260
importance to skin structure, 3, 13
in interstitial fluid, 5–6
and keloid(s), 6, 7
in lymphedema affected skin, 226
and MMPs, 11–12
molecular structure, 4–6
mRNA, 6
D–penicillamine, 10
reduction of, 266
retinoid studies, 234
in scars, 258
in Scleredema of Buschke affected skin, 219
skin thickening, 6, 7
and skin thickness, 9–11
and stiffness, 6, 157–158
in *Striae distensae*, 7
in systemic sclerosis affected skin, 125
and tensile strength, 3, 20
types, *see* individual Type of collagen
Ultraviolet (UV) radiation, 10–11
Collagen fibers
animal *in vitro* (*ex vivo*) relaxation studies, 22
animal *in vitro* (*ex vivo*) thermocontraction studies, 26–27
animals *in vitro* (*ex vivo*) anisotropy studies, 21
in dermis, 231–232
and distensibility, 42, 211, 217
ECM structural component, 3
and hydration, 6, 10, 200
Langer's lines, 45
proteolytic enzymes, effect on, 6
and rigidity, 217
schematic of structure, 5
in scleroderma affected skin, 208–210
and skin thickness, 217
thickness of, 10
viscoelastic-to-elastic ratio, impact on, 189
and viscosity, 91
Color
blanching of skin under test, 167
human *in vivo* Colorimeter studies, 267
lipodermatosclerosis, 142
pinkness, 234
of scleroderma affected skin, 111
Colorimeter, 267
Compression test
age of skin, impact on, 232
Cutech extensometer, using, 83
human *in vivo* deformation force studies, 163
and stiffness, 87
Conditioning of connective tissue, *see* Preconditioning
Conductance, skin, 202, 249
Congenital skin diseases
anetoderma, 9, 10
cutis laxa, 8, 10
elastic fiber abnormalities, 8–9
elastoderma, 9, 10
Marfan syndrome, 8–9, 10
pseudoxanthoma elasticum, 9, 10, 45
Contractubex, 258–259, 266
COR, *see* Coefficient of Restitution (COR)
Corneocytes
AHAs, impact of, 234
and dry skin, 241–242
plasticity of, 200
in *stratum corneum*, 199
Corneometer
capacitance of skin, 201
human *in vivo* hydration studies, 251
human *in vivo* moisturizer studies, 253–254
Corticoid treatment of keloids, *see* Triamcinolone acetonide
Corticosteroids, *see also* Glucocorticosteroids
animal *in vitro* (*ex vivo*) relaxation studies, 22
animal *in vitro* (*ex vivo*) repeated strain studies, 24
animal *in vitro* (*ex vivo*) stress–strain studies, 19–20
animal *in vitro* (*ex vivo*) wound studies, 26
antikeloid treatment, 258–259
atrophy, skin, 45
and collagen synthesis, 10
modeling to assay atrophogenicity, 20
and strain rate, 20
ultrasound image analysis of treated skin, 265–266
Cortisol acetate, 19
Coulomb modulus, 79
Cracking, of dry skin, 241–243
Creep(ing), *see also* Hysteresis
age of skin, impact on studies, 232–233
animal *in vitro* (*ex vivo*) stress–strain studies, 20, 21
animal *in vitro* (*ex vivo*) studies, 24, 25
animal *in vitro* (*ex vivo*) tissue expansion studies, 21
in animal tissue, 140
bonding between skin and tester, 85
Dermaflex A, 113
fast, *see* Delayed distension
human *in vivo* deformation studies, 64

human *in vivo* Dermaflex A studies, 111–112, 245
human *in vivo* Durometer studies, 163
human *in vivo* elasticity studies, 42
human *in vivo* extensometer studies, 81, 83, 86
human *in vivo* hydration studies, 203
human *in vivo* skin friction studies, 54
human *in vivo* stress–strain studies, 111–112, 114, 245
human *in vivo* stress studies, 64
human *in vivo* tensile strength studies, 209–210
human *in vivo* Twistometer studies, 67–68
modeling of, 18, 68–70
parameter, Dermaflex A, 113
phenomenon related to deformation, 216
stationary, 69, 71
transient, *see* Delayed distension
Croton oil, 201
Cryotherapy, 258
Cutaneous Absorption Coefficient (CAC), 152–159
Cutaneous edema, 210
Cutech extensometer, 83–85
Cutis laxa, 8, 10
Cutometer
biological elasticity, 94, 180, 217
comparison with Dermaflex A, 188
comparison with Dermagraph, 134–135
comparison with Sclerimeter, 125
delayed distension, 94, 180, 181, 217
disadvantages, 79, 124, 263–264
final distension, 94, 180, 217
final retraction, 94
gross elasticity, 94, 180, 217
hardware and mechanical principles, 91–93, 244
human *in vivo* antiaging treatment, fish polysaccharides study, 236
human *in vivo* EDS studies, 218–219
human *in vivo* erysipelas studies, 224–225, 227
human *in vivo* hydration studies, 250, 253
human *in vivo* keloid treatment studies, 259, 262–264
human *in vivo* lymphedema studies, 226–227
human *in vivo* psoriasis studies, 222–224, 227
human *in vivo* SB studies, 219–222, 226–227
human *in vivo* scleroderma studies, 219–222, 226–227
hysteresis parameter, 217
immediate distension, 94, 180, 181, 217
immediate retraction, 94, 180, 181, 217
measurements and analysis of data, 181, 182, 187–196, 216–218
net elasticity, 180, 217
parameters for skin measurement, 180, 181, 246–248
probe, impact on measurement, 91–92
residual skin elevation, 180
skin thickness measurement, 94
subject factors, 183
test area size, 92, 188, 217
test axis, 217
total recovery deformation at end of stress–off period, 180, 181
viscoelastic-to-elastic ratio, 94, 180, 217
viscopart, 217
Young's modulus, 79
Cysteine, 11
Cytoindenter, 164
Cytokines, 11

D

D–squame tapes, 202
Decartelization of collagen, 266
Deformation
age of skin, impact on studies, 232–237
angular, human *in vivo* DMT studies, 65
delayed-to-immediate ratio, *see* Viscoelastic-to-elastic ratio
delayed viscoelastic, *see* Delayed distension
elastic, *see* Immediate distension
glycosaminoglycans, impact on, 232
human *in vivo* Twistometer studies, 67
interstitial fluid, displacement of, 233
mechanical property of skin, 217
modeling of, 18, 53, 55–57, 64
modeling of, animal *in vitro* studies, 18
modeling of, human *in vivo* skin friction studies, 53, 55–57
modeling of, human *in vivo* stress studies, 64
parameter for skin measurement, 245–246
time, as a factor in experiments, 64, 245–246
Degree of extension, 22
Dehiscence of a wound, 26
DEJ, *see* Dermal–epidermal junction (DEJ)
Delayed distension
human *in vivo* Cutometer studies, 188–190
human *in vivo* DTM studies, 74
human *in vivo* EDS studies, 218
human *in vivo* elasticity studies, 188–190
human *in vivo* erysipelas studies, 224–225, 227
human *in vivo* hydration studies, 250–251

human *in vivo* keloid treatment studies, 259–260
human *in vivo* lymphedema studies, 226
human *in vivo* psoriasis studies, 222–223, 227
human *in vivo* SB studies, 220, 227
human *in vivo* scleroderma studies, 220, 227
modeling of, 71
parameter, Cutometer, 94, 180, 181, 217
parameter, DTM, 180
parameter, time–strain, 246–248
parameter, Twistometer, 217
rheological modeling of, 68–72
time–strain measurement, 246–248
Twistometer, 217
Delayed viscoelastic deformation, *see* Delayed distension
Delipidization with diethyl ether, 164
Dermaflex A
capacitance, skin, 245
comparison with Cutometer, 188
creep(ing), 113
distensibility, 113
elasticity studies, 42, 113
hardware and mechanical principles, 111–113, 245
human *in vivo* biological elasticity studies, 189
human *in vivo* histamine wheals studies, 201
human *in vivo* hydration studies, 203–204, 250, 253
human *in vivo* psoriasis studies, 222
hysteresis parameter, 113, 217
and the Langer's lines, 113
measurements and analysis of data, 113–114, 216–218
parameters for skin measurement, 246–248
probe, impact on measurement, 113
Relative Elastic Retraction (RER), 217
Resilient Distension (RD), 217
skin thickness measurement, 111
Tensile Distensibility (TD), 217
test area size, 188, 217
test axis, 217
Dermagraph
background, 124–125
comparison with Cutometer, 134–135
hardware and mechanical principles, 126–127
measurements and analysis of data, 126–135
residual distensibility, 128
skin tethering, 124
Dermal chondroitine sulfates, 10
Dermal–epidermal junction (DEJ), 12
Dermal hypoplasia, focal, 7
Dermal sclerosis, *see* Sclerosis, systemic
Dermal Torque Meter (DTM)
biological elasticity, 180
burns, study of the treatment of, 181
contact pressure variable, 182
delayed distension, 180
development of, 64
environmental factors, 182–183
final distension, 180
gross elasticity, 180
hardware and mechanical principles, 65–67, 244
human *in vivo* hydration studies, 181, 182–183, 250–251, 253
immediate distension, 180
immediate retraction, 180
Langer's lines, 64
measurements and analysis of data, 72–75, 181
modeling of, 70–72, 180–182
net elasticity, 180
parameters for skin measurement, 180, 181, 246–248
probe, impact on measurement, 65, 68–69, 72–75
rheological modeling of, 66, 70–72
scars, study of, 181
skin thickness measurement, 72
test area size, 64
total recovery deformation at end of stress-off period, 180
viscoelastic-to-elastic ratio, 180
Young's modulus, 72
DermaLab
disadvantages, 120–121
elast, 117, 119–120
hardware and mechanical principles, 117–119
human *in vivo* hydration studies, 254
measurements and analysis of data, 119–121
modeling of, 119–120
parameters for skin measurement, 119
Young's modulus, 117, 119–121
Dermatitis
fissures follow orientation of joints, 45, 163
human *in vivo* skin friction studies, 49
relationship to *stratum corneum* health, 162
Dermatology, 64, 77
Dermatosparaxis in cattle, 45
Dermis
composition of, 3, 187, 231–232
structure of, 210
thickness, 199
Desmosines, 8
Desmosomes, 234
Desmotropic drugs
animal *in vitro* (*ex vivo*) isorheological point studies, 24

animal *in vitro* (*ex vivo*) stress–strain studies, 19, 20
animal *in vivo* repeated strain studies, 29, 30
animal *in vivo* stress–strain studies, 28
Desquamation, 163, 233–234, 241–242
Diabetes
collagen, increased synthesis of, 6, 7
foot ulceration, 143–145
glycosilation of collagen, 6
skin thickening, 6, 7
stiffness of skin, 20
streptozotocin-induced, 20
ulcers, 143–145
VPT study, 145
Differential scanning calorimetry, 27
Directional differences, *see* Anisotropic properties of skin
Displacement, amplitude or angular (AMP), 151–159, *see also* Hysteresis
Displacement/force tests, *see* Hysteresis
Displacement, instantaneous, 104
Displacement, peak, 103–104
Distensibility
age of skin, impact on studies, 232–233
and collagen fibers, 42, 211, 217
and elastic fibers, 217
final-to-delayed distension ratio, 72
final-to-immediate distension ratio, 72
human *in vivo* Dermaflex A studies, 112, 114, 250
human *in vivo* Dermagraph studies, 124–125, 128–135
human *in vivo* DermaLab studies, 117
human *in vivo* elasticity studies, 44
human *in vivo* hydration studies, 250
human *in vivo* stress–strain studies, 112, 114, 117
and hydration, 112, 202–204
and joint motility, 45
mechanical property of skin, 217
parameter, Dermaflex A, 113
parameter, elasticity, 42
retinoid studies, 234–235
and skin thickness, 44
skin thinning in, 44
in systemic sclerosis affected skin, 124
tensile, *see* Tensile Distensibility (TD)
Distension, delayed, *see* Delayed distension
Distortion, angular, *see* Torsion, test method
Dithranol, 224
DNA, 11, 258
Dogs, 20–21, 26
DPM Nova, *see* Nova Dermal Phase Meter 9003
Drug treatments
ACTH, 20
amino–acetronitrile, 20
androgens, 20
biosynthetic human growth hormone, 20
corticosteroids, *see* Corticosteroids
cortisol acetate, 19
desmotropic, *see* Desmotropic drugs
estrogens, 20, 157–158
fish polysaccharides, 235–236
gestagens, 20
glucocorticoid, *see* Glucocorticoids
glucocorticosteroids, 19–20
lathyrogenic compounds, 20
nonsteroidal anti–inflammatory, 20
D-penicillamine, *see* D-penicillamine
prednisolone, *see* Prednisolone
prednisolone acetate, *see* Prednisolone acetate
streptozotocin, 20
thyroid hormones, 20
Dry skin
comparison, clinical assessment and test, 249–250
definition, 241–242
DSR, *see* Dynamic spring rate (DSR)
DTM, *see* Dermal Torque Meter (DTM)
Durometer
background, 139–140
hardware and mechanical principles, 140
human *in vitro* keratin studies, 163
human *in vivo* scleroderma studies, 163
measurements and analysis of data, 141–144
and pitting edema affected skin, 143
Dynamic Spring Rate (DSR)
human *in vivo* GBE studies, 252
human *in vivo* LSR studies, 108
measurements and analysis of data, 100
of the *stratum corneum*, 103
Dyshidrosis, autonomic, 143

E

EB, *see* Epidermolysis bullosa (EB)
Eccrine glands, 3
Echogenicity, *see* Ultrasound image analysis
ECM, *see* Extracellular Matrix (ECM)
Eczema, 200
Edema
animal hydration studies, 201–202
animal SWP studies, 201
caused by suction chamber device, 120, 121
caused by UV exposure, 233
cutaneous, 210
and erysipelas, 224–225
of the head, 46
histamine wheal, 112

in lymphedema affected skin, 226
mucinous, 112
pitting, tested with Durometer, 143
in psoriasis affected skin, 223
retinoid studies, 234
and scleroderma, 209, 211
Ultraviolet (UV) radiation, 233
viscoelastic-to-elastic ratio, 211
viscoelasticity in affected skin, 201
EDS, *see* Ehlers–Danlos Syndrome (EDS)
Ehlers–Danlos Syndrome (EDS)
collagen, mutations in, 7, 8, 9, 219
elastic fibers, impact on, 219
elasticity, 218
fibroblasts in, 9
final distension measurements, 218
human *in vivo* Cutometer studies, 96, 218–219
human *in vivo* Twistometer studies, 218
hyperextensibility in, 7, 8, 9, 218–219
increased joint and skin motility, 45
lysyl oxidase in cells, 7
and modulus of elasticity, 218
and net elasticity, 218
procollagen, appearance of, 7, 8
skin fragility in, 7
skin thickening, 218–219
skin thickness, effect on, 9, 218–219
skin thinning in, 7, 8
and Tensile Distensibility (TD), 218
Type I collagen in, 7, 8
Type III collagen in, 7, 8, 9
Type III procollagen in, 8
Type V collagen in, 7, 8
Elast
parameter, DermaLab, 117, 119–120
Young's modulus, 117
Elastic deformation, *see also* Immediate distension
Elastic deformation recovery, *see* Immediate retraction
Elastic fibers
in actinic elastosis affected skin, 11
age of skin, impact on, 189, 193
anetoderma, 9
and biological elasticity, 193
composition of, 8
congenital skin diseases, 8–9
in dermis, 231–232
ECM structural component, 3
in EDS affected skin, 219
and elasticity, 91, 217
elastoderma, 9
in erysipelas affected skin, 225
and hydration, 260
in lymphedema affected skin, 226
retraction, relationship to, 42
and rigidity, 217
in scleroderma affected skin, 208
and skin thickness, 217
and tensile strength, 208–210
thickness of, 11
Elastic properties of skin, *see* Elasticity
Elastic recovery, *see* Net elasticity
Elastic recovery-to-total deformation ratio, *see* Biological elasticity
Elastic response, *see* Immediate distension
Elastic retraction, *see* Recovery
Elasticity, *see also* Distensibility
age of skin, impact on studies, 11, 232–233
AHAs, impact on, 234
AMP, 151–159
animal *in vitro* (*ex vivo*) hysteresis studies, 21–23
animal *in vitro* (*ex vivo*) relaxation studies, 22
animal *in vitro* (*ex vivo*) stress–strain studies, 20–21
animal *in vitro* (*ex vivo*) tissue expansion studies, 20–21
CAC, 152–159
COR, 152–158
Dermaflex A studies, 42
distensibility, 42
and EDS, 218
and elastic fibers, 91, 217
and elastin, 4
in elastoderma affected skin, 9
fish polysaccharides, 235–236
human *in vivo* and *ex vivo* studies, 187
human *in vivo* Cutometer studies, 259
human *in vivo* Dermaflex A studies, 114, 250
human *in vivo* Dermagraph studies, 130–134
human *in vivo* DermaLab studies, 117
human *in vivo* extensometer studies, 83
human *in vivo* hydration studies, 250
human *in vivo* keloid treatment studies, 259
human *in vivo* skin friction studies, 50
human *in vivo* stress–strain studies, 117
mechanical property of skin, 216, 217, 232
modeling of, 42–43, 74–75, 182, 203
of papillar dermis, 41
parameter, Dermaflex A, 113
parameter, time–strain, 248
parameters for skin measurement, 42–43
recovery, 42
of reticular dermis, 41
retinoid studies, 234–235
rheological modeling of impact of hydration, 203
in scars, 266
in Scleredema of Buschke affected skin, 219

skin tethering, scoring of, 126, 209
and softness, 252
of *stratum corneum*, 41, 249
time, as a factor in experiments, 42
time–strain measurement, 248
Ultraviolet (UV) radiation, 189, 233, 236–237
Elasticity, modulus of, *see* Modulus of elasticity; Young's modulus
Elasticity ratio, *see* Net elasticity
Elastin
in actinic elastosis affected skin, 10
alterations in synthesis, 10–11
in cutis laxa affected skin, 8
degradation of, 11–12
and elasticity, 4, 187
fibrillin 1 gene, 8–9
fibrillogenesis, abnormal, 9
and hydration, 200
importance to skin structure, 3, 13
maintenance of elasticity and resilience, 4
Marfan syndrome, 8–9
and MMPs, 11–12
molecular structure, 8–10
mRNA levels in photoaged skin, 11
pseudoxanthoma elasticum, 9
tretinoin, impact of, 12
Elastoderma, 9, 10
Elastometer, *see* Ball–elastomer test method; DermaLab
Elastosis, 233, 235
Elaunin fibers, 8
Electrical field, keloid treatment, 262
Elevation, test method, 243
Emollients, 96, *see also* Oil/Water emulsions
Endoproteinases, cleavage of procollagens, 5–6
Endothelial cells, 12
Energy density formula, 24
Energy dissipation, 22
Energy input, 22
Entactin, 12
Enzymes
glycolytic, 258
hydrocylysine, 4, 6
lysyl hydroxylase, 4
lysyl oxidase, 6, 8
prolyl–3–hydroxylase, 4
prolyl–4–hydroxylase, 4
proteolytic, effect on collagen fibers, 6
in scars, 258
Epidermis
atrophic skin treatment, 234
composition of, 3, 187
thickness, 199
Epidermolysis bullosa (EB), 4, 12
Epithelial cells, 12
Erysipelas
and edema, 224–225
elastic fibers, appearance of, 225
human *in vivo* Cutometer studies, 218, 224–225, 227
interstitial fluid in, 225
skin thickening, 224–225
softening of the skin in, 225
Erythematous scaly plaques, 222
Estrogens, 20, 157–158
Ethanol, 203, 250
Extactum capae gel, 266
Extensibility
animal *in vitro* (*ex vivo*) anisotropy studies, 21
animal *in vitro* (*ex vivo*) hysteresis studies, 22
animal *in vitro* (*ex vivo*) stress–strain studies, 20
and anisotropy, 187
and glucocorticoid treatment, 20
Gunner portable hand–held extensometer, 80
human *in vivo* lymphedema studies, 226
immediate, 68
and induration, 220, 222
and solar irradiation, 233
of *stratum corneum*, 249
time–strain measurement, 247
Extensometer
disadvantages, 86
modeling for, 86–88
rheological modeling of, 80, 86–88
test methods, 79
types of, 80–85
use of, 85–86, 89
Young's modulus, 87
Extracellular Matrix (ECM)
animal *in vitro* (*ex vivo*) relaxation studies, 22
collagen fibers, 3
composition of, 3–4, 231–232
Contractubex, keloid treatment, 266
elastic fibers in, 3
fibrillar network, 3
and ground substance, relationship, 3
protease degradation of components, 11
UV radiation, effect on, 11, 236–237
Extrinsic aging, *see* Photoaging
Eyelid, 94

F

Face
elastic recovery differences due to race, 183
human *in vivo* Dermagraph studies, 124, 129–134
human *in vivo* systemic sclerosis studies, 124

retinoid studies, 234
wrinkle formation, 10
Fast creep, *see* Delayed distension
Fe^{2+}, 4
Feet
human *in vivo* Dermagraph studies, 124, 129–134
human *in vivo* systemic sclerosis studies, 124
neuropathic, associated with diabetes, 143–145
Fibril(s), 4
Fibrillar network, 3
Fibrillin, 8
Fibrillin 1 gene, 8–9, 10
Fibrillogenesis, abnormal, 8, 9
Fibroblasts
and the basement membrane, 12
collagen synthesis in, 4, 157–158
in EDS affected skin, 9
human *in vivo* keloid treatment studies, 266
in scars, 258
tretinoin, impact of, 234
UV radiation, effect on, 11
Fibrocytes, alignment of nuclei, 267
Fibronectin deposits, 258
Fibroplasia, 258
Fibrosis, 125
Final deviation length, *see* Final distension
Final distension, *see also* Tensile Distensibility (TD)
and EDS, 218
final-to-delayed distension ratio, 72
final-to-immediate distension ratio, 72
human *in vivo* Cutometer studies, 188–190, 250–252, 259–260
human *in vivo* Dermagraph studies, 128, 130–135
human *in vivo* elasticity studies, 188–190
human *in vivo* erysipelas studies, 224–225, 226–227
human *in vivo* hydration studies, 250–252
human *in vivo* keloid treatment studies, 259–260, 262–264
human *in vivo* lymphedema studies, 226
human *in vivo* psoriasis studies, 222–224, 226–227
human *in vivo* SB studies, 219–222, 226–227
human *in vivo* scleroderma studies, 220, 226–227
parameter, Cutometer, 94, 180, 217
parameter, DTM, 180
parameter, time–strain, 246–248
time–strain measurement, 246–248
Final retraction, 94
Final skin deformation, *see* Final distension
Fingers
human *in vivo* Durometer studies, 141, 163
human *in vivo* Microindentometer studies, 165, 169–170
human *in vivo* scleroderma studies, 141
impedance of, 165
of systemic sclerosis affected patients, 126
Finite-element-based modeling
human *in vivo* extensometer studies, 83, 87–88
mechanical properties of the skin, 18, 64
of skin subjected to stress, 68
Fish polysaccharides, 235–236
Fissuring, of dry skin, 241–243
FIT equations, 68–69
Flavonoids, 235
Flumethasone pivalate, 224
Focal dermal hypoplasia, 6, 7
Follicular horns, 234
Force/displacement tests, *see* Hysteresis
Force, instantaneous, 104
Force, peak, 103–104
Forearm. *see also* Arms
elastic recovery differences due to race, 183
human *in vitro* indentation studies, 164
human *in vivo* acroscleroderma studies, 211
human *in vivo* Cutometer studies, 94–95, 183, 187–195, 251
human *in vivo* Dermaflex A studies, 250
human *in vivo* DTM studies, 72–74
human *in vivo* Durometer studies, 142, 163
human *in vivo* EDS studies, 9, 218
human *in vivo* elasticity studies, 187–196
human *in vivo* hydration studies, 202, 204, 249–251
human *in vivo* IDRA studies, 154–157
human *in vivo* LSR studies, 106
human *in vivo* Microindentometer studies, 167
human *in vivo* SB studies, 219–220
human *in vivo* skin friction studies, 50, 52–55
human *in vivo stratum corneum* studies, 106
human *in vivo* Twistometer studies, 249
retinoid studies, 234
stress–strain measurement, 247
Young's modulus, 50
Forehead
human *in vivo* Cutometer studies, 187–195
human *in vivo* Durometer studies, 141–142
human *in vivo* elasticity studies, 187–195
human *in vivo* hydration studies, 249–250
human *in vivo* scleroderma studies, 141–142
human *in vivo* Twistometer studies, 249–250

Formulation of hydration products
gel network, 107
oil-in-water emulsion, *see* Oil/Water emulsions
Free energy of the skin, 50
Friction force measurements
of calibrated steel plate, 51
human *in vivo* DTM studies, 72
human *in vivo* extensometer studies, 81
human *in vivo* indentation studies, 163
human *in vivo* skin friction studies, 52–53
modeling of, 55–57
moisturization effects, 52–53
of polymeric (silicone) material, 52
set-up procedure, 50
skin friction coefficient, 50
static and dynamic properties, effect on, 54–55
Young's modulus, 57
Friedmann test, 262
Fruit acids, 233

G

GAG(s), *see* Glycosaminoglycans
Gas-Bearing Electrodynamometer (GBE)
comparison with LSR, 105–106
disadvantages, 101–102, 108
history, 99–101
human *in vivo* elasticity studies, 44
human *in vivo stratum corneum* studies, 99–101
test methods, 245
Young's modulus, 252
Gas Bearing Extensometer, 252
GBE, *see* Gas-Bearing Electrodynamometer (GBE)
92–kDa gelatinase, 12
Gender, as a factor in experiments, *see* Sex, as a factor in experiments
Gene encoding, 4
Genetic defects associated with collagen
in EDS affected skin, 8
in Osteogenesis imperfecta affected skin, 8
proα1(I) or proα2(I) Type I procollagen, 8
Genodermatosis, 8–9
Gestagens, 20
Giraffes, 45, 189
Glands
apocrine, in dermis, 3
eccrine, in dermis, 3
Glucocorticoids
collagen, causing reduced amounts of, 6
and extensibility, 20
glycosaminoglycans, synthesis of, 260
treatment of rats, 20
Glucocorticosteroids, 19–20
Glucosaminoglycans, 219
Glucosteroid treatment, 258
Glycerin
human *in vivo* Cutometer studies, 251
human *in vivo* Dermaflex A studies, 250
human *in vivo* hydration studies, 250, 251
impact on skin mechanics, 203
Glycerol, 107, 249, 252
Glycine, 4, 5, 8
Glycolic acid, 233
Glycolytic enzymes, 258
Glycoproteins
basement membrane, impact on, 12
laminin-1, 12
maintenance of macromolecular matrix, 208
in microfibrils, 8
nidogen/entactin, 12
in scars, 258
Glycosaminoglycans
and aging of skin, 193
AHAs, impact on, 234
and deformation, 232
glucocorticoids, effect of, 260
and hydration, 4, 200, 232
synthesis of in keloid tissue, 260
viscosity of human skin, 193
viscosity of rat skin, 260
viscous properties of skin, impact on, 232
Glycosilation
of collagen in diabetic thick skin, 6
of hydroxylysine and asparagine residues, 5
and skin hardness, 143
Gross elasticity
human *in vivo* EDS studies, 218
human *in vivo* erysipelas studies, 224–225, 227
human *in vivo* lymphedema studies, 226
human *in vivo* psoriasis studies, 222–223, 227
human *in vivo* SB studies, 219, 220, 227
human *in vivo* scleroderma studies, 220, 227
parameter, Cutometer, 94, 180, 217
parameter, DTM, 180
Ground substance
composition of, 91, 187
ECM component, 3
and hydration, 200
protoglycans in, 91
and viscosity, 91, 193
Guinea pigs, 20, 26
Gunner portable hand-held extensometer, 80

H

Hairless mouse model, 20
Hallpot transducer(s), 149
Hand
    human *in vivo* Dermagraph studies, 124, 129–134
    human *in vivo* LSR studies, 106
    human *in vivo* systemic sclerosis studies, 124
Head edema, 46
α-helical conformation, 8
Hemidesmosome(s), 12
Hemodialysis, 202
Heparin, 266
Heparin sulfate proteoglycan perlecan, 12
Hereditary skin conditions
    collagen changes in, 8
    EDS, 8
    epidermolysis bullosa, 12
    osteogenesis imperfecta, 8
Hertzian theory, 53, 56
Hidebinding, 209, 210, 211
Histamine wheals
    creep(ing), impact on, 42
    and edema, 112
    human *in vivo* Dermaflex A studies, 111–112, 114, 201
    human *in vivo* elasticity studies, 44
Hook's law, 19
Hormones
    androgens, 20
    animal *in vitro* (*ex vivo*) stress–strain studies, 19
    ATCH, 20
    biomechanical properties of skin, effect on, 20
    biosynthetic human growth, 20
    estrogens, 20, 157–158
    gestagens, 20
    thyroid, 20, 22
Horny membrane
    biological elasticity, 251
    human *in vivo* indentation studies, 164
    and hydration, 242–243, 251–252
    stripping away, effect on hydration, 251
    viscoelastic-to-elastic ratio, impact on, 251
Hot water bath, immersion in, 249
Human *in vitro* indentation studies, 163–165
Human *in vivo* AMP studies, 155–159
Human *in vivo* CAC studies, 157–159
Human *in vivo* Colorimeter studies, 267
Human *in vivo* COR studies, 157–158
Human *in vivo* Corneometer studies, 251, 253–254
Human *in vivo* Cutometer studies
    abdomen, 187–195
    actinic elastosis, 94
    age of skin, 94, 188–194
    ankle, 187–195
    arms, 187–195
    back, 187–195
    biological elasticity, 94, 189–192, 251, 259–260
    delayed distension, 188–190
    Ehlers–Danlos Syndrome (EDS), 96, 218–219
    elasticity, 259
    erysipelas, 218, 224–225, 227
    final distension, 188–190, 250–252, 259–260
    of the forearm, 94–95, 183, 187–195, 251
    of the forehead, 187–195
    glycerin, affect on, 251
    humidity of the test area, 92
    hydration, effect on, 181, 183
    immediate distension, 188–190
    immediate retraction, 188–190, 259–260
    of legs, 250–252
    lymphedema, 218, 226
    moisturizers, effect of, 96, 181
    net elasticity measurements, 251
    oil/water emulsions, 251
    papillar dermis, 94
    paraffin oil, effect of, 251
    photoaging, 189
    preconditioning, 182
    psoriasis, 96, 218, 222–224, 227
    residual skin elevation, 250
    Resilient Distension (RD), 250
    of the reticular dermis, 94
    of SB, 218–222
    of scars, 96
    of scleroderma, 219–222, 227
    sex, as a factor in experiments, 94, 183
    of skin thickness, 189
    of stratum corneum, 250–251
    stress–strain measurement, 92, 244
    striae distensae, 96
    of systemic sclerosis affected patients, 96
    temperature of skin, 92
    test axis, 92
    test position, 188
    of the thigh, 187–195
    time, as a factor in experiments, 92
    time–strain measurement, 92, 94, 188–196, 244, 246–248
    total recovery deformation at end of stress-off period, 250–251
    Ultraviolet (UV) radiation, 96
    urea, impact on, 251
    using water, 250
    viscoelastic-to-elastic ratio, 189, 193–196, 250–251, 259–260

of viscosity, 259
of vulvar skin, 94
Human *in vivo* Dermaflex A studies, 111–114, 201–204, 218, 222–224, 245, 250
Human *in vivo* Dermagraph studies, 124–125, 128–135
Human *in vivo* DermaLab studies, 117–121
Human *in vivo* diabetes studies, 143–145
Human *in vivo* DTM studies, 64–65, 72–75, 181, 244, 250–251
Human *in vivo* Durometer studies, 139–144, 163
Human *in vivo* EDS studies, 9, 218–219
Human *in vivo* elasticity studies, 42–44, 114, 183, 187–195
Human *in vivo* erysipelas studies, 224–227
Human *in vivo* extensometer studies, 81–87
Human *in vivo* GBE studies, 100–101, 252
Human *in vivo* histamine wheals studies, 19, 201
Human *in vivo* hydration studies, 106–108, 202–204, 249–254
Human *in vivo* IDRA studies, 151–159
Human *in vivo* keloid treatment studies, 259–266
Human *in vivo* lipodermatosclerosis studies, 142–144
Human *in vivo* LSR studies, 104–108
Human *in vivo* lymphedema studies, 226–227
Human *in vivo* Microindentometer studies, 165–174
Human *in vivo* moisturizer studies, 253–254
Human *in vivo* Nova DPM 9003 studies, 106–107, 250–254
Human *in vivo* psoriasis studies, 222–227
Human *in vivo* SB studies, 219–222, 226–227
Human *in vivo* scleroderma studies, 141–142, 163, 209–211, 219–222, 226–227
Human *in vivo* skin friction studies, 49–57
Human *in vivo* stiffness studies, 155–159
Human *in vivo stratum corneum* studies, 99–101, 104, 106–108, 202
Human *in vivo* stress–strain studies, 19, 111–112, 114, 117, 202–204, 216, 245
Human *in vivo* stress studies, 64
Human *in vivo* systemic sclerosis studies, 124
Human *in vivo* tensile strength studies, 79, 207–210
Human *in vivo* Twistometer studies, 64, 67–68, 218, 249–250
Human *in vivo* VESA studies, 204
Human *in vivo* viscoelasticity studies, 94
Human retinoid studies, 234–235
Humectant compounds, 107, 242
Humidity
human *in vivo* Cutometer studies, 92
human *in vivo* LSR studies, 106
human *in vivo* skin friction studies, 52
human *in vivo stratum corneum* studies, 106, 242
human *in vivo* studies, 182
Hyaluronate, 200
Hyaluronic acid, 232, 258
Hydration, 199–204
and capacitance, skin, 202–203, 249–251
collagen fiber solubility, 6, 10
of collagen fibers, 200, 260
and conductance, skin, 202, 249
dermal chrondroitine sulfates, 10
Dermal Torque Meter (DTM), 181, 182–183
and distensibility, 112, 202–204
and elastic fibers, 260
electrical measurements, 249
gel network formulation, 107
and glycosaminoglycans, 4, 200, 232
of the ground substance, 200
of the horny membrane, 242–243, 251, 252
human *in vitro* indentation studies, 164
human *in vitro stratum corneum* studies, 164
human *in vivo* Cutometer studies, 181, 183
human *in vivo* Dermaflex A studies, 112, 114
human *in vivo* DermaLab studies, 117, 121
human *in vivo* DTM studies, 64
human *in vivo* elasticity studies, 43, 44
human *in vivo* erysipelas studies, 225
human *in vivo* LSR studies, 106–108
human *in vivo* silicone gel sheet keloid treatment studies, 262
human *in vivo* skin friction studies, 49, 53, 55, 57
human *in vivo stratum corneum* studies, 106–108
human *in vivo* stress–strain studies, 112, 114, 117
human *in vivo* studies, 181–183, 252–254
human *in vivo* tactile sensor studies, 252
human *in vivo* Twistometer studies, 64
of the intercellular ground substance, 91
keratin(s), effect on, 199, 200
maintenance of water balance, 4
oil-in-water emulsion formulation, *see* Oil/Water emulsions
of the papillar dermis, 199, 200
and plasticity, 199–204
protoglycans, impact on, 232
psoriasis, effect on, 200
of reticular dermis, 200
in scar tissue, 258
and Shear wave propagation (SWP) testing, 204
and skin impedance, 202
and skin thickness, 44, 200, 202
and skin thinning, 44

and softness, 44
of the *stratum corneum*, 199–204, 241–243
and Transepidermal Water Loss (TEWL), 202, 241–243
and viscoelastic-to-elastic ratio, 217
and viscoelasticity, 217, 243
Young's modulus, 249
Hydrocolloid occlusive dressing, 224
Hydrocylysine, 4, 6
Hydrophobic elastin domains, 8
Hydrothermal isometric tension curves, 27
Hydroxylation, 4–5
Hydroxylysine, 4, 6
Hydroxylysine residues of glycosilation, 5
3–hydroxyproline, 4
4–hydroxyproline, 4, 5
Hygroscopy, 242
Hyperextensibility
in EDS affected skin, 7, 8, 9, 218–219
in Marfan syndrome affected skin, 10
Hyperkeratosis, metatarsal, 143–144
Hyperpigmented skin, 142
Hyperplastic tissue, 234
Hypertrophic scars
definition, 257
human *in vivo* Cutometer studies, 96
pathophysiology, 258
treatments, 258–267
Hypoderma
modeling of, 88
and scleroderma, 211
Hypodermic septae, 219
Hypoplasia, focal dermal, 6, 7
Hysteresis
animal *in vitro* (*ex vivo*) studies, 21–23
bonding between skin and tester, 85
human *in vivo* Dermaflex A studies, 111–112, 203–204, 250
human *in vivo* elasticity studies, 42
human *in vivo* erysipelas studies, 224–225
human *in vivo* GBE studies, 100–101
human *in vivo* hydration studies, 202–204, 250
human *in vivo* lymphedema studies, 226
human *in vivo* psoriasis studies, 222–224
human *in vivo* skin friction studies, 56–57
human *in vivo stratum corneum* studies, 100–101
human *in vivo* stress–strain studies, 111–112, 202–204, 216
modeling of, 18
parameter, Cutometer, 217
parameter, Dermaflex A, 113, 217
parameter, stress–strain, 248
parameter, time–strain, 247–248
retinoid studies, 234–235
stress–strain measurement, 248
time–strain measurement, 247–248
and viscoelasticity, 216

I

Ichthyosis, 162
IDRA, *see* Integrated Dynamic Rebound Analyzer (IDRA)
Immediate deformation, *see* Immediate distension
Immediate distension
human *in vivo* Cutometer studies, 188–190
human *in vivo* Dermagraph studies, 128, 130–135
human *in vivo* DTM studies, 74
human *in vivo* EDS studies, 218
human *in vivo* elasticity studies, 188–190
human *in vivo* erysipelas studies, 224–225, 226–227
human *in vivo* hydration studies, 249–252
human *in vivo* keloid treatment studies, 259–260
human *in vivo* lymphedema studies, 226
human *in vivo* psoriasis studies, 222–223, 226–227
human *in vivo* SB studies, 219–220, 226–227
human *in vivo* scleroderma studies, 220, 226–227
modeling of, for DMT, 71
parameter, Cutometer, 94, 180, 181, 217
parameter, DTM, 180
parameter, time–strain, 246–248
parameter, Twistometer, 217
rheological modeling of, 68–70
of *stratum corneum*, 249
time–strain measurement, 246–248
Twistometer, 217
urea, impact on, 249
Weibull's theory, 69–70
Immediate elastic recovery, *see* Immediate retraction
Immediate elastic response, *see* Immediate distension
Immediate elasticity, *see* Immediate distension
Immediate extensibility, 68
Immediate retraction
human *in vivo* Cutometer studies, 188–190, 259–260
human *in vivo* DTM studies, 74, 250–251
human *in vivo* elasticity studies, 188–190
human *in vivo* erysipelas studies, 224–225, 227

human *in vivo* hydration studies, 249, 250–251
human *in vivo* keloid treatment studies, 259–260
human *in vivo* lymphedema studies, 226
human *in vivo* psoriasis studies, 222–223, 227
human *in vivo* SB studies, 220, 227
human *in vivo* scleroderma studies, 211, 220, 227
human *in vivo* Twistometer studies, 249
modeling of, for DMT, 71
parameter, Cutometer, 94, 180, 181, 217
parameter, DTM, 180
parameter, time–strain, 246–248
of *stratum corneum*, 249
time–strain measurement, 246–248
Impedance, skin
of human finger pulp, 165
human *in vivo* hydration studies, 250–252, 254
human *in vivo* LSR studies, 106–107
human *in vivo* Nova DPM 9003 studies, 106–107
human *in vivo stratum corneum* studies, 106–107
and hydration, 202, 249
Indentation depth, 53, 54
Indentation test instrument
Cytoindenter, 164
Durometer, 139–145, 163
Microindentometer, 161, 164–174
modeling of, 174–176
perpendicular skin measurement, 243
research on, 163–164
Induration, 219–220, 222
Infraclavicular regions, 219
Instantaneous deviation length, *see* Immediate distension
Instantaneous displacement, 104
Instantaneous force, 104
Instantaneous recovery length, *see* Immediate retraction
INSTRON test instrument
animal *in vitro* (*ex vivo*) hysteresis studies, 21–22
animal *in vitro* (*ex vivo*) isorheological point studies, 23
animal *in vitro* (*ex vivo*) relaxation studies, 22
animal *in vitro* (*ex vivo*) repeated strain studies, 24
animal *in vitro* (*ex vivo*) stress–strain studies, 18–21
animal *in vitro* (*ex vivo*) wound studies, 26
animal *in vivo* stress–strain, 27
biaxial elongation test method, 18–24, 26–27
Integrated Dynamic Rebound Analyzer (IDRA)
ballistometry, 147–159
hardware and mechanical principles, 147–149
measurements and analysis of data, 151–159
modeling of, 149–154
Integrin(s), 12
Intercellular lipid material, *see* Lipids
Intercorneocitary cohesion, 234
Interface shear stress, 56
Interfacial adhesion component of friction, 55, 56
Interferon, 258
Interstitial fluid
collagen synthesis, rate of, 5–6
displacement of, during deformation, 233
in erysipelas affected skin, 225
human *in vivo* elasticity studies, 42, 44
in psoriasis affected skin, 223
and viscoelasticity, 217
Intrinsic aging, *see* Chronoaging
Isorheological point, 23–24, 25
Isotropic properties of skin
human *in vivo* extensometer studies, 83
human *in vivo* stress studies, 64
modeling of, 78

K

Keloid(s)
age of skin, 262
back, site of tests, 262
collagen, excessive amounts of, 6, 7
corticosteroids, impact on, 258–259
and glycosaminoglycans, 260
historical record, 257
pathophysiology, 258
pressure bandaging of, 258
sex, treatment differences due to, 262
and skin thickening, 7
skin thickness during treatment, 258
surgery to remove, 258
time, as a factor in treatment of, 258–259, 262
treatments, 258–267
ultrasound image analysis, 258, 262
viscoelasticity, impact on treatment of, 266
Kelvin model, 18, 71, 165
Keratin(s)
basement membrane, impact on, 12
cytoskeleton linkage to laminin–5 in basement membrane, 12
in epidermis, 187
in epidermolysis bullosa affected skin, 12
and hydration, 199, 200
in *stratum corneum*, 41, 107
Keratinization, abnormal, 162, 163

Keratinocytes
atypical and dysplastic, 234
in epidermis, 3
in epidermolysis bullosa affected skin, 12
Ultraviolet (UV) radiation, effect of, 11
Keratoses, actinic, 234
Kinase, 234

L

Lactic acid, 233, 242, 249
Lamellar lipid structures, 242, 243
Lamina densa, 12
Lamina lucida, 12
Laminin(s), 12
Langerhans cells, 3
Langer's lines
animal *in vitro* (*ex vivo*) studies, 21
anisotropic properties of skin, 21
collagen fibers, 45
and the Dermaflex A, 113
and the DTM, 64
orientation in human test subjects, 21, 43, 45
and Shear wave propagation (SWP) testing, 211
and the Twistometer, 64
Lanugo hairs, 113
Laser Doppler perfusion imager, 142
Laser treatment, 96
Lathyrogenic drugs, 20
Legs
human *in vivo* Cutometer hydration studies, 250–252
human *in vivo* Dermaflex A hydration studies, 250
human *in vivo* DTM hydration studies, 250–251
human *in vivo* Durometer studies, 142–143
human *in vivo* hydration studies, 249
human *in vivo* lipodermatosclerosis studies, 142–143
human *in vivo* Twistometer studies, 249
Lesions, 7, 10
Levarometry, test method, 245
Lichenification furrows in atopic dermatitis, 45
Ligaments and Type I collagen, 8
Linear Skin Rheometer (LSR)
calibration, 104–105
comparison with GBE, 105–106
data processing, 105
hardware and mechanical principles, 102–103
history, 99–100
human *in vivo stratum corneum* studies, 99–100, 106–108
measurements and analysis of data, 105–106
modeling, 103–104
Linear Variable Differential Transformer (LVDT) Sensor
Baker, Bader, and Hopewell extensometer, 83
GBE, 100, 102
Gunner portable hand-held extensometer, 80
human *in vitro* indentation studies, 164, 165
human *in vivo* extensometer studies, 83
Magnetic extensometer, 81
Lipidic films, occlusive, 107
Lipids
hydration retention of the *stratum corneum*, 242–243
hydrophobic component of skin barrier, 199
in moisturizers, 203
structure of, 200
Lipodermatosclerosis, 142–144
Liposuction, 96
LSR, *see* Linear Skin Rheometer (LSR)
LVDT, *see* Linear Variable Differential Transformer (LVDT) Sensor
Lymphedema
collagen, appearance of, 226
elastic fibers, appearance of, 226
human *in vivo* Cutometer studies, 218, 226
human *in vivo* lower limb studies, 226
in skin thickening, 226
viscoelasticity reduction postmastectomy, 201
Lymphocytes, 258
Lymphokines, 11
Lymphovenous insufficiency, development of, 225
Lysine, 4, 6, 8
Lysyl hydroxylase
collagen, catalyzing of, 4
in EDS affected skin, mutation of gene, 7
Lysyl oxidase
collagen cross–linking, 8
and collagen synthesis, 6
in cutis laxa affected skin, 8
in EDS affected skin, 7

M

Macromolecular matrix, 208
Magnetic extensometer, 81
Magnetic Resonance Imaging (MRI)
human *in vivo* hydration studies, 202
human *in vivo* scleroderma studies, 141
human *in vivo* tensile strength studies, 210
Malpighii cellular stratum, 41
Manschot extensometer, *see* Magnetic extensometer

Marfan syndrome, 8–9, 10
Massages, connective tissue, 258
Mast cells, 258, 262
Mathematical modeling, mechanical properties of skin, 17
Matrix metalloproteinases (MMPs), 236–237
  in actinic elastosis affected skin, 10, 11–12
  collagen, regulation of degradation of, 11–12
  elastin, regulation of degradation of, 11–12
  Type I collagen, 11
  Ultraviolet (UV) radiation, effect of, 11–12
Maximal deformation, *see* Final distension
Maximal distensibility, *see* Final distension
Maxwell model, 18, 71
Mechanical Recovery, *see* Recovery
Medial malleoli, *see* Ankle
Melanocytes, 3
Merkel cells, 3
Mesenchymal connective tissue, 12
Metalloproteinases, *see* Matrix metalloproteinases (MMPs)
Metastasis, 11
Methotrexate, 142
Mice
  hairless mouse model, 20
  inbred, 20
  laboratory, 26
  tight-skin mutant, 21
  Ultraviolet (UV) radiation, effect of, 20
Microcirculation, 235
Microfibrils, 4, 8
Microindentometer
  background, 161–162
  hardware and mechanical principles, 164–168, 174–175
  measurements and analysis of data, 169–174
MMPs, *see* Matrix metalloproteinases (MMPs)
Modeling
  anisotropic properties of skin, 64, 78, 87, 120
  to assay atrophogenicity of corticosteroids, 20
  Burger's rheological model, 68–69, 71
  contact pressure variable, 181–182
  of creep(ing), 18, 68–70
  of deformation, 18, 53, 55–57, 64
  of delayed distension, 71
  for the DermaLab, 120
  for the DTM, 70–71, 180–182
  of elasticity, 74–75, 203
  elasticity parameters, 42–43, 182
  environmental factors, 182–183
  for an extensometer, 86–87
  finite–element–based, *see* Finite–element–based modeling
  of friction force measurements, 55–57
  hairless mouse, 20
  history, 17–18, 63–64
  human *in vivo* skin friction studies, 53–57
  of hypoderma, 88
  of hysteresis, 18
  immediate distension, 71
  immediate retraction, 71
  of indentation test instrument, 174–176
  instrumentation factors, 180–183
  of Integrated Dynamic Rebound Analyzer (IDRA), 149–154
  of isotropic material, 78
  of the LSR, 103–104
  mathematical, 17–18, 71
  mechanical, 18, 78, 165
  modulus of elasticity, *see* Modulus of elasticity; Young's modulus
  net elasticity, 71–72
  parameters for skin measurement, 180
  preconditioning or pretension, 182
  of recovery, 18, 69–70
  of relaxation, 18
  rheological, *see* Rheological modeling
  rigidity, 78
  of stress–strain measurement, 18
  subject factors, 183–184
  tensile strength, 87
  of the viscous properties of skin, 64
  Young's modulus, *see* Young's modulus
Modified Skin Score (MSS), 141, 142
Modulus of elasticity, *see also* Young's modulus
  animal *in vitro* (*ex vivo*) hysteresis studies, 21–23
  animal *in vitro* (*ex vivo*) isorheological point studies, 24
  animal *in vitro* (*ex vivo*) stress–strain studies, 19, 20
  animal *in vivo* recovery studies, 30
  animal *in vivo* repeated strain studies, 28–29
  and EDS, 218
  human *in vivo* hydration studies, 202
  human *in vivo* keloid treatment studies, 262–263
  human *in vivo* skin friction studies, 53, 54, 55
  in modeling distension, 72
  and strain rate, 20
  Ultraviolet (UV) radiation, effect of, 20
Moisturizers
  friction force measurements, 52–53
  human *in vivo* Cutometer studies, 96, 181
  human *in vivo* DTM studies, 72, 181
  human *in vivo* GBE studies, 101
  human *in vivo* hydration studies, 203–204
  human *in vivo* LSR studies, 106–108
  human *in vivo* skin friction studies, 52–54

human *in vivo stratum corneum* studies, 101, 106–108
retinoids, containing, 201
Monoaxial test method
GBE, 44, 99–102, 108
LSR, 102–108
tribometer, 50–55
Monomelic scleroderma, 209
Morphea, Pasini–Pierini atrophic, 44
Morphea, scleroderma, 141–142, 209
MRI, *see* Magnetic Resonance Imaging (MRI)
mRNA, 6, 11
MSS, *see* Modified Skin Score (MSS)
Mucinous edema, 112
Mucopolysaccharides, 235
Muscle tissue studies, 163–164
Muscular fibers, impact on anisotropy, 21
Myeloma, associated with SB, 219
Myofibroblast, 258

N

NAMAS calibration standards, 104
Nape of the neck, location of test site, 262
Natural Moisturizing Factors (NMF), 242
Net elasticity
and EDS, 218
human *in vivo* Cutometer studies, 251
human *in vivo* hydration studies, 249, 251
human *in vivo* SB studies, 219
human *in vivo* Twistometer studies, 249
modeling of, for DMT, 71–72
parameter, Cutometer, 180, 217
parameter, DTM, 180
parameter, Twistometer, 217
preconditioning or pretension, effect on, 182
stratum corneum, after removal of, 249
Neuropathic foot, 143–145
Nidogen, 12
Normalization, *see* Preconditioning
Nova Dermal Phase Meter 9003
human *in vivo* hydration studies, 250–252, 254
human *in vivo* LSR studies, 106–107
human *in vivo stratum corneum* studies, 106–107

O

O/W, *see* Oil/Water emulsions
Occlusion, glove patch, 163
Occlusive therapy of plaque psoriasis, 224
OG, *see* Operating Gap (OG) in DMT probe
Oil/Water emulsions
human *in vivo* Cutometer studies, 251
human *in vivo* GBE studies, 252
human *in vivo* hydration studies, 203, 249, 251, 252
human *in vivo* LSR studies, 107
human *in vivo stratum corneum* studies, 107
human *in vivo* Twistometer studies, 249
Onion extract, 266
Operating Gap (OG) in DMT probe, 65
Orthostatic fluid accumulation, 45–46, 114, 189
Osteogenesis imperfecta, 7, 8
Overall elasticity, *see* Gross elasticity
Oxidative damage, 237
2–oxoglutarate, 4
Oxygen, 4, 11
Oxytalan fibers, 8, 10, 11

P

Palm, site of elasticity test, 187–195
Papillar dermis
of actinically damaged skin, 193
elasticity of, 41
human *in vivo* Cutometer studies, 94
and hydration, 199, 200
retinoid studies, 234
vasodilation in, 223
Paraffin oil
human *in vivo* Cutometer studies, 251
human *in vivo* Dermaflex A studies, 250
human *in vivo* hydration studies, 250, 251
impact on skin mechanics, 203
Paraproteinemia, 220
Pasini–Pierini atrophic morphea, 44, 114
Peak displacement, 103–104
Peak force, 103–104
Peelings, chemical, 96
D-penicillamine
animal *in vitro* (*ex vivo*) creep(ing) studies, 24
animal *in vitro* (*ex vivo*) hysteresis studies, 22
animal *in vitro* (*ex vivo*) isorheological point studies, 24
animal *in vitro* (*ex vivo*) relaxation studies, 22
animal *in vitro* (*ex vivo*) repeated strain studies, 24
animal *in vitro* (*ex vivo*) stress–strain studies, 20
animal *in vivo* recovery studies, 30
and collagen synthesis, 10
human *in vivo* elasticity studies, 45
skin fragility during treatment, 45
Perivascular cellular infiltration, 258
Perspiratio insensibilis, 199–200

Petrolatum, 249, 251
PEXP curve–fit tool, 70–71
Photoaging, 231–237, *see also* Age of skin, impact on
- ballistometry, 234
- human *in vivo* Cutometer studies, 189
- human *in vivo* DTM studies, 64
- human *in vivo* elasticity studies, 189
- human *in vivo* Twistometer studies, 64
- tretinoin, impact of, 234–235
- Ultraviolet (UV) radiation, 11

Photodynamic therapy, 142
Physicochemical nature of skin, 50
Pigs, 20, 26, 28, 83
PIIINP, 5, 9
PINP, 5
Pitting edema, 143
Plasma cells, 258
Plastic occlusive dressing, 224
Plastic surgery, 114
Plasticity
- age of skin, 22
- animal *in vitro* (*ex vivo*) creep(ing) studies, 24
- animal *in vitro* (*ex vivo*) relaxation studies, 22
- animal *in vitro* (*ex vivo*) repeated strain studies, 24
- of corneocytes, 200
- human *in vivo* Dermagraph studies, 128
- human *in vivo* Durometer studies, 140
- human *in vivo* LSR studies, 106–108
- human *in vivo stratum corneum* studies, 106–108
- and hydration, 199–204
- parameter, time–strain, 247

Plectin, 12
Poisson's coefficient, 53, 78
Polyalanine sequences, 8
Polymerization of collagen, 266
Polypeptide chains, 4, 6, 8
Polysaccharides, fish, 235–236
Postauricular, site of elasticity test, 187–195
Preconditioning
- animal *in vivo* repeated strain studies, 29
- biological elasticity, 182
- human *in vivo* Cutometer studies, 182
- human *in vivo* DTM studies, 74–75
- human *in vivo* skin friction studies, 52
- modeling of, 182
- net elasticity, effect on, 182

Prednisolone
- animal *in vitro* (*ex vivo*) creep(ing) studies, 24
- animal *in vitro* (*ex vivo*) hysteresis studies, 22
- animal *in vitro* (*ex vivo*) repeated strain studies, 24
- animal *in vitro* (*ex vivo*) stress–strain studies, 19

Prednisolone acetate
- animal *in vitro* (*ex vivo*) isorheological point studies, 24
- animal *in vitro* (*ex vivo*) stress–strain studies, 19
- animal *in vivo* recovery studies, 30

Pressure sores, 164
Prestressed, *see* Preconditioning
Pretension, 45–46
proα1(I) or proα2(I) Type I procollagen, 8
Probe, impact on measurement, 217
- Cutometer, 91–92
- Dermaflex A, 113
- DTM, 65, 68–69, 72–75
- human *in vivo* DermaLab studies, 118–120
- human *in vivo* elasticity studies, 43–44
- human *in vivo* scleroderma studies, 211
- human *in vivo* tensile properties studies, 211
- of psoriasis affected skin, 223
- of scleredema affected skin, 220
- of Scleredema of Buschke affected skin, 220, 222
- Twistometer, 65

Procollagen
- collagen synthesis, 5–6
- in EDS affected skin, 7, 8
- type I, 8, 10
- type III, 8, 10

Proline, 4, 5, 8
Prolyl–3–hydroxylase, 4
Prolyl–4–hydroxylase, 4
Protease, 11
Proteolytic, 6
Protoglycans
- and hydration, 232
- in intercellular ground substance, 91
- maintenance of macromolecular matrix, 208
- reduction of, 266
- in scar formation, 258, 260
- and scleroderma, 211
- viscoelastic-to-elastic ratio, impact on, 211

Pseudo-Young's modulus, *see* Coefficient of elasticity
Pseudoscars, *see also* Scars
Pseudoxanthoma elasticum
- elastin, abnormal accumulation of, 10
- elastin fibrillogenesis, abnormal, 9
- increased joint and skin motility, 45

Psoriasis Vulgaris
- edema in, 223
- human *in vivo* Cutometer studies, 96, 218, 222–224, 227

human *in vivo* Dermaflex A studies, 111–112, 114, 218, 222–224
and hydration, 200
interstitial fluid in, 223
occlusive therapy of plaque, 224
probe, impact on measurement of, 223
skin thickening, 222–224
softening of the skin in, 222–224
and stiffness, 222–224
Pure elasticity, *see* Net elasticity
Pyrrolidone carboxylic acid, 242

R

Rabbits, 20, 26, 201
Race of test subjects, 183, 188
Radiation therapy, 20
Rash, 234
Rats
laboratory, 18–24, 26–30
Sprague–Dawley male, 25–26
Raynaud's syndrome, 209
RD, *see* Resilient Distension (RD)
Reconstructive dermal substitutions, 96
Recovery
age of skin, impact on studies, 11, 232
animal *in vitro* (*ex vivo*) isorheological point studies, 23
animal *in vitro* (*ex vivo*) relaxation studies, 22, 23, 25
animal *in vitro* (*ex vivo*) stress–strain studies, 20
animal *in vivo* studies, 29–30
elastic, *see* Net elasticity
in elastoderma affected skin, 9
human *in vivo* elasticity studies, 44
human *in vivo* extensometer studies, 80, 83
human *in vivo* hydration studies, 202–203
human *in vivo* Twistometer studies, 67–68
modeling of, 18, 69–70
parameter, elasticity, 42
Reflex algodystrophy, 135
Relative Elastic Retraction (RER)
human *in vivo* EDS studies, 218
human *in vivo* erysipelas studies, 224–225, 227
human *in vivo* lymphedema studies, 226
human *in vivo* psoriasis studies, 222–223, 227
human *in vivo* SB studies, 219, 220, 227
human *in vivo* scleroderma studies, 220, 227
parameter, Dermaflex A, 217
parameter, time–strain, 248
Relaxation
age of skin, impact on studies, 232–233
animal *in vitro* (*ex vivo*) studies, 20–23, 25
animal *in vitro* (*ex vivo*) thermocontraction studies, 27
Gunner portable hand-held extensometer, 80
human *in vivo* deformation studies, 64
human *in vivo* Dermagraph studies, 128, 130–133
human *in vivo* extensometer studies, 86
human *in vivo* stress studies, 64
modeling of, 18
Repeated strain
animal *in vitro* (*ex vivo*) studies, 24
animal *in vivo* studies, 28–29
RER, *see* Relative Elastic Retraction (RER)
Residual deformation, *see* Residual skin elevation
Residual distensibility, *see also* Residual skin elevation
human *in vivo* Dermagraph studies, 130–135
parameter, Dermagraph, 128
Residual extension, 22
Residual skin elevation, *see also* Resilient Distension (RD)
human *in vivo* Cutometer studies, 250
human *in vivo* erysipelas studies, 224–225
human *in vivo* lymphedema studies, 226
human *in vivo* psoriasis studies, 222–224
parameter, Cutometer, 180
parameter, time–strain, 247–248
Residual stress, 22
Resilience, 4
Resilient Distension (RD)
human *in vivo* Cutometer studies, 250
human *in vivo* Dermaflex A studies, 250
human *in vivo* erysipelas studies, 224–225
human *in vivo* hydration studies, 250
human *in vivo* lymphedema studies, 226
human *in vivo* psoriasis studies, 222–224
parameter, Dermaflex A, 217
parameter, time–strain, 247–248
Resonance frequency, test method, 245
Response, elastic, *see* Delayed distension
restitutio ad integrum, 29–30
Reticular dermis
human *in vivo* Cutometer studies, 94
elasticity of, 41
and hydration, 200
mechanical properties, 187
in scleroderma affected skin, 208, 211
Retinoids, 201, 234–235
Retraction
and elastic fibers, 42
immediate retraction-to-total deformation ration, *see* Biological elasticity
total retraction-to-total deformation ratio, *see* Relative Elastic Retraction (RER)

Rex durometer, 139–140
Rheological modeling
  Burger's, 68–69
  of delayed distension, 68–72
  for the DTM, 66, 70–72
  of elasticity, impact of hydration, 203
  for the extensometer, 80, 86–88
  of immediate distension, 68–70
  of mechanical behavior of a material, 78
  parameters, DTM, 71
  of skin subjected to stress, 64, 68–70
Rigidity
  age of skin, impact on studies, 232
  collagen fibers, appearance of, 217
  and EDS, 218
  elastic fibers, appearance of, 217
  modeling term, 78
  and solar irradiation, 233
Rodnan total skin thickness score, 124, 209
Rotary Variable Differential Transformer, 149
Roughness, 202, 234
Ruffini and Meissner end organs, 46

S

Salicylic acid, 224
Sample preparation, 18, 21
SB, *see* Scleredema of Buschke
Scaling, 202, 241–243
Scapula
  human *in vivo* elasticity studies, 187–195
  human *in vivo* keloid treatment studies, 262
Scars
  antikeloidal treatment of, 257–267
  collagen, appearance of, 258
  DMT studies, 181
  and elasticity, 266
  and enzyme activity, 258
  fibroblasts in, 258
  glycoproteins in, 258
  human *in vivo* Cutometer studies, 96
  human *in vivo* extensometer studies, 87
  hydration, effect on, 258
  hypertrophic, 96, 257, 258–267
  mechanical properties study, 181
  protoglycans, impact on, 258, 260
  viscoelasticity, impact on, 266
Scleredema
  collagen, appearance of, 6
  probe, impact on measurement, 220
Scleredema of Buschke
  collagen, appearance of, 219
  human *in vivo* Cutometer studies, 218–222
  mechanical properties, summary, 227
  probe, impact on measurement, 220, 222
  and skin hardness, 219
  skin thickening, 219
Sclerimeter, 124, 125, 133–134
Scleroderma
  acroscleroderma, 45, 209, 211, 212
  collagen fibers, appearance of, 208–210
  collagen, increased synthesis of, 6, 7
  color of affected skin, 111
  diffuse, 209
  edema in, 209, 211
  elastic fibers, appearance of, 208
  human *in vivo* Cutometer studies, 219–222, 227
  human *in vivo* Dermaflex A studies, 111–112, 114
  human *in vivo* DermaLab studies, 119–121
  human *in vivo* Durometer studies, 141–142, 163
  human *in vivo* elasticity studies, 44
  human *in vivo* stress–strain studies, 111–112, 114
  hypoderma in affected skin, 211
  induration, 209–211
  monomelic, 209
  morphea, 141–142, 209
  patient concerns, 46
  protoglycans, impact on, 211
  reticular dermis, 208, 211
  sclerotic plaque of morphea, 112
  and shear modulus, 210
  and Shear wave propagation (SWP) testing, 211
  and skin tethering, 210, 211
  skin thickening, 7, 210
  and stiffness, 111, 114, 211–212
  and tensile strength, 207–212
  Young's modulus, 210
Sclerosis, systemic
  clinical assessment of, 209
  dermal, 45
  distensibility in, 124
  fibrosis, 125
  in fingers, 126
  human *in vivo* Cutometer studies, 96
  human *in vivo* Dermagraph studies, 124–126, 129–135
  human *in vivo* Durometer studies, 141–142
  skin thickening, 124
  and tethering, 124
  Type I collagen in, 125
  Type III collagen in, 125
Sclerotic plaque of morphea, 112
Sebocytes, 210
Serial stresses, *see* Preconditioning

Serine, 11
Sex, as a factor in experiments
human *in vivo* Cutometer studies, 94, 183
human *in vivo* Dermaflex A studies, 114
human *in vivo* Dermagraph studies, 131–134
human *in vivo* Durometer studies, 142
human *in vivo* tensile strength studies, 207, 209
impact on skin tests, 44, 183
keloid treatment assessment, 262
Shear modulus
calculation for human skin, 78
formula for, 72
human *in vivo* skin friction studies, 54
and scleroderma, 210
Shear wave propagation (SWP) testing
animal *in vivo* hydration studies, 201
human *in vivo* stress studies, 64
and hydration, 204
and Langer's lines, 211
and scleroderma, 211
Silicone gel sheet, 258, 260, 262–263, 265–266
Simpson's formula, 28
Skicon, 254
Skin atrophy
age–related, 6, 7
animal *in vitro* (*ex vivo*) stress–strain studies, 19, 20
collagen, appearance of, 10
collagen, reduced amounts of, 6, 7
corticosteroids, impact on, 45
human *in vivo* extensometer studies, 82
steroid-induced, 6, 7
thinning of the skin, 7
Types of collagen in, 7
wrinkle formation, 10
Skin cancer, 11
Skin color, *see* Color; Race
Skin distensibility, *see* Immediate distension
Skin elevation, residual, *see* Residual skin elevation
Skin extensibility, *see* Immediate distension
Skin extension, *see* Immediate distension
Skin fragility
in EDS affected skin, 7
in Osteogenesis imperfecta affected skin, 7
during penicillamine treatment, 45
Skin friction coefficient, 50
Skin friction force studies, 49–57
Skin furrows, 50
Skin hardness
and glycosilation, 143
human *in vivo* diabetes studies, 143–145
human *in vivo* Durometer studies, 141–144, 163
human *in vivo* IDRA studies, 151–159
human *in vivo* lipodermatosclerosis studies, 142–144
human *in vivo* Rex durometer studies, 139
human *in vivo* scleroderma studies, 141–142, 209
and Scleredema of Buschke, 219
and stiffness, 162
of the *stratum corneum*, 162
and tensile strength, 162
Skin induration
human *in vivo* Durometer studies, 141–143
and scleroderma, 209–211
Skin–Score, for systemic sclerosis affected skin, 126
Skin stiffening, *see* Stiffness
Skin tautness, 7
Skin tethering
human *in vivo* Dermagraph studies, 126, 128–129
parameter, Dermagraph, 124
and scleroderma, 210, 211
scoring of skin elasticity, 126, 209
in systemic sclerosis affected skin, 124
Skin thickening
in actinic damaged skin, 233
collagen, metabolic disturbance of, 6, 7
in diabetes, 6, 7
in EDS affected skin, 218–219
in erysipelas affected skin, 224–225
in keloids, 7
in lymphedema affected skin, 226
oral treatment, fish polysaccharides, 235
in psoriasis affected skin, 222–224
in SB affected skin, 219
in scleroderma affected skin, 7, 210
in systemic sclerosis affected skin, 124
ultraviolet (UV) radiation, effect of, 20
Skin thickness
age dependence, 10, 11, 19
AHAs, impact on, 234
animal *in vitro* (*ex vivo*) stress–strain studies, 19, 20, 21
and collagen fibers, 217
collagen, impact on, 9–11
in DermaLab modeling, 119–120
and distensibility, 44
drug treatments, effect of, 19–21
in EDS affected skin, 9, 218–219
and elastic fibers, 217
human *in vivo* Cutometer studies, 189
human *in vivo* DermaLab studies, 120, 121
human *in vivo* elasticity studies, 44, 189
human *in vivo* keloid treatment studies, 262, 266

and hydration, 44, 200, 202
index, 209
keloid treatment assessment, 258
measurement of, by Dermaflex A, 111
measurement of, by ultrasound, prior to Cutometer studies, 94
measurement of, prior to DTM studies, 72
measurement of, with calipers, 18, 27, 28, 29
and measurement parameters, 217, 247–248
Rodnan total score, 124, 209
and solar irradiation, 233
standard, 120
and tensile strength, 208–209
ultrasound image analysis, 94
Ultraviolet (UV) radiation, 11
ultraviolet (UV) radiation, effect of, 20
Skin thinning
in atrophy, 7
and distensibility, 44
in EDS affected skin, 7, 8
in focal dermal hypoplasia, 7
and hydration, 44
oral treatment, fish polysaccharides, 235
in Osteogenesis imperfecta, 7
in Pasini–Pierini atrophic morphea affected skin, 44
Skin tightness, 101
Skin *turgor*, 44, 114
Sodium pyrrolidone carboxylate, 249
Softening of the skin
in erysipelas affected skin, 225
human *in vivo* LSR studies, 106–108
human *in vivo stratum corneum* studies, 106–108
in psoriasis affected skin, 222–224
Softness, *see also* Dynamic spring rate (DSR)
and elasticity, 252
human *in vivo* DermaLab studies, 119, 120, 121
human *in vivo* GBE studies, 101
human *in vivo* hydration studies, 252
human *in vivo stratum corneum* studies, 101
and hydration, 44
Solar elastosis, 210
Solar irradiation, 233, *see also* Ultraviolet (UV) radiation
Spring and dashpot test instrument
Kelvin parallel element, 18
Maxwell series element, 18
muscle tissue hardness, study of, 163
Voigt parallel element, 18
Static limit friction force, 54
Stationary creep, 69, 71
"Step" phenomenon, 21
Sternum, location of test site, 262
Steroid-induced skin atrophy, 6, 7
"Stick–slip" motion pattern, 55
Stiffness, *see also* Skin hardness
animal hydration studies, 201
animal *in vitro* (*ex vivo*) stress–strain studies, 20
animal *in vivo* extensometer studies, 83
animal SWP studies, 201–202
collagen, appearance of, 157–158
collagen, excessive accumulation of, 6
compression test, 87
in diabetes affected rat skin, 20
in the early postpartum period, rat skin, 20
human *in vivo* Dermaflex A studies, 111, 114
human *in vivo* DermaLab studies, 119, 120, 121
human *in vivo* IDRA studies, 152–159
human *in vivo* indentation studies, 162, 164
human *in vivo* skin friction studies, 55
lateral, of skin, 54
mechanical property of skin, 216
orthostatic fluid accumulation, 45, 189
parameter, time–strain, 247
of psoriasis affected skin, 222–224
of scleroderma affected skin, 111, 114, 211–212
and skin hardness, 162
of the *stratum corneum*, 162, 164, 200
and tensile strength, 162
Storiform pattern, 258
Strain rate, 20
Strain–time measurement, *see* Time–strain measurement
Stratum corneum
AHAs effect upon, 233–234
corneocytes in, 199
and dermatitis, 162
and the Dynamic Spring Rate (DSR), 103
elasticity of, 41, 249
extensibility of, 249
human *in vivo* Cutometer studies, 250–251
human *in vivo* DTM studies, 64, 250–251
human *in vivo* GBE studies, 100, 252
human *in vivo* hydration studies, 106–108, 249, 250–251, 252
human *in vivo* indentation studies, 161–166, 169–176
human *in vivo* LSR studies, 106–108
human *in vivo* mechanical properties studies, 100
human *in vivo* Microindentometer studies, 164–166, 169–174
human *in vivo* skin friction studies, 57
human *in vivo* Twistometer studies, 64, 249
and hydration, 199–204, 241–243

immediate retraction, measurement of, 249
keratin(s) in, 41, 107
net elasticity measurements, 249
skin hardness, 162
stiffness, 162, 164, 200
tensile strength, 162
ultrasound and suction studies, 181
Streptococcal infection of the dermis, *see* Erysipelas
Streptozotocin-induced diabetes, 20
Stress–strain measurement, *see also* Distensibility
animal *in vitro* (*ex vivo*) anisotropy studies, 21
animal *in vitro* (*ex vivo*) hysteresis studies, 21–23
animal *in vitro* (*ex vivo*) studies, 18–21
animal *in vitro* (*ex vivo*) wound studies, 26
animal *in vivo* studies, 27–30
of the forearm, 247
human *in vivo* anisotropy studies, 21
human *in vivo* Cutometer studies, 92, 244
human *in vivo* Dermaflex A studies, 112, 114
human *in vivo* DTM studies, 244
human *in vivo* GBE studies, 252
human *in vivo* IDRA studies, 151–159
hysteresis, 248
modeling of, 18
parameters for skin measurement, 248
Stretching, test method
of planar skin mechanics, 243
uniaxial stretching devices, 245
Striae distensae
collagen, reduced synthesis in, 7
human *in vivo* Cutometer studies, 96
in Marfan syndrome affected skin, 10
Stripping
removal of horny membrane, 251
removal of *stratum corneum*, 249
Stromelysin, 12
Subcutaneous fat, 181, 211
Suction, test method
animal models, *in vivo* studies, 28
BTC-2000, 244
causing edema, 120, 121
Cutometer, 91–96, 187–196, 217, 243–244
Dermaflex A, 111–114, 201, 217, 243
Dermagraph, 124–135
DermaLab, 117–121, 244
human *in vivo* elasticity studies, 43
and hydration, 203, 252–254
of perpendicular skin mechanics, 243
and scleroderma, 211
Sugar, 4
Sun, exposure to, *see* Solar irradiation; Ultraviolet (UV) radiation
Suppleness, 78
SWP, *see* Shear wave propagation (SWP) testing

T

Tactile sensor, 252
TD, *see* Tensile distensibility (TD)
Teleangectasia, 234, 237
Temperature of skin
animal *in vitro* (*ex vivo*) studies, 18
human *in vivo* Cutometer studies, 92
human *in vivo* Dermagraph studies, 129
human *in vivo* Durometer studies, 140
human *in vivo* elasticity studies, 43
human *in vivo* LSR studies, 106
human *in vivo* skin friction studies, 52
human *in vivo stratum corneum* studies, 106
human *in vivo* studies, 182
Temperature, shrinkage, 26–27
Temple, site of test, 94–95
Tensile Distensibility (TD), *see also* Final distension
and EDS, 218
human *in vivo* DTM hydration studies, 250–251
human *in vivo* erysipelas studies, 224–225, 226–227
human *in vivo* keloid treatment studies, 262–264
human *in vivo* lymphedema studies, 226
human *in vivo* psoriasis studies, 222–224, 226–227
human *in vivo* SB studies, 219–222, 226–227
human *in vivo* scleroderma studies, 220, 226–227
parameter, Dermaflex A, 217
parameter, time–strain, 246–248
Tensile force, 87
Tensile strength
animal *in vitro* (*ex vivo*) stress–strain studies, 19, 20
animal *in vitro* (*ex vivo*) wound studies, 25–26
animal *in vivo* studies, 27–30
calculation for skin, 19
circadian variations of, 20
from collagen, 3, 20
and drug treatments, animal studies, 19–21
and elastic fibers, 208–210
human *in vivo* DMT studies, 65
human *in vivo* extensometer studies, 83–85
modeling *in vivo* tests, 87
of scleroderma affected skin, 207–212
and skin hardness, 162
and skin thickness, 208–209

and stiffness, 162
of the *stratum corneum*, 162
Ultraviolet (UV) radiation, effect of, 20
Tension, half-life of, 24
Tension test, 83–85
Terminal hairs, 113
Test area size
Cutometer, 92, 188, 217
Dermaflex A, 188, 217
DMT, 64
human *in vivo* Durometer studies, 142
human *in vivo* scleroderma studies, 211
Twistometer, 64, 217
Test axis
animal *in vitro* (*ex vivo*) anisotropy studies, 21
human *in vivo* Cutometer studies, 92, 217
human *in vivo* Dermaflex A studies, 113, 217
human *in vivo* Dermagraph studies, 132–133
human *in vivo* DTM studies, 64, 72, 73
human *in vivo* Durometer studies, 140
human *in vivo* elasticity studies, 43
human *in vivo* GBE studies, 100
human *in vivo* IDRA studies, 154–155
human *in vivo* Microindentometer studies, 167, 173–174
human *in vivo* skin friction studies, 50–55
human *in vivo* Twistometer studies, 217
human skin mechanical behavior studies, 78–79
Test position
human *in vivo* Cutometer studies, 188
human *in vivo* Dermaflex A studies, 113
human *in vivo* Dermagraph studies, 132–133
human *in vivo* DermaLab studies, 120
human *in vivo* DTM studies, 72–74
human *in vivo* Durometer studies, 140
human *in vivo* elasticity studies, 43
human *in vivo* extensometer studies, 86
human *in vivo* IDRA studies, 154–155
human *in vivo* LSR studies, 106
human *in vivo* Microindentometer studies, 167, 173–174
human *in vivo* studies, 182
human *in vivo* tensile strength studies, 209
Test standards, lack of, 43–44
Test subject, animal
adrenalectomized, 20
dogs, 20–21, 26
giraffe, 45, 189
guinea pigs, 20, 26
mice, hairless, 20
mice, inbred, 20
mice, laboratory, 26
mice, tight-skin, 21
pigs, 20, 26, 28, 83
rabbits, 20, 26, 201
rats, laboratory, 18–24, 26–30
rats, male Sprague–Dawley, 25–26
TEWL, *see* Transepidermal Water Loss (TEWL)
Thenar eminence, 169–173, *see also* Palm, site of elasticity test
Thermocontraction, 26–27
Thigh
human *in vitro* indentation studies, 164
human *in vivo* Cutometer studies, 187–195
human *in vivo* Dermagraph studies, 132–134
human *in vivo* Durometer studies, 163
human *in vivo* elasticity studies, 187–195
Thyroid hormones, 20, 22
Time, as a factor in experiments
deformation, 64, 245–246
elasticity, 42
human *in vivo* Cutometer studies, 92
human *in vivo* DermaLab studies, 117
human *in vivo* tensile strength studies, 209–210
keloid treatment assessment, 258–259, 262
sample preparation, animal *in vitro* studies, 18
viscoelasticity, 216
Time, as a factor in modeling, 69, 72
Time–strain measurement
delayed distension, 246–248
of elasticity, 248
extensibility, 247
final distension, 246–248
human *in vivo* Cutometer studies, 92, 94, 188–196, 244, 246–248
human *in vivo* Dermaflex A studies, 245
human *in vivo* DTM studies, 244
human *in vivo* tensile strength studies, 209–210
hysteresis, 247–248
immediate retraction, 246–248
plasticity, 247
Relative Elastic Retraction (RER), 248
Residual skin elevation, 247–248
Resilient Distension (RD), 247–248
stiffness, 247
Tensile Distensibility (TD), 246–248
total recovery deformation at end of stress-off period, 246–248
viscoelasticity, 216
TIMPs, *see* Tissue inhibitors of matrix metalloproteinases (TIMPs)
Tissue expansion, 20–21, 114
Tissue inhibitors of matrix metalloproteinases (TIMPs)
basement membrane, impact on, 12
collagen, impact on, 11
MMPs, impact on, 12, 236–237

Tocoferoxil, 237
Tonicity, 247, *see also* Immediate retraction
Torsion, test method
  age of skin, 232
  DTM, 64–75, 244
  human *in vivo* elasticity studies, 43–44
  hydration studies, 244, 252–254
  of planar skin mechanics, 243
  Twistometer, 64, 66–67, 217, 244
Torso, 132–134
Total extensibility, *see* Final distension
Total recovery deformation at end of stress-off period, 180
  human *in vivo* Cutometer studies, 250–251
  human *in vivo* DTM studies, 250–251
  human *in vivo* hydration studies, 250–251
  parameter, Cutometer, 180, 181
  parameter, DTM, 180
  parameter, time–strain, 246–248
Touch, sense of, 46
Transcutaneous oxygen pressure ($TcpO_2$), 142–144
Transcutaneous oxygen tension measurements, 141
Transepidermal Water Loss (TEWL)
  guidelines for, 179
  human *in vivo* DermaLab studies, 117
  human *in vivo* stress–strain studies, 117
  and hydration, 202, 241–243
  oral treatment, fish polysaccharides, 235
Transferase, 234
Transient creep, *see* Delayed distension
Translucent skin, 9
Treatment site, impact on skin tests, 183
Tretinoin
  animal photoaging studies, 234
  elastin, impact on, 12
  fibroblasts, affect on, 234
  human photoaging studies, 234–235
Triamcinolone acetonide, 258–259, 261, 262
Tribometer, 50
Tropoelastin, 8
Tumorlike thickening of the skin, 7
Tumors, 11
Twistometer, *see also* Dermal Torque Meter (DTM)
  delayed distension, 217
  development of, 64
  disadvantages, 79
  human *in vivo* EDS studies, 218
  human *in vivo* hydration studies, 249–250
  immediate distension, 217
  measurements and analysis of data, 216–218
  net elasticity parameter, 217
  probe, impact on measurement, 65
  test area size, 64, 217
  test axis, 217
  Young's modulus, 249
Type I collagen
  abundance in skin, 4
  in age–related atrophy, 7
  degradation by MMPs, 11
  in EDS affected skin, 7, 8
  in Osteogenesis imperfecta affected skin, 7, 8
  ratio of Type I to Type III, 6
  in steroid–induced atrophy, 7
  in systemic sclerosis affected skin, 125
Type I procollagen, 8, 258
Type II collagen, 11
Type III collagen
  abundance in skin, 4
  age of skin, amount in, 6
  atrophy, skin, 7
  degradation by MMPs, 11
  in EDS affected skin, 7, 8, 9
  ratio of Type I to Type III, 6
  in systemic sclerosis affected skin, 125
Type III procollagen, 8
Type IV collagen
  abundance in skin, 4
  in basement membrane, 12
  degradation by MMPs, 11
Type V collagen
  abundance in skin, 4
  in EDS affected skin, 7, 8
Type VII collagen
  abundance in skin, 4
  in epidermolysis bullosa affected skin, 12
  laminin–5 linkage to, in *lamina densa*, 12
Type XVII collagen, 12

U

Ulcers
  from diabetes, 143–145
  lipodermatosclerosis, associated with, 142–143
  neuropathic, associated with diabetes, 143
  venous, 45–46, 142–143
Ultimate extension, 21, 24
Ultimate load, 19–21
Ultimate modulus of elasticity, *see* Modulus of elasticity
Ultimate strain, 19–21, 24
Ultrasound image analysis
  animal *in vitro* (*ex vivo*) studies, 20
  animal *in vivo* stiffness studies, 83
  echogenicity of upper dermis, decrease in, 11

of corticosteroids treatment, 265–266
human *in vivo* keloid treatment studies, 265–266
human *in vivo* psoriasis studies, 222
human *in vivo* scleroderma studies, 141
human *in vivo* tensile strength studies, 210
intradermal water and structural texture interactions, 44
keloid treatment assessment, 258, 262
orthostatic fluid accumulation in human legs, 46
skin thickness measurement for Cutometer, 94
of the *stratum corneum*, 181
Ultraviolet (UV) radiation
actinic elastosis, 11, 233
animal *in vitro* (*ex vivo*) stress–strain studies, 20
collagen synthesis, effect on, 10–11
damage to DNA, 11
ECM, effect on, 11, 236–237
and edema, 233
elasticity studies, 189, 233, 236–237
fibroblasts, affect on, 11
human *in vivo* Cutometer studies, 96
human *in vivo* DTM studies, 64
human *in vivo* tensile strength studies, 209
human *in vivo* Twistometer studies, 64
keratinocytes, number after treatment with, 11
mice, effect on, 20
MMPs, effect on, 11–12
modulus of elasticity of, 20
molecular structure of skin, effect on, 11–12, 236–237
photoaging, 11
and skin thickening, 20
skin thickness, effect on, 11, 20, 233
tensile strength, 20
tretinoin studies, 234
and ultimate load, 20
and ultimate strain, 20
UVA definition, 11
UVB definition, 11
Uniassial tests, 232
Uniaxial tension, test method, 64, 79–85, 267
Unmyelinated axons, 3
Urea
human *in vivo* Cutometer studies, 251
human *in vivo* hydration studies, 251
human *in vivo stratum corneum* studies, 202
immediate distension, impact on, 249
natural moisturizing factor, 242
UV, *see* Ultraviolet (UV) radiation

V

Valine, 8
Vascular dilation in the dermis, 225
Vasodilatation in the papillary dermis, 223
Venous leg ulcers, 45–46
Verhoeff–van Gieson stain, 10
Vertical vector of skin elasticity, *see* Orthostatic fluid accumulation
VESA, *see* Viscoelasticity Skin Analyzer (VESA)
Vibration perception threshold (VPT), 144, 145
Viscoelastic and viscous deviation length, *see* Delayed distension
Viscoelastic creep, *see* Delayed distension
Viscoelastic ratio, *see* Viscoelastic-to-elastic ratio
Viscoelastic-to-elastic ratio
age of skin, 189, 193–196
and edema, 211
horny membrane, impact on, 251
human *in vivo* Cutometer studies, 189, 193–196, 250–251, 259–260
human *in vivo* DTM studies, 72
human *in vivo* EDS studies, 218
human *in vivo* elasticity studies, 189, 193–196
human *in vivo* erysipelas studies, 224–225, 227
human *in vivo* hydration studies, 250–251
human *in vivo* keloid treatment studies, 259–260, 262–264
human *in vivo* lymphedema studies, 226
human *in vivo* psoriasis studies, 222–224, 227
human *in vivo* SB studies, 219, 221–222, 227
human *in vivo* scleroderma studies, 211, 220, 227
and hydration, 217
parameter, Cutometer, 94, 180, 217
parameter, DTM, 180
protoglycans, impact on, 211
Viscoelasticity
age of skin, impact on studies, 232
animal *in vitro* (*ex vivo*) isorheological studies, 23
animal *in vitro* (*ex vivo*) relaxation studies, 22
animal *in vitro* (*ex vivo*) stress–strain studies, 20
in edematous skin, 201
human *in vivo* Dermaflex A studies, 114, 250
human *in vivo* elasticity parameters, 41, 42
human *in vivo* GBE studies, 252
human *in vivo* hydration studies, 250
human *in vivo* skin friction studies, 55–57
human *in vivo* VESA studies, 204
and hydration, 217, 243
hypertrophic scar treatment assessment, 266
hysteresis, 216
interstitial fluid, 217

keloid treatment assessment, 266
mechanical property of skin, 216, 245–246
oral treatment, fish polysaccharides, 235–236
retinoid studies, 234–235
time, as a factor in experiments, 216
time–strain measurement, 216
Viscoelasticity Skin Analyzer (VESA), 201, 204
Viscopart
human *in vivo* erysipelas studies, 224–225, 227
human *in vivo* lymphedema studies, 226
human *in vivo* psoriasis studies, 222–223, 227
human *in vivo* SB studies, 219, 220, 227
human *in vivo* scleroderma studies, 220, 227
parameter, Cutometer, 217
Viscosity
animal *in vitro* (*ex vivo*) creep(ing) studies, 24
animal *in vitro* (*ex vivo*) isorheological point studies, 24
of collagen fibers, 91
Glycosaminoglycans, 193
of the ground substance, 91, 193
human *in vivo* keloid treatment Cutometer studies, 259
Viscous deformation, *see* Delayed distension
Viscous extension, *see* Creep(ing)
Viscous properties of skin
animal *in vitro* (*ex vivo*) creep(ing) studies, 24
animal *in vitro* (*ex vivo*) relaxation studies, 22
animal *in vitro* (*ex vivo*) repeated strain studies, 24
glycosaminoglycans, impact on, 232
human *in vivo* LSR studies, 104
human *in vivo* *stratum corneum* studies, 104
human *in vivo* Twistometer studies, 67–68
modeling of, 64
Vitamin E, 237
Voigt model, 18, 69, 71, 165
VPT, *see* Vibration perception threshold (VPT)
Vulvar skin, 94

W

W/O, *see* Oil/Water emulsions
Water, accumulation in dermis, *see* Hydration
Water, hot, impact on elasticity, 249
Water/Oil emulsions, *see* Oil/Water emulsions
Water, tap
human *in vivo* Cutometer studies, 250
human *in vivo* Dermaflex A studies, 250
human *in vivo* DTM studies, 181
human *in vivo* GBE studies, 252
human *in vivo* Twistometer studies, 249
impact on skin mechanics, 202–203
Wave propagation, test method, 243, *see also* Shear wave propagation (SWP) testing
Weibull's theory, 69–70
Wijn extensometer, *see* Magnetic extensometer
Wounds, healing of, *see also* Scars
animal *in vitro* (*ex vivo*) studies, 25–26
proteases, affect on, 11
pseudoscars, 45
traction or tangential, in aged skin, 45
Wrinkles
atrophy, skin, 10
chronoaging, 10
in cutis laxa affected skin, 8
on the face, 10
MMP inhibitors, effect on, 237
oral treatment, fish polysaccharides, 235
pseudoxanthoma elasticum, 9, 10
retinoid studies, 234

X

X rays, 28, 258
Xerosis, 241–242

Y

Young's modulus, *see also* Modulus of elasticity
age of skin, impact on, 232
in anisotrophic skin modeling, 47, 120
in Cutometer modeling, 79
in DermaLab modeling, 117, 119–121
in DTM modeling, 72
elast, 117
in extensometer modeling, 87
in friction force modeling, 57
in GBE modeling, 252
for human forearm, 50
in hydration modeling, 249
in isotropic material modeling, 78, 120
measurement standardization, 179–180
and scleroderma, 210
of the skin, 53, 54
in stress–strain modeling, 19, 20, 248
in Twistometer modeling, 249